Minerals in Africa

Minerals in Africa

Opportunities for the Continent's Industrialisation

Francis P. Gudyanga
Zimbabwe Academy of Sciences

CRC Press
Taylor & Francis Group
Boca Raton London New York

CRC Press is an imprint of the
Taylor & Francis Group, an **informa** business

AN A K PETERS BOOK

CRC Press/Balkema is an imprint of the Taylor & Francis Group, an informa business

© 2020 Taylor & Francis Group, London, UK

Typeset by Apex CoVantage, LLC

Library of Congress Cataloging-in-Publication data

Applied for

Published by: CRC Press/Balkema
                 Schipholweg 107C, 2316 XC Leiden, The Netherlands
                 e-mail: Pub.NL@taylorandfrancis.com
                 www.crcpress.com – www.taylorandfrancis.com

ISBN: 978-0-367-42015-4 (Hbk)
ISBN: 978-0-367-81740-4 (eBook)

DOI: 10.1201/9780367817404
DOI: https://doi.org/10.1201//9780367817404

Contents

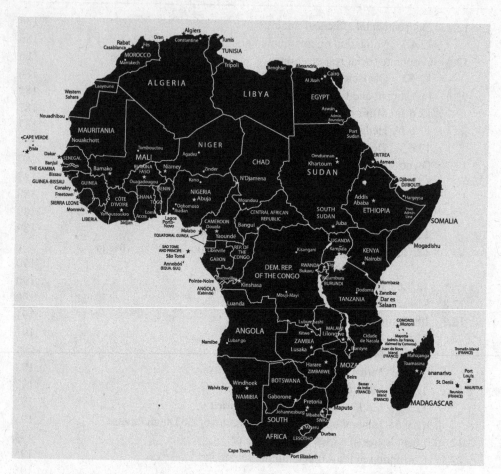

Figure 0.1 Map of Africa with countries and their capital and major cities

Source: www.shutterstock.com/search/africa+map

Foreword

Modern industry and enterprises rely on a variety of materials for functional economies, jobs and for improving the quality of life. Reliance on access to a growing number of materials has been necessitated by rapid technological progress which requires hi-tech goods. The origins of these materials include metallic ores as well as industrial minerals which anchor downstream industries. Thus, products derived from minerals are intrinsically linked to most industries across all supply chain stages, and consequently are essential to our societies. These products are growing in complexity and sophistication with a concomitant increase in the number and breadth of raw materials used in their production.

The African continent is richly endowed with minerals from which these industrial products are derived. Thus, Africa's export-oriented mining and quarrying is driven primarily by the commodity hunger of the world's largest economies. There is incessant concern that Africa is not receiving fair prices for its minerals on the global market. It has now become obvious that the greatest loss to Africa has resulted from the exportation of these minerals as raw ores or concentrates. To appreciate this fact, one has to look at the low status of the African economies, and contrast this with the contribution to world economies by minerals from Africa and the extent of colonial plunder of these minerals.

The increasing global need for minerals as industrial inputs has led to a sustained production of primary metals worldwide, as well as greater interest in mineral exploration in Africa. Ore deposits are being obtained with increasingly complex mineralogy and metallurgical properties as the more easily treatable resources are depleted. However, Africa is not exploiting to its own benefit the full potentials represented by this mineral wealth. The exploitation of minerals from Africa has been used to achieve advanced economic developments in the rest of the world, reflecting how they can, and should, be used for Africa's industrialisation.

The continent of Africa is today facing a NEW (North East West) colonial invasion, which is just as devastating in scale and impact as that from which it suffered during the nineteenth century. As before, the new colonialism is driven by a determination to plunder the natural resources of Africa, especially its strategic energy and mineral resources. There is a long-term involvement of the governments of the industrialised countries, through their trade and investment policies, to manoeuvre their foreign companies' access to African raw materials.

Therefore, the continent's minerals present a unique opportunity for cooperation between African countries when negotiating for favourable trade terms and prices with foreign industrialised countries. If Africa is to leapfrog into the technological advancement of the twenty-first century, the continent needs to develop, in concert, an adequate number of indigenous materials scientists, engineers and related professionals who will actively contribute to finding solutions to continental trading problems.

Africa's greatest need in the minerals industry is value addition. This has been recognised by the African leaders who have made pronouncements on the matter at various national, regional and continental fora. It is fortunate that Professor Gudyanga, an African metallurgist and materials scientist of notable expertise and breadth of technological skills has been prevailed upon to write such a book. We believe that his book will prove to be of great value to applied and practising metallurgists and materials scientists who wish to become engaged in the beneficiation/extraction and value addition of minerals found in Africa. This book will be useful also for those primarily responsible for policy issues pertinent to mining in Africa, who wish to know more about the formulation and implementation of suitable policies for the furtherance of Africa's industrialisation. This was envisaged by the African Mining Vision of 2009 in which were expressed aims of having the continent pursue "a transparent, equitable, and optimal exploitation of mineral resources to underpin its broad-based sustainable growth and socio-economic development".

The bridging of the gap between the so-called 'technical' and 'policy' dimensions of mining in Africa is essential for the rapid industrialisation of the continent. Professor Gudyanga has done a great service in helping to break down this wholly artificial barrier in one area at least. This is in sync with Agenda 2063 which is the blueprint and master plan for transforming Africa into the global powerhouse of the future, founded on the African Union (AU) Vision of "an integrated, prosperous, and peaceful Africa, driven by its own citizens and representing a dynamic force in the international arena".

The writing of this book was inspired by the desire to promote the continent's mineral-driven industrialisation by integrating information on African minerals, their beneficiation/extraction and value addition with mineral policy-oriented issues. It was also informed by concepts and continent-wide deliberations that resulted in the establishment, in 2016, of the Pan African Minerals University of Science and Technology (PAMUST), which is co-located with the Scientific and Industrial Research & Development Centre (SIRDC) in Harare, Zimbabwe. The aims and objectives of that university is to provide advanced and appropriate training in geosciences, metallurgy, mining engineering, materials science and engineering and mineral-enabling ICTs, as well as mineral economics. It is our considered view that the ideas expressed in this book are very relevant to PAMUST and SIRDC, and indeed to other institutions on the continent with similar programmes.

It gives me great pleasure to recommend this book to all who are interested in the mineral-driven industrialisation of Africa.

Professor Christopher J. Chetsanga

Founding Fellow and Inaugural President of the
Zimbabwe Academy of Sciences (ZAS)

Fellow of both the African Academy of Sciences (AAS) and
The World Academy of Sciences (TWAS)

Preface

Africa's dire need to industrialise is acknowledged continent-wide and it is evident that the continent's vast mineral resources can catalyse that industrialisation. This requires Africa's promotion of local beneficiation and value addition of minerals to yield materials on which modern Africa's industry and society can rely. This book is, therefore, about transforming Africa's comparative advantages in mineral wealth into the continent's competitive edge regarding materials. Mineral beneficiation and value addition form the basis and provide opportunities for mineral-driven industrialisation of Africa. The scope of this book is threefold with interconnected relationships: information, technical, and policy oriented issues. For this reason, the book is organised into three parts as follows.

Part I: Information about minerals in Africa

This part is comprised of the first chapter of the book, Chapter 1, which presents a panoramic overview of minerals found on the African continent. The information about the minerals in Africa is summarised in Table 1.1 and Table 1.2. Table 1.1 lists the minerals found in each African country. Conversely, Table 1.2 lists the African countries in which a specific mineral is found. Underground mineral deposits may be found to stretch across territorial boundaries that are recognised on the surface. The map of Africa, Figure 0.1, in conjunction with Tables 1.1 and 1.2, provide information of geological and economic interest for neighbouring countries. The presence of a mineral in one country is a pointer to the likely possibility of that mineral being present in the neighbouring country, if it has not been identified already.

The chapter also serves as an introduction to the whole book, pointing out the paradox of the vastness of the mineral resources on the African continent and the poverty of its citizens. It is observed that these African mineral resources have been critical ingredients to the industrialisation of countries other than the continent itself. Reference is made to the efforts by African leaders to adopt and embrace appropriate strategies that can add value to mineral resources and resultantly improve the lives of citizens in African countries. It is the view of the current African leaders that the mining sector is a potential driving engine for the continent's development and industrialisation.

Part II: Mineral beneficiation and value addition

This part details, in eight chapters, the technical exploitation of about 120 minerals in different African countries. These 120 minerals are clustered into the following respective eight chapters: (2) precious metals, (3) base metals, (4) radioactive materials, (5) industrial minerals, (6) precious stones, (7) semi-precious stones, (8) dimension stones, and (9) energy sources.

For each mineral a general brief *overview* is given covering its origins and geological occurrences on the earth's crust. In the case of metallic ores, the form in which the metal is found is given. Some metals like gold may exist as native metal, perhaps alloyed to a greater or lesser extent with other metals. For other metallic ores, the chemical compositions in which a metal exists with other elements is described.

The overview on the geological aspect of the mineral is followed by a description of its beneficiation. The term *beneficiation* in this book is used to refer to the entire processing of mineral-bearing ores to produce commercial end-products. In this case beneficiation starts with the process of removing gangue materials from the mineral ore to produce a higher-grade product (concentrate) and a waste stream (tailings). The sequential stages are: particle size reduction, separation of particle sizes, solid/liquid separation and concentration techniques which may be gravity separation, froth flotation, electrostatic and magnetic separation processes, as well as optical sensors. The beneficiation process ends with the recovery of the valuable product.

In some textbooks on extractive metallurgy the term beneficiation refers strictly only to the aforementioned processes. This is followed, for metallic ores, by *extraction* processes to recover the metals from concentrates produced by the mineral processing referred to prior. The extraction processes may be pyrometallurgical or hydrometallurgical and final recovery. Pyrometallurgy consists of the thermal treatment of minerals, metallurgical ores and concentrates to enable recovery of valuable metals in pure form or as intermediate compounds or alloys, suitable as feed for further processing. Hydrometallurgy is a technique involving the use of aqueous chemistry for the recovery of metals from ores, concentrates and recycled or residual materials. Hydrometallurgy is typically divided into three general areas: leaching, solution concentration and purification, as well as metal recovery which is achieved by electrowinning, precipitation, electrorefining, electroplating or electroforming.

The basic beneficiation and extraction processes in this section can be further learnt from textbooks on extractive metallurgy. In this Part II the essential ideas are described concisely in order to present the theory in the form in which it is to be used, and here a moderate knowledge of chemistry is assumed. Depending on the nature of the mineral, such as its mineralogy, all or some of the previously mentioned stages may be involved. In this book suitable techniques and technologies in each stage are described for the beneficiation of each mineral.

Lastly, the production of individual mineral concentrates/metals and fabrication of finished products, as well as the uses of each mineral and/or materials derived from them are described. This section is collectively termed *value addition*. The words *beneficiation* and *value addition* are often used interchangeably. In this book, a very subtle but key differentiation is made; beneficiation refers to the processing of mineral-bearing ores to produce concentrates or intermediate products, while value addition is the production of individual metals and fabrication of finished products, which have a much larger multiplier effect on the economy. Value addition therefore feeds-off and is subsequent to beneficiation/extraction; in other words, without beneficiation/extraction there can be no value addition.

Part III: Policy issues about mining in Africa

The last three chapters in Part III of the book discuss the contexts under which these minerals are, or can be, exploited: (10) mineral-driven Africa's industrialisation, (11) science and technology and (12) mineral economics.

Chapter 10 discusses the role that minerals can play in Africa's industrialisation. Observation is made on how the minerals from Africa have been successfully exploited by other countries for the benefit of their economies. Value addition is identified as an imperative for Africa's industrialisation. This issue leads to the discussion in Chapter 11 on the role of science and technology coupled to the need to the training of the critical mass of scientists, engineers, and the broad issue of the role of higher education for socio-economic development of the continent. Chapter 12 deals with mineral economics in the context of historical plunder of African minerals in colonial and neocolonial periods, dynamics of global trade, environmental issues, current corporate ownership of African minerals, taxations, investments and desirable governance policies. The purpose of these last three chapters is to create the conditions in which the advancement of Africa's industrialisation can take place notwithstanding the glaring underdevelopment of the continent in the context of current global challenges and opportunities.

The intended readership of the book is intentionally varied in an attempt to address the aspirations of stakeholders in and beyond the mining industry including, but not limited to: governments (who through their role as regulators, play a crucial role in enabling mining companies to maximise their contribution), investors, contractors and suppliers, service providers, mining-affected communities, civil society organisations, organised labour and academia and research institutions as well as downstream users. It is intended to inform the generality of the African public with policy and decision makers as a special target by providing basic information about what and how minerals can best contribute to the industrialisation of individual countries in Africa. Mineral economists can advise stakeholders on the trade and market opportunities for the mineral value-added products.

It is hoped that the book will be a useful reference material for mining undergraduate students on beneficiation and value addition of each of the minerals found in Africa. The book, while presenting a broad overview of beneficiation and value addition of Africa's minerals, provides crucial starting material for postgraduate research students and R&D institutions who wish to delve into more advanced methods of beneficiation/extraction and utilisation of mineral-derived materials that are in Africa for the purpose of industrialisation of the continent.

The overall desired contribution of the book is the promotion of full value addition of Africa's minerals, the production of mineral products that fetch premiums in their markets, create jobs, grow the economy and ultimately lead to better qualities of life for most people in Africa. The aspirations of the African citizens can only be met when the continent's minerals have been transformed locally into materials desired by the rest of the global society. This will be realised when Africa has transitioned from a resource-based economy into acknowledge-based economy. For Africa to transition from a resource-based economy to a knowledge-based economy demands significant investments in science and technology, at levels unprecedented in our continent's history.

Francis P. Gudyanga
2020

Acknowledgements

I am indebted to colleagues in academia, industry and government with whom I interacted in the course of my studies and working career for contributing so much to my further knowledge of metallurgy, materials science and engineering that is used in writing Part II of this book. I particularly wish to acknowledge the seminal contributions of Dr. Josephat Zimba, a past President of the African Minerals Research Society (AMRS), during the gestation of the book.

I am very grateful to the former President of the Republic of Zimbabwe, Cde. R.G. Mugabe, for appointing me (during the years 2007–2017) Permanent Secretary of the Government of Zimbabwe in the following three Ministries: Science and Technology Development; Mines and Mining Development; and Higher and Tertiary Education Science and Technology Development. Those appointments afforded me opportunities for gaining policy insights that crystallised into some of the ideas which have found expression in Part I and Part III of the book.

I am also grateful to Professor C.J. Chetsanga, Inaugural President of the Zimbabwe Academy of Sciences, Ambrose Made, a geoscientist and remote sensing expert, and Emeritus Professor Erik Navara, a colleague metallurgist at the University of Zimbabwe and SIRDC, for their valuable input and editorial corrections to the manuscript of the book. None of them, of course, is in anyway responsible for errors, misrepresentations and misinterpretations that may remain; these are mine alone.

I wish to thank Rukudzo Gudyanga for offering his valuable legal opinion on matters related to the production of this book. I similarly appreciate the assistance of Moses Chibharo, Takumurwa Gudyanga, Samson Katikiti and Jana Mindlová by providing IT solutions where and when they were necessary.

Last but not least, I wish to thank Mrs. Passion Tawasika, a principal executive assistant in the government of Zimbabwe, for her supportive secretarial services in the course of my research for the manuscript of the book.

About the author

Francis P. Gudyanga

Professor Francis Gudyanga has a BSc (Hons) degree in *Applied Chemistry* from Hatfield Polytechnic (now University of Hertfordshire) (UK), MSc in *Analytical Chemistry* from Chelsea College, University of London, MPhil in *Metallurgy* from Brunel University, Uxbridge (UK). He obtained a PhD in *Mineral Technology* from the Royal School of Mines, Imperial College of Science, Technology and Medicine, University of London, concurrently with a DIC in *Electrochemical Engineering*. He is a metallurgist and applied chemist with over 30 years' experience in the minerals industry covering plant operations, management, R&D at institutions in the UK, South Africa and Zimbabwe, teaching university undergraduates and supervision of doctoral research students. His research interests are in extractive metallurgy, materials technology, and S&T policy. He serves on several national, continental and international committees and boards. He was a member of the Executive Board of the International Council for Science (ICSU), 2002–2008. He held senior positions in the Government of Zimbabwe including serving as Permanent Secretary for ten years in three ministries, namely: Science and Technology Development, Mines and Mining Development, and Higher and Tertiary Education Science & Technology Development. Professor Gudyanga is a past (2016–2019) President of the Zimbabwe Academy of Sciences.

Acronyms

2IE	International Institute for Water and Environmental Engineering
AAS	African Academy of Sciences
AAS	atomic absorption spectroscopy
ACS	American Chemical Society
ADPA	Association of African Diamond Producing Countries
AIME	American Institute of Mining, Metallurgical, and Petroleum Engineers
AISTs	African Institutes of Science and Technology
AMASA	Annual Meetings of African Science Academies
AMD	acid mine drainage
AMRS	African Materials Research Society
AMV	African Mining Vision
APFO	ammonium pentadecafluoro-octanoate
APT	ammonium paratungstate
ASTM	American Society for Testing and Materials (now ASTM International)
ATL	amine-treated lignite
AU	African Union
AUC	African Union Commission
AUST	African University of Science and Technology
BFS	bitumen feedstock
BGS	British Geological Survey
BMR	base metal refinery
BTU	British Thermal Units
CBM	coalbed methane
CCA	chromated copper arsenate
CCM	caustic calcined magnesite
CE	circular economy
CEPR	Centre for Economic Policy Research
ICSU	International Council for Science
CIGS	copper indium gallium (di)selenide
CIMFR	Central Institute of Mining & Fuel Research
CSEM	Case Studies in Environmental Medicine
CSIRO	Council for Scientific and Industrial Research Organisation
CSIR	Council for Scientific and Industrial Research
CSR	coke strength after reaction
CSR	corporate social responsibility

CTL	coal to liquid
CUTS	Consumer Unity & Trust Society
CVD	chemical vapour deposition
CZT	cadmium zinc telluride
DBM	dead burned magnesite
DC	direct current
DCL/ICL	direct/indirect coal liquefaction
DIC	Diploma of Imperial College
DME	dimethyl aether
DSA	dimensionally stable anodes
DU	depleted uranium
EASAC	European Academies Science Advisory Council
ECA	Economic Commission for Africa
EEC	European Economic Community
Eh-pH diagrams	Pourbaix diagrams
EIA	environmental impact assessment
EoL	end-of-life
FIR (4IR)	Fourth Industrial Revolution (Industry 4.0)
FM	fused magnesia
FPDs	flat panel devices
GAN	gallium nitride
GDP	gross domestic product
GIA	Gemological Institute of America
GIS	Geographical Information System
GPS	global positioning system
GTAW	gas tungsten arc welding
GTL	gas to liquids
HAZ	heat affected zone
HCC	hard coking coal
HDI	Human Development Index
HPHT	high-pressure high-temperature
HREE	heavy rare earth elements
HSLA	high-strength, low-alloy
HSS	high-speed tool steels
IAC	InterAcademy Council
IAP	InterAcademy Partnership
ICMM	International Council on Mining and Metals
ICP	Inductively coupled plasma
ICs	integrated circuits
ICTs	information and communication technologies
IFC	International Finance Corporation
IFG	insulation fibreglass
IGCC	integrated gasification combined cycle
IHERD	Innovation, Higher Education and Research for Development
IJERA	International Journal of Engineering Research and Application
IPCS	International Programme on Chemical Safety
IR	infrared

ISAMPE	International Symposium on Advanced Mechanical and Power Engineering
ISG	International Study Group
ISO	International Standards Organisation
ITO	indium tin oxide
ITS	international temperature scale
IUGS	International Union of Geological Sciences
IUPAC	International Union of Pure and Applied Chemistry
KPCS	Kimberley Process Certification Scheme
LCD	liquid crystal diode
LCT	Li – Cs – Ta pegmatite
LD	lethal dose
LEDs	light-emitting diodes
LME	London Metals Exchange
LNG	liquefied natural gas
MARID	mica-amphibole-rutile-ilmenite-diopside
MG-Si	metallurgical grade silicon
MIC	microwave integrated circuits
MINTEKA	metallurgical R&D institute in South Africa
MIS	Management Information System
MMO	mixed-metal oxide
MOVPE	metalorganic vapour phase epitaxy
MSA	material system analysis
MTG	methanol to gasoline process
NASAC	Network of African Science Academies
NdYAG	neodymium yttrium aluminium garnet
NGL	natural gas liquids
NMAIST	Nelson Mandela African Institute of Science and Technology
NiMHB	nickel metal hydride batteries
NMI	The Nelson Mandela Institution
NIST	National Institute of Standards and Technology
OECD	Organisation for Economic Cooperation and Development
OEMs	original equipment manufacturers
OPEC	Organisation of petroleum exporting countries
PAMUST	Pan African Minerals University of Science and Technology
PCI	low pulverised coal
PFOA	pentadecafluoro-octanoic acid
PGE	platinum group elements
PGMs	platinum group metals
PHCC	premium hard coking coal
PMR	precious metal refinery
PPPs	public-private partnerships
PTFE	polytetrafluoroethylene
PV	photovoltaics
PVC	polyvinyl chloride
PWR	pressurised water reactor
R&D	Research and Development

RECs	Regional Economic Communities
REE	rare earth elements
REMs	rare earth metals
REOs	rare earth oxides
RF	radio-frequency
RFID	radio-frequency identification
ROI	return-on-investments
ROM	run of mine
RSC	Royal Society of Chemistry
S&E	Science and Engineering
SADC	Southern Africa Development Community
SAPS	Structural Adjustment Programmes
SAXS	small-angle X-ray scattering
SCEPS	stored chemical energy propulsion system
SCF	standard cubic feet
SCM	Supply Chain Management
SDGs	Sustainable Development Goals
SEM	scanning electron microscopy
SEP	standard electrode potential
SERF	spin-exchange relaxation-free
SET	Science, Engineering and Technology
SHE	standard hydrogen electrode
SHG	special high grade
SIRDC	Scientific and Industrial Research & Development Centre (Zimbabwe)
SNG	substitute natural gas
SOE	state owned enterprise
SOS	silicon on sapphire
SPA	super purity aluminium
SSCC	semi-soft coking coal
S&T	Science and Technology
STG+	syngas to gasoline plus process
STI	Science, Technology and Innovation
SVHC	substances of very high concern
SX/EW	solvent extraction and electrowinning
SYNGAS	synthesis gas (mixture of CO and H_2)
TEM	transmission electron microscopy
TFG	textile fibreglass
TLD	thermoluminescent radiation dosimetry
TMI	trimethylindium
TWAS	The World Academy of Sciences
UEMOA	Common Mining Policy and "Code Miniere Communautaire"
UG	under ground
UNCTD	United Nations Conference on Trade and Development
UNEP	United Nations Environmental Programme
UNFCCC	United Nations Framework Convention on Climate Change
UNIDO	United Nations Industrial and Development Organisation
USA or US	United States of America

USGS	U.S. Geological Survey
UV	ultraviolet
VAM	vacuum arc melting
VATs	value added taxes
VUV	vacuum ultraviolet
WD	World Bank
WEEE	water electrical and electronic equipment
XPD	X-ray photoelectron diffractometry
XRF	x-ray fluorescence
XRT	X-ray transmission
YAG	yttrium aluminium garnet, $Y_3Al_2(AlO_4)_3$,
YSZ	yttria-stabilized zirconia
ZAS	Zimbabwe Academy of Sciences

Minerals and their chemical formulae

abuyelite	Li_2CO_3
albite	$NaAlSi_3O_8$
allanite	$(Ca,Fe^2)_2(R,Al,Fe^3)_3Si_3O_{12}OH$; (R = rare elements)
alumina	Al_2O_3
amphibole	complex Mg,Fe silicate, $RSiO_3$ with R = Ca, Mg, or Fe
andalusite	Al_2SiO_5
anglesite	$PbSO_4$
anorthite	$CaAl_2Si_2O_8$
(apatite) chlorapatite	$Ca_{10}(PO_4)_6(Cl)_2$
(apatite) fluorapatite	$Ca_{10}(PO_4)_6(F)_2$
(apatite) hydroxylapatite	$Ca_{10}(PO_4)_6(OH)_2$
argentite	Ag_2S
arsenopyrite	$FeAsS$
autunite	$Ca(UO_2)_2(PO_4)_2 \cdot 10H_2O$
azurite	$Cu_3(CO_3)_2(OH)$ or $CuCO_3 \cdot Cu(OH)_2$
baryte or barite	$BaSO_4$
bauxite (alumina)	Al_2O_3
bertrandite	$BeOAl_2O_6SiO_2$
beryl	$Be_3Al_2(SiO_3)_6$ or $Be_4Si_2O_7(OH)_2$ or $Be_3Al_2Si_6O_{18}$.
bikitaite	$|Li(H_2O)|[AlSi_2O_6]$
bismite (bismutite)	$(BiO)_2CO_3$
bismuthite (bismuthinite)	Bi_2S_3
boehmite	γ-$AlO(OH)$
borax	$Na_2B_4O_7 \cdot 10H_2O$
braunite	$(Mn^{2+}Mn_6^{3+})(SiO_{12})$
brucite	$Mg(OH)_2$
cinnabar or vermillion	HgS
carnotite	$K_2(UO_2)_2(VO_4)_2 \cdot 3H_2O$
cassiterite	SnO_2
castorite	$LiAlSi_4O_{10}$
celestine	$SrSO_4$.
cerussite	$PbCO_3$

chalcocite	Cu_2S
chalcopyrite	$CuFeS_2$
chlorargyrite	$AgCl$
chromite	$FeCr_2O_4$
chrysoberyl	$BeAl_2O_4$
cinnabar (or vermillion)	HgS
clinoptilolite	$(Na,K)AlSi_5O_{111}H_2O$
colemanite	$[Ca_2B_6O_{11}.5H_2O$
columbite	$[(Fe,Mn,Mg)(Nb,Ta)_2O_6]$
covellite	CuS
cristobalite	$3Al_2O_3 \cdot 2SiO_2$
crocoite	$PbCrO_4$
cryolite	Na_3AlF_6
cuprite	Cu_2O
diaspore	$\alpha\text{-}AlO(OH)$
dolomite	$CaMg(CO_3)_2$
dravite	$NaMg_3(Al,Mg)_6B_3Si_6O_{27}(OH)$
epidote	$Ca_2(Al,Fe)_3Si_3O_{12}OH$
feldspars	$(KAlSi_3O_8 - NaAlSi_3O_8 - CaAl_2Si_2O_8)$
ferberite	$FeWO_4$
fluorspar (or fluorite)	CaF_2
galena	PbS
gallite	$CuGaS_2$
garnierite	$(Ni,Mg)_3Si_2O_5(OH)_4$
geikielite	$MgTiO_3$
germanite	GeS_2
gibbsite	$Al(OH)_3$
goethite	$FeO(OH)$
greenockite	CdS
gypsum	$CaSO_4 \cdot 2H_2O$
hafnia	HfO_2
haematite	Fe_2O_3
howlite	$Ca_2B_5SiO_9(OH)_5$
hübnerite	$MnWO_4$
hydroxylbastnasite	$NdCO_3OH$
ilmenite	$FeTiO_3$
johachidolite	$CaAlB_3O_7$
kaolinite (kaolin)	$Al_2Si_2O_5(OH)_4$ or $Al_2O_3 \cdot 2SiO_2 \cdot 2H_2O$
knorringite	$Mg_3Cr_2(SiO_4)_3$
kyabbnite	$Al_2O_3SiO_2$
kyanite (disthene, rhaeticite and cyanite).	$Al_2O_3 \cdot SiO_2$
lepidolite	$K(Li,Al,Rb)_3(Al,Si)_4O_{10}(F,OH)_2$
lewisite	$ClCH{=}CHAsCl_2,$
limestone	$CaCO_3$
limonite	$(Fe,Ni)O(OH)$
magnesia	MgO
magnesite	$MgCO_3$

magnetite	Fe_3O_4
malachite	$Cu_2(CO_3)(OH)_2$ or $CuCO_3.Cu(OH)_2$
metakaolin	$Al_2Si_2O_7$
millerite	NiS
molybdenite	MoS_2
mullite	$Al_2O_3.2SiO_2$
nahcolite	$NaHCO_3$
natron	$Na_2CO_3 \cdot 10H_2O$
nepheline syenite	$(Na_2O + K_2O)/SiO_2$ and $(Na_2O + K_2O)/Al_2O_3$
olivine	$[(Mg,Fe)_2SiO_4]$ with $Mg > Fe$
orpiment	As_2S_3
parasite	$Ca(Ce, La, Nd)_2(CO_3)_3F_2$
patronite	VS_4
periclase	MgO
patronite	VS_4
pentlandite	$(Ni,Fe)_9S_8$
perovskite	$CaTiO_3.$
petalite (castorite)	$LiAlSi_4O_{10}$
pitchblende (uraninite)	UO_2
pollucite	$(Cs,Na)_2Al_2Si_4O_{12} \cdot 2H_2O$
potash	KCl
potassium feldspar	(K-spar) $KAlSi_3O_8$
powellite	$CaMoO_4$
psilomelane	$Ba,H_2O)_2Mn_5O_{10}$
pyrargyrite	Ag_3SbS_3
pyrite (marchasite)	FeS_2
pyrochlore	$[(Ca,Na,Ce)(Nb,Ti,Ta)_2(O,OH,F)_7]$
pyrolusite	$MnO2$
pyrope	$Mg_3Al_2(SiO_4)_3$
pyrophanite	$MnTiO_3$
pyroxene	$RSiO_3$ (R = Ca, Mg, Fe, Al, etc.)
pyrrhotite	FeS
quartz	SiO_2
realgar	As_4S_4
rhodochrosite	$MnCO_3$
rutile	TiO_2
salt	$NaCl$
scheelite	$CaWO_4$
serpentine	$(Mg,Fe)_3Si_2O_5(OH)_4$
silica	SiO_2
soda	ash Na_2CO_3
sphalerite	$(Zn,Fe)S$
spodumene	$LiAl(SiO_3)_2$
steatite	$(MgO)_3(SiO_2)_4.$
stibnite	Sb_2S_3
suanite	$Mg_2B_2O_5$
tanzanite	$Ca_2Al_3(SiO_4)_3(OH)$

talc	$H_2Mg_3(SiO_3)_4$ or $Mg_3Si_4O_{10}(OH)_2$
tantalite	$[(Fe,Mn)Ta_2O_6]$
thermonatrite	$NaHCO_3$
thorianite	ThO_2
thorite	$ThSiO_4$
titania	TiO_2.
topaz	$Al_{2Si}O_4(F,OH)_2$
torbernite	$Cu(UO_2)_2(PO_4)_2.8H_2O$
tourmaline	$NaMg_3(Al,Mg)_6B_3Si_6O_{27}(OH)$
uraninite (pitchblende)	UO_2
uranophane	$Ca(UO_2)2Si_2O_7 \cdot 6H_2O$
vanadinite	$Pb_5(VO_4)_3Cl$
vermillion (cinnabar)	HgS
wolframite	$(Fe,Mn)WO_4$
wollastonite	$CaSiO_3$
wulfenite	$PbMoO_4$
yttrium aluminium garnet	$(YAG)\ Y_3Al_2(AlO_4)_3$
zirconia	$ZrSiO_4$

Figures

Tables

Part I

Information about minerals in Africa

Chapter 1

Africa's mineral resources

1.1 Introduction

The abundant mineral ore resources in Africa [1–4] constitute a major input into the industrial sector throughout the world. They therefore have a great potential of contributing significantly to the industrialisation of many African countries, most of which are blessed with these vast mineral resources. These mineral resources have been critical ingredients to the industrialisation of countries other than those on the continent itself [3,5–7]. Unfortunately, even with this abundance of mineral resources, citizens in most – if not all – countries in the continent are still poor, prompting the African leaders and development partners to meet on many occasions [8–14] seeking to adopt and embrace appropriate strategies that can add value to mineral resources and effectively improve the lives of citizens in the African countries. The African Mining Vision of 2009 [5] read together with the International Study Group Report on Africa's Mineral Regimes [3] is indicative of the seriousness with which African leaders view the mining sector as a potential driving engine for the continent's industrialisation and development [5,15] through mineral beneficiation and value addition.

In Tables 1.1 and 1.2, this book lists the known mineral deposits in the various African countries [1–3,6]. This listing is not exhaustive as more minerals are being discovered with ongoing exploration activities. Besides, the lists are also only within the bounds of the information that was accessed at the time of writing. In the subsequent eight chapters, current technical processes for the mineral beneficiation and value addition for about 120 minerals [16] estimated to be in economic quantities [3] in different African countries are concisely described. For each mineral a brief *overview* covering its origins and geological occurrences is given. This is followed by a description of its *beneficiation*, outlining the essential chemical reactions involved and the conditions under which these reactions take place. Lastly the use of each mineral and/or material(s) derived from it is discussed in terms of materials science and materials technology, collectively termed *value addition*. Attempt has been made in the last three chapters of the book to situate these beneficiation and value addition processes in the contexts of the role that is played by African mineral-derived materials [4,7,17–21] in the global economy, developments in science and technology [22–33], and relevant issues in mineral economics [35–38].

This book draws attention to the prospect of transforming Africa's comparative advantages in minerals into the continent's competitive edge regarding materials. Mineral beneficiation and value addition form the basis, and provide opportunities, for the mineral-driven industrialisation of Africa.

1.2 Occurrences of minerals in Africa

Mining in Africa dates back to ancient times as demonstrated by old mineral workings found in various parts of the continent such as Ngwenya Mine in Eswatini [1247]. This mine is considered to be the world's oldest, having been worked on 41,000 to 43,000 years ago for the extraction of red ochre from haematite (Fe_2O_3) ore deposits. These ancient workings have often been sites of major mines especially for gold, copper and iron ore. The systematic exploration and documentation of the geosocial locations of mineral deposits in African countries [1–3,6] which started with the European colonial settlers is still low, with large areas still unmapped. Many African countries have not geologically mapped their territories sufficiently and therefore have no full economic evaluation of the minerals and appreciation of their mineral resources. It is from these and other varied sources that Tables 1.1 and 1.2 were compiled. This information was substantially valid at the time of writing, having been sourced from and crosschecked with updated databases. Table 1.1 lists the minerals/metals found in each African country while Table 1.2 lists the African countries where a specific mineral/metal is found.

From Table 1.1 each African country is able to have holistic information about its mineral wealth. This is valuable for economic planning for the country and for marketing to potential investors.

Table 1.2, which lists all the African countries in which a specific mineral is found, is useful to suppliers and consumers who would be able to target specific countries for a particular mineral of interest. The table provides information that forms the basis for collaboration and joint initiatives between neighbouring countries in research for beneficiation and value addition. The groupings of countries that produce the same minerals constitute potential memberships of African mineral cartels for negotiating and determining favourable marketing and price conditions for those minerals.

Underground, the mineral deposits may be found to stretch across territorial boundaries that are recognised on the surface. As such, a mineral found in one country is likely to be present in a neighbouring country. Such cases can be confirmed by inspecting the map of Africa, Figure 0.1 in conjunction with Tables 1.1 and 1.2 which list the minerals in each African country and the African countries in which specific minerals are found, respectively. This is of geological and economic interest to neighbouring countries. The presence of a mineral in one country is a pointer to the likely possibility of that mineral being present in the neighbouring country.

Table 1.1 Lists of minerals in African countries

Country (former colonial power(s)) (countries it shares land border with)	Minerals of the country
Algeria (France) (Libya, Mali, Mauritania, Morocco, Niger, Tunisia)	barite, bentonite, clays, crushed stone, gold, gravel, gypsum, helium, iron, lead, limestone, marble, nitrogen fertiliser, petroleum (oil), phosphate rock, phosphates, pozzolan, quartz, salt, sand, silver, uranium, zinc
Angola (Portugal) (DR Congo, Namibia, Zambia)	asphalt, bauxite, copper, diamonds, feldspar, fluorite, gold, granite, gypsum, iron, kaolin, lead, manganese, marble, mica, petroleum (oil), phosphates, quartz, silver, sulphur, talc, tin, uranium, wolframite, zinc
Benin (France) (Burkina Faso, Niger, Nigeria, Togo)	clay, gold, iron, limestone, marble, petroleum (oil), phosphate, silica sand, titanium
Botswana (Britain) (Namibia, South Africa, Zimbabwe)	coal, copper, diamonds, gold, iron, nickel, potash, salt, silver, soda ash
Burkina Faso (France) (Benin, Ghana, Ivory Coast, Mali, Niger, Togo)	bauxite, copper, gold, lead, limestone, manganese, marble, nickel, phosphates, pumice, salt, zinc
Burundi (Germany, Belgium) (DR Congo, Rwanda, Tanzania)	cobalt, copper, dolomites, gold, iron, kaolin, limestone, marble, nickel, niobium, peat, phosphates, platinum group metals, rare earth elements, tantalum, tin, tungsten, uranium, vanadium
Cameroon (Germany, France, Britain) (Chad, Congo, Central African Republic, Equatorial Guinea, Gabon, Nigeria)	bauxite, cobalt, gold, granite, iron, nepheline syenite, nickel, petroleum (oil), rutile
Cape Verde (Portugal)	basalt rock, clay, gypsum, iron, kaolin, limestone, pozzolana, salt
Central African Republic (France) (Cameroon, Chad, Congo, DR Congo, South Sudan, Sudan)	copper, diamonds, gold, graphite, ilmenite, iron, kaolin, kyanite, lignite, limestone, manganese, monazite, petroleum (oil), quartz, rutile, salt, tin, uranium
Chad (France) (Cameroon, Central African Republic, Libya, Niger, Nigeria, Sudan)	gold, gravel, kaolin, limestone, natron, petroleum (oil), salt, sand, soda ash, uranium
Comoros (France)	clay, crushed stone, gravel, sand
Congo (France) (Central African Republic, DR Congo, Gabon)	copper, diamonds, gold, iron, lead, magnesium, petroleum (oil), phosphates, potash, uranium, zinc
DR Congo (Belgium) (sharing land border with Angola, Burundi, Central African Republic, Congo, South Sudan, Rwanda, Tanzania, Uganda, Zambia)	coal, cobalt, copper, diamonds, gold, manganese, niobium, petroleum (oil), silver, tantalum, tin, tungsten, uranium, zinc
Djibouti (France) (Ethiopia, Eritrea, Sudan)	clay, diatomite, gold, granite, gypsum, limestone, marble, perlite, petroleum (oil), pumice, salt
Egypt (Britain, France) (Libya, Sudan)	aluminium, asbestos, barite, basalt, bentonite, coal, coke, dolomite, feldspar, fluorspar, gold, granite, gravel, gypsum, ilmenite, iron, iron, kaolin, lead, limestone, manganese, marble, petroleum (oil), phosphates, quartz, rare earth elements, salt, sand, sandstone, talc, zinc

(Continued)

Table 1.1 (Continued)

Country (former colonial power(s)) (countries it shares land border with)	Minerals of the country
Equatorial Guinea (Spain) (Cameroon, Gabon)	bauxite, clay, columbite, diamonds, gold, gravel, iron, molybdenite, ornamental stone, petroleum (oil), sand, tantalum, tin, tungsten
Eritrea (Italy) (Djibouti, Ethiopia, Sudan)	ceramics, copper, gold, granite, gypsum, iron, limestone, marble, natural gas, petroleum (oil), potash, salt, zinc
Eswatini (Britain) (Mozambique, South Africa)	arsenic, asbestos, clay, coal, copper, diamonds, gold, kaolin, manganese, nickel, quarry stone, silica, talc, tin
Ethiopia (occupied by Italy 1935–1941) (Djibouti, Eritrea, Kenya, Somalia, South Sudan, Sudan)	coal, copper, crushed stone, dimension stone, gemstones, gold, iron, limestone, manganese, marble, molybdenum, natural gas, nickel, niobium, phosphorus, platinum, potash, salt, soda ash, tantalum
Gabon (France) (Cameroon, Congo, Equatorial Guinea)	diamonds, gold, iron, lead, manganese, marble, niobium, petroleum (oil), phosphate, potash, uranium, zinc
Gambia (Britain) (Senegal)	clay, Ilmenite, rutile, silica, tin, titanium, zircon
Ghana (Britain) (Burkina Faso, Ivory Coast, Togo)	bauxite, diamonds, gold, limestone, manganese, natural gas, petroleum, salt, petroleum (oil), silver
Guinea (France) (Guinea-Bissau, Ivory Coast, Mali, Liberia, Senegal)	bauxite, diamonds, gold, iron, petroleum (oil), uranium
Guinea-Bissau (Portugal) (Guinea, Senegal)	bauxite, clays, diamond, gold, granite, gravel, limestone, petroleum (oil), phosphate rock, sand
Ivory Coast (France) (Burkina Faso, Ghana, Guinea, Liberia, Mali)	bauxite, cobalt, colombo-tantalite, copper, diamonds, gold, ilmenite, iron, manganese, nickel, petroleum (oil)
Kenya (Britain) (Ethiopia, Somalia, South Sudan, Tanzania, Uganda)	diatomite, fluorspar, gemstones, gemstones, petroleum (oil), gold, gypsum, iron, kyanite, lead, limestone, manganese, silica, soda ash, titanium, vermiculite, zinc
Lesotho (Britain) (South Africa)	building stone, clay, diamonds, dimension stone, gravel, uranium, agate, sand, sand
Liberia (France) (Guinea, Ivory Coast, Sierra Leone)	bauxite, beryl, chromite, columbite-tantalite, copper, diamonds, gold, ilmenite, iron, kyanite, lead, molybdenum, monazite, nickel, petroleum (oil), phosphates, rare earth elements, rutile, silica, tin, uranium, zinc, zircon
Libya (Italy) (Algeria, Chad, Egypt, Niger, Sudan, Tunisia)	diamonds gold, gypsum, iron, iron, lime, limestone, magnesium, magnetite, methanol, natural gas, petroleum (oil), phosphate rock, potassium, salt, sulphur
Madagascar (France)	bauxite, chromite, coal, cobalt, graphite, ilmenite, mica, nickel, petroleum (oil), quartz, rare earth elements, rutile, semi-precious stones, tar sands, vanadium, zircon

Malawi (Britain) *(Mozambique, Tanzania, Zambia)*
bauxite, coal, copper, diamonds, limestone, niobium, petroleum (oil), phosphate, rubies, sapphires, uranium

Mali (France) *(Algeria, Burkina Faso, Guinea, Ivory Coast, Mauritania, Niger, Senegal)*
amethyst, epidote, garnet, gold, granite, gypsum, kaolin, limestone, phosphates, prehnite, quartzs, uranium

Mauritania (France) *(Algeria, Mali, Senegal, Western Sahara)*
copper, diamonds, gold, gypsum, iron, limestone, petroleum (oil), phosphate, quartz, salt

Mauritius (France)
basalt, lime from coral, salt

Morocco (Spain) *(Algeria, Western Sahara)*
anthracite, antimony, barite, cobalt, copper, fluorspar, gold, iron, lead, lead, limestone, manganese, phosphates, salt, silver, tin, titanium, tungsten, zinc

Mozambique (Portugal) *(Eswatini (Swaziland), Malawi, South Africa, Tanzania, Zambia, Zimbabwe)*
aluminium, clays, coal, gemstones, gold, graphite, iron, limestone, natural gas, niobium, petroleum (oil), tantalum, titanium, uranium, zirconium

Namibia (Germany, South Africa) *(Angola, Botswana, South Africa, Zambia)*
cadmium, copper, diamonds, fluorspar, gold, lead, limestone, lithium, manganese, petroleum (oil), silver, tin, tungsten, uranium, wollastonite, zinc

Niger (France) *(Benin, Burkina Faso, Chad, Libya, Mali, Nigeria)*
coal, gold, gypsum, iron, limestone, molybdenum, petroleum (oil), phosphates, salt, silver, tin, uranium

Nigeria (Britain) *(Benin, Cameroon, Chad, Niger)*
bitumen, coal, coking coal, gold, iron, lead, lignite, limestone, niobium, petroleum (oil), tantalite, tin, uranium, wolframite, zinc

Rwanda (Germany, Belgium) *(Burundi, DR Congo, Tanzania, Uganda)*
amblygonite, beryl, coltan (niobo-tantalite), gold, monazite, tin, tungsten

Sao Tome and Principe (Portugal)
clay, limestone, natural gas, petroleum (oil), rare earths minerals, volcanic rock

Senegal (France) *(Gambia, Guinea, Guinea-Bissau, Mali, Mauritania)*
barite, chromium, copper, decorative stones, gold, iron, limestone, nickel, petroleum (oil), phosphates, platinum group metals, salts, titanium, zircon

Seychelles (Britain)
clay, coral, sand, stone

Sierra Leone (Britain) *(Guinea, Liberia)*
bauxite, diamonds, gold, iron, limonite, rutile

Somalia (Italy) *(Djibouti, Ethiopia, Kenya)*
feldspar, gemstones, gypsum, iron, kaolin, limestone, natural gas, petroleum (oil), quartz, salt, silica, tantalum, tin, uranium

South Africa (Britain) *(Botswana, Eswatini (Swaziland), Lesotho, Mozambique, Namibia, Zimbabwe)*
antimony, chromium, coal, copper, diamonds, gold, iron, lead, manganese, nickel, phosphates, platinum group metals, rare earth elements, silver, tin, uranium, vanadium, vermiculite, zinc

South Sudan (Britain) *(Central African Republic, DR Congo, Ethiopia, Kenya, Sudan, Uganda)*
asbestos, chrome, clay, copper, diamonds, dolomite, gold, iron, kaolin, lead, limestone, manganese, marble, mica, petroleum (oil), silver, tin, tungsten, zinc

Sudan (Britain) *(Chad, Central African Republic, Egypt, Eritrea, Ethiopia, Libya, South Sudan)*
chrome, copper, gold, gypsum, iron, limestone, mica, petroleum (oil), salt, silver, tungsten, zinc

(Continued)

Table 1.1 (Continued)

Country (former colonial power(s)) (countries it shares land border with)	Minerals of the country
Tanzania (Germany, Britain) (Burundi, DR Congo, Kenya, Malawi, Mozambique, Rwanda, Uganda, Zambia)	alexandrite, bentonite, coal, copper, diamonds, diatomite, emerald, garnet, gemstones, gold, gypsum, iron, kaolin, lime, nickel, phosphates, platinum, ruby, salt, sapphire, silver, tanzanite, tin, vermiculite
Togo (Germany, France) (Benin, Burkina Faso, Ghana)	bauxite, clinker, diamonds, gold, gypsum, iron, limestone, manganese, marble, phosphate, rutile, zinc
Tunisia (France) (Algeria, Libya)	iron, lead, petroleum (oil), phosphates, salt, zinc
Uganda (Britain) (DR Congo, Kenya, Rwanda, South Sudan, Tanzania)	beryl, chrome, clays, cobalt, copper, diamonds, diatomite, dimension stones, feldspar, gold, graphite, gypsum, iron, kaolin, kyanite, lead, limestone, lithium, marble, nickel, niobium, petroleum (oil), phosphates, platinum group metals, rare earth elements, salt, sand, tantalite, tungsten, uranium, vermiculite, zinc
Western Sahara (Spain) (Mauritania, Morocco)	iron, natural gas, phosphates, uranium
Zambia (Britain) (Angola, Botswana, DR Congo, Malawi, Mozambique, Namibia, Tanzania, Zimbabwe)	amethyst, aquamarine, coal, cobalt, copper, emeralds, gold, lead, silver, tourmaline, uranium, zinc
Zimbabwe (Britain) (Botswana, Mozambique, South Africa, Zambia)	agate, aluminium, amethyst, antimony, apatite, aquamarine, arsenic, asbestos, barytes, bentonite, beryl, bismuth, borates, caesium, chrome, clay, coal bed methane, coal, cobalt, copper, cordierite, corundum, diamonds, diatomite, dolomite, emeralds, feldspar, fluorite, garnet, gold, granite, graphite, gypsum, howlite, iron, jade, kaolin, kyanite, lead, limestone, limestone, lithium, magnesite, magnetite, mercury, mica, molybdenum, monazite, mtorolite, nickel, niobium, perlite, phosphate, platinum group metals, pyrite, pyrophyllite, quartz, rare earth elements, ruby, sapphire, selenium, silica, sillimanite, silver, talc, tantalite, tin, titanium, topaz, tungsten, uranium, vanadium, vermiculite, zinc, zirconium

Table 1.2 Lists of African countries in which specific minerals are found

Agate	Botswana, Lesotho, Madagascar, Malawi, Zimbabwe
Aquamarine	Benin, Madagascar, Malawi, Mozambique, Namibia, Nigeria, Zambia, Zimbabwe
Aluminium (Al)	Angola, Burkina Faso, Cameroon, Cape Verde, Chad, Congo, Egypt, Equatorial Guinea, Ghana, Guinea, Guinea-Bissau, Ivory Coast, Liberia, Libya, Madagascar, Malawi, Mali, Mozambique, Sierra Leone, Somalia, Togo, Tunisia, Zimbabwe
Antimony (Sb)	Madagascar, Morocco, Namibia, South Africa, Western Sahara, Zimbabwe
Apatite	Malawi, Zimbabwe
Arsenic (As)	Eswatini, Western Sahara, Zimbabwe
Asbestos	Botswana, Egypt, Eritrea, Eswatini, Kenya, South Sudan, Zimbabwe
Asphalt	Angola, Egypt, Madagascar
Baryte (Barite) $(BaSO_4)$	Algeria, Egypt, Eritrea, Ghana, Kenya, Morocco, Niger, Togo, Western Sahara, Zimbabwe
Basalt	Cape Verde, Egypt, Ghana, Mauritius
Bentonite	Algeria, Egypt, Malawi, Mozambique, Tanzania, Togo, Western Sahara, Zimbabwe
Beryl $(Be_3Al_2(SiO_3)_6$ or $Be_4Si_2O_7(OH)_2$ or $Be_3Al_2Si_6O_{18}$)	Angola, Egypt, Liberia, Libya, Madagascar, Mozambique, Nigeria, Rwanda, Uganda, Zimbabwe
Beryllium (Be)	Angola, Mozambique
Bismuth (Bi)	Zimbabwe
Borates	Zimbabwe
Cadmium (Cd)	Congo, Namibia, Togo
Caesium (Cs)	Zimbabwe
Chromium (Cr)	Botswana, Mali, Senegal, South Africa, South Sudan, Sudan, Zimbabwe
Chrysoberyl $(BeAl_2O_4)$	Madagascar
Clays	Algeria, Angola, Benin, Botswana, Cameroon, Cape Verde, Chad, Comoros, Djibouti, Egypt, Equatorial Guinea, Eswatini, Gambia, Ghana, Guinea-Bissau, Kenya, Lesotho, Morocco, Mozambique, Rwanda, Sao Tome and Principe, Seychelles, South Sudan, Togo, Tunisia, Uganda, Western Sahara, Zimbabwe
Coal	Botswana, Congo, DR Congo, Egypt, Eswatini, Ethiopia, Gabon, Kenya, Madagascar, Malawi, Morocco, Mozambique, Namibia, Niger, Nigeria, South Africa, Tanzania, Uganda, Zambia, Zimbabwe
Coal Bed Methane (CBM)	Zimbabwe
Cobalt (Co)	Botswana, Burundi, Cameroon, Congo, DR Congo, Ivory Coast, Liberia, Libya, Madagascar, Morocco, Niger, Tanzania, Uganda, Western Sahara, Zambia, Zimbabwe
Copper (Cu)	Angola, Botswana, Burkina Faso, Burundi, Central African Republic, Congo, DR Congo, Egypt, Eritrea, Eswatini, Ethiopia, Ivory Coast, Kenya, Liberia, Libya, Madagascar, Malawi, Mali, Mauritania, Morocco, Mozambique, Namibia, Niger, Senegal, Somalia, South Africa, South Sudan, Sudan, Tanzania, Uganda, Western Sahara, Zambia, Zimbabwe

(Continued)

Table 1.2 (Continued)

Cordierite	Madagascar, Tanzania, Zimbabwe
Corundum (Al_2O_3)	Kenya, Malawi, Zimbabwe
Diamond (C)	Angola, Benin, Botswana, Burkina Faso, Cameroon, Central African Republic, Congo, DR Congo, Equatorial Guinea, Eswatini, Ethiopia, Gabon, Ghana, Guinea, Guinea-Bissau, Ivory Coast, Lesotho, Liberia, Libya, Malawi, Mali, Mauritania, Namibia, Niger, Sierra Leone, South Africa, South Sudan, Sudan, Tanzania, Togo, Tunisia, Uganda, Zimbabwe
Diatomite	Djibouti, DR Congo, Ethiopia, Kenya, Mozambique, Rwanda, Tanzania, Uganda, Zimbabwe
Dimension Stone	Ethiopia, Kenya, Lesotho, Togo, Uganda
Dolomite $(CaMg(CO_3)_2)$	Burkina Faso, Burundi, Egypt, Ethiopia, Ghana, Madagascar, Malawi, South Sudan, Sudan, Togo, Zimbabwe
Emerald	Madagascar, Tanzania, Zambia, Zimbabwe
Epidote $(Ca_2Al_2(Fe^{3+},Al)(SiO_4)(Si_2O_7)O(OH))$	Mali
Feldspar $(KAlSi_3O_8 - NaAlSi_3O_8 - CaAl_2Si_2O_8)$	Angola, Botswana, Burundi, Egypt, Eritrea, Ethiopia, Ghana, Somalia, Uganda, Western, Sahara, Zimbabwe
Fluorspar (CaF_2)	Angola, Egypt, Kenya, Morocco, Mozambique, Namibia
Gallium (Ga)	Zimbabwe
Garnet	Kenya, Madagascar, Malawi, Mali, Mozambique, Namibia, Tanzania, Togo, Zimbabwe
Germanium (Ge)	Congo
Gold (Au)	Algeria, Angola, Benin, Botswana, Burkina Faso, Burundi, Cameroon, Central African Republic, Chad, Congo, Djibouti, DR Congo, Egypt, Equatorial Guinea, Eritrea, Eswatini, Ethiopia, Gabon, Ghana, Guinea-Bissau, Ivory Coast, Kenya, Liberia, Libya, Madagascar, Malawi, Mali, Mauritania, Morocco, Mozambique, Namibia, Niger, Nigeria, Rwanda, Senegal, Sierra Leone, Somalia, South Africa, South Sudan, Sudan, Tanzania, Togo, Uganda, Western Sahara, Zambia, Zimbabwe
Granite	Angola, Burkina Faso, Cameroon, Djibouti, Egypt, Eritrea, Ghana, Guinea-Bissau, Liberia, Libya, Mali, Mozambique, Somalia, Zimbabwe
Graphite	Botswana, Central African Republic, Guinea, Kenya, Madagascar, Malawi, Mozambique, Niger, Uganda, Zimbabwe
Gypsum $(CaSO_4 \cdot 2H_2O)$	Algeria, Angola, Botswana, Cape Verde, Djibouti, Egypt, Eritrea, Ghana, Kenya, Libya, Mali, Mauritania, Niger, Rwanda, Somalia, Sudan, Tanzania, Togo, Tunisia, Uganda, Zimbabwe
Hafnium (Hf)	Zimbabwe
Howlite $(Ca_2B_5SiO_9(OH)_5)$	Zimbabwe
Indium (In)	Zimbabwe
Iron (Fe)	Algeria, Angola, Benin, Botswana, Burundi, Cameroon, Cape Verde, Central African Republic, Congo, Egypt, Equatorial Guinea, Eritrea, Eswatini, Ethiopia, Gabon, Ghana, Guinea, Ivory Coast, Kenya, Liberia, Libya, Madagascar, Malawi, Mali, Mauritania, Morocco, Mozambique, Niger, Nigeria, Senegal, Sierra Leone, Somalia, South Africa, South Sudan, Sudan, Tanzania, Togo, Tunisia, Uganda, Western Sahara, Zambia, Zimbabwe

Jade — Zimbabwe

Kaolin (Kaolinite) $(Al_2Si_2O_5(OH)_4$ or $Al_2O_3 \cdot 2SiO_2 \cdot 2H_2O)$ — Algeria, Angola, Burundi, Cape Verde, Central African Republic, Chad, Egypt, Eswatini, Ethiopia, Ghana, Malawi, Mali, Rwanda, Somalia, South Sudan, Sudan, Tanzania, Togo, Uganda, Zimbabwe

Kyanite $(Al_2O_3 \cdot SiO_2)$ — Central African Republic, Kenya, Liberia, Libya, Malawi, Togo, Uganda, Zimbabwe

Lead (Pb) — Algeria, Angola, Burkina Faso, Congo, Egypt, Gabon, Kenya, Liberia, Libya, Mali, Morocco, Namibia, Nigeria, South Africa, South Sudan, Sudan, Tunisia, Uganda, Western Sahara, Zambia, Zimbabwe

Lithium (Li) — DR Congo, Mali, Namibia, Niger, Uganda, Zimbabwe

Magnesite $(MgCO_3)$ — Ghana, Zimbabwe

Magnesium (Mg) — Congo, Libya

Magnetite (Fe_3O_4) — Libya, Zimbabwe

Manganese (Mn) — Angola, Botswana, Burkina Faso, Central African Republic, Congo, DR Congo, Egypt, Eswatini, Ethiopia, Gabon, Ghana, Guinea, Ivory Coast, Kenya, Liberia, Libya, Madagascar, Mali, Morocco, Namibia, South Africa, South Sudan, Sudan, Togo, Western Sahara, Zambia, Zimbabwe

Marble — Algeria, Angola, Benin, Burkina Faso, Burundi, Djibouti, Egypt, Eritrea, Ethiopia, Gabon, Ghana, Kenya, Madagascar, Malawi, Mali, Mozambique, Somalia, South Sudan, Sudan, Togo, Uganda

Mercury (Hg) — Algeria, Western Sahara, Zimbabwe

Mica — Angola, Ghana, Madagascar, South Sudan, Sudan, Zimbabwe

Molybdenum (Mo) — Ethiopia, Liberia, Libya, Niger

Mtorolite (Chrome Chalcedony) — Zimbabwe

Natron $(Na_2CO_3 \cdot 10H_2O)$ — Chad

Natural Gas — Algeria, Angola, Cameroon, Congo, Egypt, Equatorial Guinea, Eritrea, Ethiopia, Gabon, Ghana, Ivory Coast, Kenya, Libya, Morocco, Mozambique, Namibia, Nigeria, Tao Tome and Principe, Senegal, Somalia, South Africa, Tanzania, Tunisia, Uganda, Western Sahara.

Nepheline Syenite $((Na_2O+K_2O)/SiO_2$ and $(Na_2O+K_2O)/Al_2O_3)$ — Cameroon

Nickel (Ni) — Angola, Botswana, Burkina Faso, Burundi, Cameroon, Central African Republic, Eswatini, Ethiopia, Guinea, Ivory Coast, Kenya, Liberia, Libya, Madagascar, Malawi, Mali, Niger, Senegal, South Africa, South Sudan, Sudan, Tanzania, Uganda, Western Sahara, Zambia, Zimbabwe

Niobium (Nb) — Burundi, DR Congo, Equatorial Guinea, Ethiopia, Gabon, Ivory Coast, Kenya, Liberia, Libya, Madagascar, Malawi, Mali, Mozambique, Niger, Nigeria, Rwanda, Somalia, Tanzania, Uganda, Zimbabwe

Opal — Zimbabwe

Pearl — Zimbabwe

Peat — Angola, Burundi, Togo

Pegmatite — Ethiopia, Madagascar, Nigeria

Petroleum (Oil) — Algeria, Angola, Benin, Cameroon, Central African Republic, Chad, Congo, Djibouti, DR Congo, Egypt, Equatorial Guinea, Eritrea, Gabon, Ghana, Guinea, Guinea-Bissau, Ivory Coast, Kenya, Liberia, Libya, Madagascar, Malawi, Mauritania, Morocco, Mozambique, Namibia, Niger, Nigeria, Sao Tome and Principe, Senegal, Somalia, South Sudan, Sudan, Tunisia, Uganda

(Continued)

Table 1.2 (Continued)

Phosphate   Algeria, Angola, Benin, Burkina Faso, Burundi, Congo, Egypt, Gabon, Ghana, Guinea-Bissau, Liberia, Libya, Malawi, Mali, Mauritania, Mauritius, Morocco, Mozambique, Niger, Senegal, South Africa, Tanzania, Togo, Tunisia, Uganda, Western Sahara, Zimbabwe

Platinum Group Metals (PMGs)   Burundi, Ethiopia, Madagascar, Mali, Namibia, Niger, Senegal, South Africa, Tanzania, Uganda, Zimbabwe

Potash (KCl)   Botswana, Congo, Eritrea, Ethiopia, Gabon,

Pozzolan   Algeria, Cameroon, Cape Verde, Rwanda, Uganda

Pumice   Burkina Faso, Djibouti

Quartz (SiO_2)   Algeria, Angola, Burundi, Central African Republic, Chad, Egypt, Ethiopia, Gambia, Madagascar, Malawi, Mali, Mauritania, Somalia, Zimbabwe

Rare Earth Elements (REE)   Angola, Burundi, Central African Republic, Egypt, Kenya, Liberia, Libya, Madagascar, Malawi, Mali, Mozambique, Namibia, Rwanda, Sao Tome and Principe, South Africa, South Sudan, Sudan, Tanzania, Zambia, Zimbabwe

Rubidium (Ru)   Zimbabwe

Ruby   Kenya, Madagascar, Malawi, Mozambique, Tanzania, Zimbabwe

Salt (NaCl)   Algeria, Angola, Botswana, Burkina Faso, Cape Verde, Central African Republic, Chad, Djibouti, Egypt, Eritrea, Ethiopia, Ghana, Guinea, Kenya, Libya, Madagascar, Mali, Mauritania, Mauritius, Morocco, Mozambique, Niger, Senegal, Somalia, South Sudan, Sudan, Tanzania, Tunisia, Uganda

Sapphire   Cameroon, Ethiopia, Madagascar, Malawi, Tanzania, Zimbabwe

Selenium (Se)   Zimbabwe

Serpentine ((Mg, $Fe)_3Si_2O_5(OH)_4$)   Zimbabwe

Silica (SiO_2)   Benin, Egypt, Eswatini, Gambia, Kenya, Liberia, Libya, Madagascar, Rwanda, Somalia, Zimbabwe

Silver (Ag)   Algeria, Angola, Botswana, Congo, DR Congo, Eritrea, Ghana, Kenya, Madagascar, Mali, Morocco, Namibia, Niger, South Africa, South Sudan, Sudan, Tanzania, Uganda. Western, Sahara, Zambia, Zimbabwe

Slate   Ghana, Zimbabwe

Soapstone   Zimbabwe

Soda Ash (Na_2CO_3)   Botswana, Chad, Egypt, Ethiopia, Gabon, Kenya

Talc $H_2Mg_3(SiO_3)_4$ or $Mg_3Si_4O_{10}(OH)_2$ Angola, Egypt, Eritrea, Eswatini, Ghana, Kenya, Mali, Morocco, Niger, Rwanda, Zimbabwe

Tantalum (Ta)   Benin, Burundi, DR Congo, Egypt, Equatorial Guinea, Ethiopia, Ivory Coast, Liberia, Libya, Madagascar, Malawi, Mozambique, Niger, Nigeria, Rwanda, Somalia, Tanzania, Uganda, Zimbabwe

Tanzanite $Ca_2Al_3(SiO_4)_3(OH)$ Tanzania

Tin (Sn)
Angola, Benin, Burundi, Cameroon, Central African Republic, Chad, Congo, DR Congo, Egypt, Equatorial Guinea, Eswatini, Ethiopia, Gambia, Liberia, Libya, Madagascar, Mali, Morocco, Namibia, Niger, Nigeria, Rwanda, Somalia, South Africa, South Sudan, Tanzania, Uganda, Zimbabwe

Titanium (Ti)
Benin, Cameroon, Central African Republic, Egypt, Gambia, Ivory Coast, Kenya, Liberia, Libya, Madagascar, Malawi, Mali, Morocco, Mozambique, Niger, Senegal, Sierra Leone, South Africa, Tanzania, Togo, Uganda, Zimbabwe

Topaz ($Al_2SiO_4(F,OH)_2$)
Madagascar, Mozambique, Namibia, Zimbabwe

Tourmaline
($NaMg_3(Al,Mg)_6B_3Si_6O_{27}(OH)$)
Benin, Madagascar, Malawi, Mozambique, Namibia, Nigeria, Tanzania, Zambia, Zimbabwe

Tungsten (W)
Angola, Burundi, Chad, DR Congo, Egypt, Equatorial Guinea, Mali, Morocco, Namibia, Nigeria, Rwanda, South Sudan, Sudan, Uganda, Zimbabwe

Uranium (U)
Algeria, Angola, Botswana, Burundi, Central African Republic, Chad, Congo, DR Congo, Gabon, Guinea, Lesotho, Liberia, Libya, Madagascar, Malawi, Mali, Mauritania, Morocco, Mozambique, Namibia, Niger, Nigeria, Somalia, South Africa, South Sudan, Tanzania, Togo, Uganda, Western Sahara, Zambia, Zimbabwe

Vermiculite
Kenya, South Africa, Tanzania, Uganda, Zimbabwe

Vanadium (V)
Angola, Burundi, Madagascar, Mozambique, South Africa, Tanzania, Zimbabwe

Wollastonite ($CaSiO_3$)
Namibia, Kenya

Zinc (Zn)
Algeria, Angola, Burkina Faso, Congo, DR Congo, Egypt, Eritrea, Gabon, Kenya, Liberia, Libya, Madagascar, Morocco, Namibia, Niger, Nigeria, South Africa, South Sudan, Sudan, Togo, Tunisia, Uganda, Western Sahara, Zambia, Zimbabwe

Zirconium (Zr)
Gambia, Kenya, Liberia, Libya, Madagascar, Malawi, Mali, Mozambique, Senegal, South Africa, Zimbabwe

Part II

Mineral beneficiation and value addition

Chapter 2

Minerals of precious metals

2.1 Gold

2.1.1 Overview of gold

Gold (Au) is chemically inert and generally occurs in an elemental form as nuggets of variable sizes, as fine grains or flakes in alluvial deposits, or as grains or microscopic particles embedded in various rocks. It may be found alloyed with silver (as electrum) or with mercury (as amalgam). Gold has a very high specific gravity such that it readily falls out of suspension in running water, usually rivers that cut through gold-bearing rocks and become concentrated alluvial gold as a constituent in 'placer' deposits of heavy *mineral sands* which include titanium minerals ilmenite and rutile, wolframite, and cassiterite. Gold may also be present sometimes as a minor constituent in a base metal (e.g. copper) concentrate from where it is recovered as a by-product during the production of the base metal such as in anode slime during the electrorefining process. Generally, gold is present in large quantities in ores which are closely associated with sulphide minerals of iron, copper and a host of other base metals.

2.1.2 Beneficiation of gold

Gold is inert and occurs generally as a native metal. As it exists in a wide range of mineralogical matrices, the choice of the recovery process depends on the mineralogy [491] of the gold-bearing ore [39].

2.1.2.1 Recovery of free gold

Direct recovery of gold 'placer' deposits of free gold is usually done by gravity concentration techniques such as gold panning, washing tables and sluicing. These are normally used by small scale or artisanal miners. In gold panning, wide, shallow pans are filled with sands and gravel that may contain gold. Shaking the submerged pan in water sorts the gold from the gravel and other material as it quickly settles to the bottom of the pan as a consequence of its greater density. In sluicing, a box with riffles set in the bottom designed to create dead zones is placed in the stream of water flow. As the gold-bearing material is placed at the top of the box carried by the current through the volt, gold settles out behind the riffles. The less dense material flows out of the box.

2.1.2.2 Gold recovery from anode slimes

Where gold is present as a minor constituent in a base metal like copper or silver concentrate it can be recovered as a by-product during the production of the base metal. That recovery is usually from the anode slime formed during the electrorefining process for the base metal in question. If the levels of the base metal(s) are high, prior leaching of the precipitate by nitric or sulphuric acid may be necessary. Alternatively, nitric acid or forced-air oven oxidation can be used to dissolve iron from the electrowinning cathodes before smelting. The iron may originate from gravity concentrates which often contain high grinding steel contents.

2.1.2.3 Recovery of sulphidic gold

Commercial quantities of gold are usually found in ores that are closely associated with sulphide minerals of iron, copper and other base metals [491,1249]. Upgrading of these base metal sulphides consequentially increase in the concentration of the gold values. The recovery [1246] of gold from its ores, therefore, has its genesis with comminution (crushing and grinding) of the ore followed by other mineral processing stages [451,1245] to prepare them for the extraction of the gold.

Cyanidation leaching [487,1255] is the most used method for the recovery of gold. In this process, cyanide salts solution such as sodium cyanide are added to the finely ground sulphide concentrates containing gold. The gold is separated from the ground rock as it dissolves in the cyanide solution. The gold is subsequently recovered by the Merrill-Crowe process as sludge from the solution by addition of zinc metal [34]. The sludge is smelted into ingots for final refinery into pure metal. Alternatively, recovery from the solution can be done by adsorption of gold on activated carbon [1139] followed by solution concentration or stripping and/or electrowinning.

2.1.2.4 Recovery of refractory gold

Gold refractoriness occurs in several forms. Gold cyanidation [499,1255], in which dissolved oxygen and cyanide ions participate, is the most used leaching process for gold recovery from its ores. For this to happen, both the oxygen and cyanide ions must access the gold particles. A substantial proportion of gold particles remain locked up as submicroscopic inclusion or in solid solution with other minerals subsequent to grinding of the ore meant to liberate the gold. Some host minerals, which are unstable in cyanide solutions, interfere with gold dissolution if they decompose into cyanicides which are oxygen and/or cyanide consuming species. Gangue includes minerals such as pyrrhote (FeS), covellite (CuS), chalcocite (Cu_2S), malaachite ($(CuCO_3) \cdot Cu(OH)_2$) and azurite ($2CuCO_3 \cdot Cu(OH)_2$) [453]. Similarly, antimony [483,1250] and arsenic-bearing minerals [483,485] decompose in alkaline solution giving antimonites, thioantimonites [489], arsenites and thioaarsenites [489], respectively, which attach onto gold surfaces thereby inhibiting further interaction between gold and the reagents. When gold is in alloy with other minerals that are insoluble in cyanide, leaching of the gold is inhibited. Charcoa, activated carbon, and some clays can adsorb gold cyanide complexes from solution prior to recovery thereby reducing gold extraction. They are referred to as "preg-robbin" carbonaceous materials [450,462,480]. Whichever the case may be, the gold is considered refractory.

Refractory gold may be rendered amenable to cyanide leaching by suitable pre-treatment to decompose the sulphides and the associated host minerals [454]. The oxidative treatments used are either thermal (roasting [467] or pressure leaching [456,459,1256]), biological (bioleaching) [1254] or chemical (oxidation or chlorination) depending on the ore mineralogy [39].

Roasting plants [467], such as Edward Roasters, have been used to treat refractory gold ore concentrates with gold recoveries averaging 75%. The remaining gold is accounted for in the calcine residue as very small unliberated particles, occluded in over-roasted sintered particles and glassy phases, and coated with substances that render the gold particles insoluble.

Pressure leaching [456,459,1256] carried out in autoclaves involves oxidation of the sulphide minerals in solution with air or oxygen at high temperatures (180–225° C) and pressure (700–2200 kPa) in acid or alkaline environment [1257]. The oxidation reactions are carried out in autoclaves in either an acid or alkaline environment. The process is fast, results in higher gold recoveries, and there is better control over impurities that are likely to cause environmental degradation [1256].

In bioleaching [476,1254], oxidative leaching of refractory gold can be catalysed by bacteria *thiobacilli ferro-oxidans* and *thio-oxidans* under carefully controlled temperature and pH.

When liberation of gold particles from the surrounding mineral matrix is the primary refractory characteristic, ultrafine grinding may be used. Sulphide minerals and carbonaceous [450,462,480] and organic material that are present in refractory ores can be oxidised by aqueous chlorine.

To meet stringent health and environmental regulations search for novel processes are required. An electrohydrometallurgical route to gold (and other metals) recovery has been investigated for the aqueous reductive decomposition of sulphides [466,472,493,1244]. The sulphides are reduced either to the metals or lower oxidation state species with hydrogen sulphide, H_2S, evolution. The hydrogen sulphide can be oxidised to elemental sulphur which is stable and can be stored indefinitely, or it can be burnt to SO_2 and used to neutralise cyanide effluent from gold leaching processes.

2.1.3 Gold value addition

Table 2.1 Uses of gold and its compounds

Gold attributes and properties	inert; extremely malleable; conducts electricity; does not tarnish; alloys well with other metals; easy to work into wires and sheets; brilliant lustre and glossy shine; does not rust; does not corrode; does not turn skin green; looks beautiful
Currency	investment bars, gold coins, gold bullion [7]; central banks bullion; universally used in political and economic transactions
Jewellery [8,9]	watches; rings; bracelets; anklets; eyebrow rings; nose rings; wire granules; alloys for sheets, wire, granules, solders, discs
Catalyst [470]	gold alloys in production of paints and glue; remove nitrogen oxide from diesel engine exhausts; improve air quality in underground mining environments; remove odours from public toilets
Reflective ability [40]	space shuttle cover; gold sunglasses to protect eyes; cockpit windows [41]; adorn buildings; aerospace industry; lubricant for mechanical parts; coats insides of space vehicles; infrared radiation protection; windows for thermal control
Electronics [42–46]	microwaves; washing machines; TV, calculators; mobile phones; DVDs and CD-ROMs; automotive, electronics, missiles, spacecraft; GPS units; desktop and laptop computers
Automotive	car electronics for ignition, control electrics, anti-lock braking system, electronic fuel injectors, crash sensors for air bags
Biomedical [47–50]	gold injections to alleviate arthritis pain; inner ear implants; wires to pacemakers; gold dust tablets to relieve fatigue and depression; drug-delivery microchips; gold isotopes in certain radiation treatments and diagnosis

(Continued)

Table 2.1 (Continued)

Dentistry	false teeth; caps or crowns; fillings; crowns; bridges and orthodontic appliances
Religious artefacts	Tutankhamen of Egypt; gold leaf in paintings to decorate glass and fine china
Gold plating	trophies; crowns; awards; religious statues; pens; watches; bathroom fittings; spectacle frames

2.2 Platinum Group Metals (PGMs)

2.2.0.1 Overview of PGMs

Iridium (Ir), osmium (Os), palladium (Pd), platinum (Pt), rhodium (Rh) and ruthenium (Ru) are the six precious metals [67] collectively referred to as platinum group metals (PGMs), and sometimes precious group elements (PGEs). They occur together having similar properties, such as catalytic activity, high density and high melting point [51] exhibiting variations with respect to each metal's abundance and applications [52]. Various levels of gold, arsenic, nickel, copper, cobalt, iron, lead, tellurium and selenium are normally found in the PGM ores.

2.2.0.2 Beneficiation of Platinum Group Metals (PGMs)

A schematic diagram, Figure 2.1, for the extraction of PGMs summarises the basic beneficiation processes for the PGMs which start with underground or open-pit mining followed by concentration and smelting resulting in successively higher-grade filter-cake for subsequent series of intricate chemical processes carried out during the refining phase. The grade and type of deposit [53,54] determine the specific processing routes thereafter.

The conventional comminution (crushing and grinding) is followed by making a fine slurry of the powdered ore and water to which chemicals are added. Potassium amyl xanthate renders the sulphide minerals hydrophobic while carboxymethylcellulose confers hydrophilicity to the waste minerals such as talc.

After the addition of frothing agents, the slurry is injected into large tanks from where the sulphide minerals clinging to the surface bubbles are collected as PGM concentrate. The concentrate is smelted to drive off sulphur, iron and other metals resulting in a converter matte which is highly enriched in Ni, Cu and PGMs. The converter matte is then ground and leached with acid, under a judicious control of oxygen levels and temperature, to dissolve nickel, iron, copper, tellurium and selenium. The final "cake" consisting of the PGM metals is leached with specialised organic acids to separate the different platinum group metals.

Alternative hydrometallurgical routes, such as the Kell Process, are being developed for the extraction of platinum group metals and base metals from flotation concentrates [1258].

2.2.0.2.1 Base Metal Refinery (BMR) of PGMs

At the base metal refinery, nickel, copper and cobalt are leached out with nickel and copper being further recovered as cathodes using the electrowinning process. Metathesis reactions in a first leaching stage separate copper and nickel from each other. Further purification of the solution removes lead, arsenic, iron, cobalt and other metals. The residual nickel is either crystallised, reduced to powder using hydrogen, or electrowon. The residue from the first stage contains mainly copper which can be dissolved in a second leaching stage. Subsequently, the copper solution is purified and copper is recovered electrolytically.

2.2.0.2.2 Precious Metal Refinery (PMR) of PGMs

The PMR is designed to isolate different metals and involves pressure leaching, solvent extraction and PGM precipitation. Pressure leaching in *aqua regia* dissolves palladium, platinum and gold leaving the other PGMs and base metal components as solids in the leach residue. Gold is segregated from the solution through solvent extraction after which platinum is precipitated as ammonium chloroplatinate on treatment with ammonium chloride. The resultant crude platinum salt is then heated to a fine powder which is then re-dissolved in *aqua regia*, precipitated once more with ammonium chloride, and calcined to pure metal.

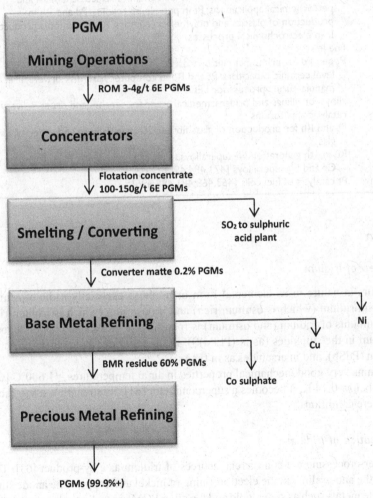

Figure 2.1 Schematic diagram for the beneficiation and extraction of PGMs

The palladium, still left in solution, is similarly subjected to processes of precipitation, re-dissolution, filtration to form a high-grade palladium salt whose powdered product is converted to metallic form through chemical reduction involving formic acid.

The preceding series of solvent extraction, evaporation and precipitation achieves separation in the following order: gold, palladium and platinum. The remaining rhodium, rhuthenium, iridium and osmium are produced as a mixed saleable cake.

2.2.0.3 PGMs value addition

Table 2.2 Uses of PGMs and their compounds [53]

PGM attributes and properties	high melting points; strength; resistance to corrosion; high resistant to wear and tarnish
Autocatalysis	Pt, Pd and Rh used for the automotive industry to reduce emission from petrol and diesel engines; all PGMs in chemical, electrochemical and petrochemical applications; Pt in petroleum refining; Pd and Rh in production of plastics and polymer precursors; Ru in ammonia production; Ir in electrochemical processes
Jewellery	fine jewellery
Electronics	Pt and Pd for printed circuit boards; Pd in mobile phones and multi-level ceramic capacitors; Pt and Ru in computer hard disk drivers; Ir in manufacturing process for LEDs
Medical and dental	alloys for fillings and bridges; medical scanners, sensors and drugs delivery
Chemical reagents	catalytic applications
Glass production	Pt and Rh for production of glass fibre, LCD manufacture and other types of glass
Alloys	Ru in 4th generation Ni superalloys in jet engine turbines; Pt-Al together with Cr and Ni superalloys [471,484,495]
Hydrogen economy	Pt catalysis of fuel cells [463,468,481,488,496]

2.2.1 Iridium

2.2.1.1 Overview of iridium

Iridium (Ir) occurs in nature in its elemental form or in alloys especially iridium-osmium alloys such as osmiridium (which is osmium rich) and iridosmium (which is iridium rich) [55,62]. Small amounts of iridium (and osmium) is found in the nickel and copper deposits replacing platinum in the sulphides (as in (Pt,Pd)S), in the tellurides as in (PtBiTe), in the antimonides as in (PdSb), and in arsenides as in (PtAs$_2$).

Iridium maintains very good mechanical properties in air at temperatures $\geq 1,600°C$ [60]. At temperatures below 0.14 K, it becomes a superconductor [61]. Its strength is a key factor in extremely severe conditions.

2.2.1.2 Beneficiation of iridium

Nickel and copper processing are commercial sources of iridium as a by-product [63]. This would be part of the anode slime in the electrorefining of nickel and copper. The anode slime also contains noble metals such as silver, gold and the other PGMs as well as selenium and tellurium. The extraction of iridium from the rest of the metals start at this point [64] necessitating the dissolution of the anode slime. One method involves fusing the slime anode with sodium peroxide followed by dissolution in *aqua regia*, and dissolution in a mixture of chlorine with hydrochloric acid [65,66]. Iridium is then separated from the other PGMs by precipitating it with ammonium to form ammonium hexachloroiridate ((NH_4)$_2$IrCl$_6$) or by extracting IrCl$_2$$^{6-}$ with organic amines [76]. This is subsequently reduced by hydrogen to yield the metal as a powder or sponge that can be treated using powder metallurgy techniques [68,69].

2.2.1.3 Iridium value addition

Table 2.3 Uses of iridium and its compounds

Iridium attributes and properties	very hard; brittle; very hard to machine, form or work; very dense; corrosion resistant even at high temperatures; very high melting point; does not dissolve in acids, even *aqua regia*, molten metals or silicates at high temperatures; can be attacked by some molten salts such as sodium cyanide or potassium cyanide [56], oxygen and halogens particularly fluorine [57] at high temperatures[58]
Iridium metal	powder metallurgy commonly employed [30,59]; employed in high-performance spark plugs [70,71], crucibles [64,72] for recrystallisation of semiconductors at high temperature; as electrodes for the production of chlorine in the chloralkali process; multi-pored spinnerets, compass bearing and balances [73,74]; high-temperature crucibles to produce oxide single-crystals (gadolinium gallium garnet and yttrium gallium garnet) [60] for use in computer memory devices and solid state lasers [70,75]; in particle physics for the production of antiprotons
Iridium compounds	forms important salts and acids; organometallic compounds used in industrial catalysis for production of acetic acid [76]
Nuclear application	iridium radioisotopes [465,1252] used in some radioisotope thermoelectric generators
Alloys	long-life aircraft engine parts; deep-water pipes; hardening agent in platinum alloys
Optics	X-ray telescopes [77]

2.2.2 Osmium

2.2.2.1 Overview of osmium

Osmium (Os) occurs in nature in its elemental form or alloyed as osmiridium (osmium rich), or iridosmium (iridium rich) [55]. It can also be found in alloys with nickel and copper. Small amounts of iridium and osmium may exchange with platinum in the nickel or copper deposits where PGMs occur as sulphides (e.g. (Pt,Pd)S), tellurides (e.g. PtBiTe), antimonides (e.g. PdSb) and arsenides (e.g. PtAs$_2$).

2.2.2.2 Beneficiation of osmium

Nickel and copper electrorefining processes are the main commercial source of osmium where it is obtained as a by-product [83] and as a component of the *anode slime* together with noble metals such as silver, gold, the PGMs, as well as non-metallic elements selenium and tellurium. The extraction of osmium, as indeed for the other PGMs, start with this anode slime [62,84].

Separation of the metals is possible after they have been brought into solution. One method involves fusion with sodium peroxide followed by dissolution in *aqua regia*, and dissolution in a mixture of chlorine with hydrochloric acid [85,86]. Osmium, ruthenium, rhodium and iridium can be separated from platinum, gold and base metals because they are insoluble in *aqua regia*. This leaves a solid residue from which rhodium can be separated by treatment with molten sodium bisulphate. The remaining insoluble residue, containing ruthenium, osmium, and iridium, is treated with sodium oxide, in which iridium remain insoluble, producing water-soluble ruthenium and osmium salts. Volatile oxides can at this stage be oxidised after which RuO_4 is separated from OsO_4 by precipitation of $(NH4)_3RuCl_6$ with ammonium chloride.

Alternatively, osmium can be separated from the other PGMs, after they have been dissolved, by distillation. Extraction of the osmium from the rest of PGMs can also be achieved with organic solvent of the volatile osmium tetroxide [87]. The final metallic product can be obtained as powder [88] or *sponge* by reduction with hydrogen.

2.2.2.3 Osmium value addition

Table 2.4 Uses of osmium and its compounds

Osmium attributes and properties	hard [79–82]; brittle; bluish-white; lustrous; low compressibility; densest element [78]; low vapour pressure; high melting point; difficult to machine, form or work
Osmium oxide	alloyed with other PGMs for high-wear applications such as fountain pen nibs, electrical contacts, and instruments pivots [89]
Osmium tetroxide	in fingerprint detection [90]; staining fatty tissue for optical and electron microscopy; fixes biological membranes in place in tissue samples and simultaneously stains them; osmium staining enhances image contrasts in transmission electron microscopy (TEM) [91]
Osmium ferricyanide (OsFeCN)	fixing and staining action [92]

2.2.3 Palladium

2.2.3.1 Overview of palladium

The two most common PGMs are palladium and platinum. Palladium has the largest production of the PGMs. It occurs together with other PGMs as well as with various levels of gold, arsenic, nickel, copper, cobalt, iron, lead, tellurium and selenium. Recovery from recycling from jewellery and coins is estimated to be above 90%.

2.2.3.2 Beneficiation of palladium

The conventional route of crushing, grinding, flotation and smelting applies in the extraction of palladium. The Ni, Cu and PGMs enriched converter matte is ground from which nickel, iron, copper, tellurium and selenium are leached out with acid under judicious control of oxygen levels and temperature. The remaining PGM powder cake is subsequently leached out with specialised organic acid to separate the different PGMs as described elsewhere in this chapter.

2.2.3.3 Palladium value addition

Table 2.5 Uses of palladium and its compounds

Palladium attributes and properties	lowest melting point and least dense of the PGMs
Palladium metal	autocatalyst for petrol engines [93,94]; soldering material; dental uses [94,96,97]
Electronics [95]	Pd (and Pd/Ag alloy) anodes manufacture in multilayer ceramic capacitors; connector platings in consumer electronics; plating of electronics components
Chemical catalyst [98,99]	speeds up hydrogenation and dehydrogenation reactions; petroleum cracking
Jewellery [94, 100–104]	watch making, blood sugar test strips, aircraft spark plugs, surgical instruments; transverse flutes; bullion has ISO currency codes of XPD; manufacture of white gold
Electrochemical studies	palladium-hydrogen electrode
Gas detections	palladium(II) chloride in detection of carbon monoxide; Pd absorbs H_2 at room temperature forming PdH_x
Printing process	platinotype printing [105] for fine-art black-and-white prints; alternative to silver

2.2.4 Platinum

2.2.4.1 Overview of platinum

Platinum (Pt) and palladium (Pd) are the most common PGMs that occur together with other PGMs as well as with various levels of gold, arsenic, nickel, copper, cobalt, iron, lead, tellurium and selenium.

2.2.4.2 Beneficiation of platinum

The conventional route of crushing, grinding, flotation and smelting applies in the extraction of platinum [107]. The Ni, Cu and PGMs enriched converter matte is ground from which nickel, iron, copper, tellurium and selenium are leached out with acid under judicious control of oxygen levels and temperature. The remaining PGM powder cake is subsequently leached out with specialised organic acid to separate the different PGMs as described elsewhere in this chapter. Platinum may also be obtained commercially as a by-product from the nickel and copper electrorefining process as described in detail earlier.

Where pure platinum occurs in placer deposits, it may be isolated by floating away lighter impurities in a liquid because platinum is significantly denser than many of its impurities. Another method of isolating platinum from a mixture with nickel and iron is by running an electromagnet over the mixture. Nickel and iron being ferromagnetic, whereas platinum is paramagnetic, will be removed from the mixture. As platinum has a higher melting point than most other substances, many impurities can be burned or melted away leaving the platinum unmelted. The insolubility of platinum in hydrochloric and sulphuric acid can be exploited by dissolving and removing the other substances by stirring the mixture in these acids and recovering the remaining platinum [108].

Purification of raw platinum material which contains gold and other PGMs may be done through processing it with *aqua regia*. Platinum, gold and palladium dissolve leaving osmium, iridium, ruthenium and rhodium unreacted. Addition of ferrous chloride precipitates the gold. After filtering the gold precipitate the platinum can in turn be precipitated as ammonium chloroplatinate by addition of ammonium chloride. The precipitate ammonium chloroplatinate can then be converted to platinum by heating. The unprecipitated hexachloroplatinate(IV) may be reductively processed to yield platinum metal by the addition of elemental zinc [88,110,111].

2.2.4.3 Platinum value addition

Table 2.6 Uses of platinum and its compounds

Platinum attributes and properties	paramagnetic; dense; malleable [106]; ductile; highly unreactive; resistant to corrosion; stable at high temperature; stable electrical properties; reacts with oxygen slowly at very high temperatures; reacts vigorously with fluorine at 500° C to form tetrafluoride; attacked by Cl, Br, I, and S; insoluble in HCl and nitric acid, but dissolves in hot *aqua regia* to form chloroplatinic acid H_2PtCl_6; resistant to wear and tarnish; chemical stability
Platinum metal	catalytic converters; laboratory equipment; electrical contacts; electrodes; platinum resistance thermometers; dentistry equipment; jewellery
Platinum compounds	cisplatin, oxliplatin and carboplatin used in chemotherapy against certain types of cancer [112–115]
Hexachloroplatinic acid	precursor to many other platinum compounds; used in photography, zinc etchings, indelible ink, plating, mirrors, porcelain colouring, and as a catalyst
Alloys	fine wires; noncorrosive laboratory containers; medical instruments; dental prostheses; electrical contacts; thermocouples; strong permanent magnets (of Pt-Co) alloys; ships; pipelines; steel piers use platinum-based anodes; Pt-based superalloys [473,484,495]
Jewellery	as a 90–95 alloy; watches
Catalyst	catalytic reforming of straight-run naphthas into higher-octane petroleum that becomes rich in aromatic compounds; PtO_2 (Adam's catalyst) used as a hydrogenation catalyst, especially for vegetable oils; catalyst for the decomposition of hydrogen peroxide into water and oxygen
Electrodes	platinum wire; standard hydrogen electrode (SHE) [116]; Pt pans and supports used in thermogravimetric analysis

2.2.5 Rhodium

2.2.5.1 Overview of rhodium

Rhodium can be considered to be a by-product of platinum and palladium with which it is normally found as a free metal or alloy. It is rarely found as a chemical compound in minerals such as bowieite and rhodplumsite. An important source of rhodium is from recycling of industrial used products, estimated at 90%.

Rhodium reduces nitrogen oxides to nitrogen and oxygen [119]:

$$2NO_x \rightarrow xO_2 + N_2$$

(1)

2.2.5.2 Beneficiation of rhodium

Extraction of rhodium from platinum involves dissolution of the ore in *aqua regia* and neutralising the acid with sodium hydroxide [120]. Addition of ammonium chloride results in the precipitation of ammonium chloroplatiname. Subsequently the other metals like copper, lead, palladium and rhodium can be precipitated with elemental zinc. Dilute nitric acid dissolves all except palladium and rhodium, which are dissolved in *aqua regia*. Rhodium is precipitated as $Na_3[RhCl_6] \cdot nH_2O$ by the addition of sodium chloride. The precipitate can be reductively decomposed to rhodium metal by reacting it with zinc metal [121].

2.2.5.3 Rhodium value addition

Table 2.7 Uses of rhodium and its compounds

Rhodium attributes and properties	rare; hard; chemically inert; noble metal; high melting point; poor malleability; resistant to corrosion; durable; high reflectance; does not form oxide even when heated [117]; absorbs oxygen only at melting point and release it at solidification [118]; resistant to acid attack; completely insoluble in nitric acid and dissolves slightly in *aqua regia*
Rhodium metal	used in three-way catalytic converters to reduce vehicle emissions [122] and especially NO_x emissions [123–125] which is not possible with other catalysts; electrical contacts [131]
Chemical catalyst	production of fibreglass [122]; catalytic carbonylation of methanol to produce acetic acid by Monsanto process [126]; catalyse addition of hydrosilanes to molecular double bonds in the manufacture or rubbers [127]; reduce benzene to cyclohexane [128]
Alloys	with Pt and Pd in high-temperature and corrosion-resistive coatings [117]; used in furnace windings, bushings for glass fibre production; thermocouple elements; electrodes for aircraft spark plugs and laboratory crucibles [120]
Nuclear application	rhodium detectors to measure neutron flux level;
Jewellery	decorations and high price [129]; electroplated on white gold and platinum to give it reflective (Ag_2S); white surface (flashing); coating sterling silver to protect against tarnish
Optics	plated rhodium by electroplating or evaporation [132]
Mammography	Rh produces characteristic X-rays [133]
Nuclear application	Rh neutron detectors used in combustion engineering nuclear [465,1252] reactors to measure neutron flux levels [134]
Rhodium complexes	median lethal dose (LD50) for rats is 198 mg of rhodium chloride ($RhCl_3$) per kg of body weight [135]; metal is harmless [136]

2.2.6 Ruthenium

2.2.6.1 Overview of ruthenium

Ruthenium (Ru), which belongs to the platinum group of metals, is usually found as a minor component of platinum ores [137–139]. Significant amounts of ruthenium in fission products of uranium-115 may be a possible source of ruthenium. However, there are some complications in both the extraction and storage for several half-lives of the decaying isotopes [147–149].

2.2.6.2 Beneficiation of ruthenium

The usual commercial source of ruthenium is as a by-product from nickel and copper mining and processing as well as by processing the PGM ores. The *anode slime* formed during the electrorefining process for copper and nickel is the starting point for the extraction of ruthenium [140,141]. Separation of the metals may involve fusion with sodium peroxide followed by dissolution in *aqua regia* and further dissolution in a mixture of chlorine and hydrochloric acid [142,143]. Because they are insoluble in *aqua regia*, osmium, ruthenium, rhodium and iridium can be separated from platinum, gold and base metals which dissolve. From the residue, rhodium can be separated by treating it with molten sodium bisulphate. The remaining ruthenium, osmium and iridium can now be treated with sodium oxide to produce water-soluble ruthenium and osmium salts which can be oxidised to their respective volatile oxides RuO_4 and OsO_4. Addition of ammonium chloride precipitates out and separates $(NH_4)_3RuCl_6$. Hydrogen is used to reduce ammonium ruthenium chloride yielding a metal powder or sponge [145] that can be treated using powder metallurgy techniques or by argon-arc welding [146]. Alternatively, the separation can be effected by distillation or extraction with organic solvents of the volatile osmium tetroxide [144].

2.2.6.3 Ruthenium value addition

Table 2.8 Uses of ruthenium and its compounds

Ruthenium attributes and properties	polyvalent; hard; inert to most other chemicals
Ruthenium metal	wear-resistant electrical contacts; production of thick-film resistors
Catalyst	splitting of H_2S using CdS particles loaded with ruthenium dioxide; used for the removal of hydrogen sulphide in oil refineries; organometallic ruthenium carbene and alkylidene complexes as catalysts for olefin metathesis for pharmaceutical chemistry [166]; Ru-promoted Co catalysts are used in the Fischer-Tropsch synthesis
Ruthenium tetroxide RuO_4	a strong oxidising agent; a potent fixative and stain for electron microscopy or organic materials; used to reveal the structure of polymer samples [150]; dipotassium ruthenate (K_2RuO_4) and potassium perruthenate ($KRuO_4$) are also known [151] compounds of ruthenium
Alloys	Pt and Pd alloys are wear-resistant electrical contacts; plating material like Rd [152] for electrical contacts [130,153]; thin coatings by electroplating [154] or spattering [155]; titanium alloy with high corrosion resistance [160]; advanced high-temperature single-crystal superalloys used in turbine blades in jet engines; fountain pen nibs
Chip resistors	thin-film chip resistors of ruthenium dioxide or lead and bismuth ruthenates [156–159]
Mixed-metal oxide (MMO)	contain Ru; used for the protection of underground and submerged structures and for electrolytic cells for chemical processes such as generating chlorine from salt water [161]; Ru complexes used as opcode sensors for oxygen [162] e.g. Ruthenium red [$(NH_3)_5Ru-O-Ru(NH_3)_4-O-Ru(NH_3)_5^{6+}$] as a biological stain to stain polyanionic molecules in microscopy [163]; in radiotherapy of eye tumours [164]; fingerprints
Ru-based oxides	Unusual properties such as quantum critical point behaviour [167], exotic superconductivity [168], and high-temperature ferromagnetism [169]

2.3 Silver

2.3.1 Overview of silver

Silver (Ag) exists on the earth's crust in three forms: as a native element, in alloys such as with gold (electrum), and in ores containing sulphur, arsenic, antimony or chlorine including argentite (Ag_2S), chlorargyrite (AgCl) and pyrargyrite (Ag_3SbS_3). Ores of copper, copper-nickel, lead and lead-zinc are the principal sources of silver.

2.3.2 Beneficiation of silver

The process of extraction of silver from its argentite ore Ag_2S involves cyanidation leaching [475]. The crushed ore is concentrated by froth flotation and then treated with sodium cyanide solution resulting in the formation of sodium argentocyanide $Na[Ag(CN)_2]$.

$$Ag_2S + 4NaCN \rightleftharpoons 2Na[Ag(CN)_2] + Na_2S \tag{2}$$

The solution of sodium argentocyanide combines with zinc dust forming sodium tetra cyanozincate and precipitated *spongy* silver.

$$Zn + 2Na[Ag(CN)_2] \rightarrow Na_2[Zn(CN)_4] + 2Ag \tag{3}$$

Pure silver is obtained by fusing the *spongy* silver with potassium nitrate. Subsequently, the silver obtained is purified by an electrolytic process.

Silver can be extracted pyrometallurgically by smelting its ore. It can also be produced as a by-product during the electrolytic refining of copper and by the application of the Parkes process on lead metal obtained from lead ores that contain small amounts of silver.

Silver ores can be dissolved in halides such as chlorine and iodide [1260]. The effect of oxidant and lead(II) ions in the leaching and electrochemistry of gold, silver and gold-silver alloys in cyanide solutions has been investigated [1255].

2.3.3 Silver value addition

Table 2.9 Uses of silver and its compounds

Silver attributes and properties	soft; high electrical conductivity; thermal conductivity; high reflectivity [171,172]; ductile; malleable; brilliant white metallic lustre [170]
Silver metal	coinage; solar panels; water filtration; jewellery and ornaments; high-value tableware and utensils; investment in forms of coins and bullion; electrical contacts and conductors; in specialised mirrors; window coatings; catalysis of chemical reactions
Silver compounds	photographic film and X-rays; dilute silver nitrate solutions used as disinfectant and microbiocides added to bandages and wound dressings, catheters and other medical instruments; silver halides are photosensitive [173]
Silver plating	improve electrical conductivity of parts and wires; thermal or infrared telescopes [179,180]; reflective telescopic mirrors [181]

(Continued)

Table 2.9 (Continued)

Jewellery	plating giving shiny silver finish (flashing)
Silver alloys	constituent of coloured carat gold alloys and carat gold solders [174]; alloyed with gold or tin and mercury to make amalgams for dental filling; solder an brazing alloys on bearing surfaces to increase galling resistance and reduce wear under heavy load
Photovoltaics	Crystalline solar photovoltaic panels [175]; plasmonic solar cells [1136]; refléctive coating for concentrated solar power reflectors [176]
Catalytic properties	water purifiers; prevents bacteria and algae in filters; sanitises water and eliminates need for chlorine [177] in hospitals, community water systems, pools and spas; oxidation reactions e.g. production of formaldehyde from methanol, converts ethylene to ethylene oxide (later hydrolysed to ethylene glycol for making polyesters; used in Oddy tests to detect reduced S compounds and carbonyl sulphides)
Electrical conductivity	electronic products – audio cables, speaker wires, power cables, hearing aids, watches [173,178]; silver oxide batteries, high capacity silver-zinc and silver-cadmium batteries
Sputtering	glass coating; insulated glazing; tinted windows; silver-coated polyester sheets [177]
Musical wind instruments	tubes [182]; flutes; brass instruments, such as trumpets and baritones [183]
Nuclear application	in control rods [465] to regulate the fission chain reaction in pressurised water nuclear reactors [1252]
Biological applications	silver stains to increase contrast and visibility of cells and organelles in microscopy; stain proteins in gel electrophoresis and polyacrylamide gels to enhance visibility and contrast of colloidal [1259] gold stain [184]; inhibits growth of bacteria and fungi on clothing, such as socks; silver nanoparticles into yarn polymer [188,189]
Medical applications	wound dressings containing silver sulphadiazine or silver nanomaterials to treat external infections; antibiotic coating in medical devices; in urinary catheters and endotracheal breathing tubes; ventilator-associated pneumonia [185,186]; silver ion (Ag$^+$) is bioactive and kills bacteria *in vitro;* silver and silver nanoparticles as antimicrobial in industrial, healthcare and cosmetic applications [187]

Chapter 3

Minerals of base metals

3.1 Aluminium

3.1.1 Overview of aluminium

Aluminium is very chemically reactive to preclude existence of native specimens except in extreme reducing environments. Aluminium is found in combination with many other elements in over 270 different minerals, the principal of which is bauxite ore [190,191] which occurs as a weathering product of low iron and silica bedrock. Because of its malleability aluminium metal is easily machined, cast, drawn and extruded with yield strength of 7–11 MPa and one-third the density and stiffness of steel.

3.1.2 Beneficiation of aluminium

Bauxite ore ($AlOx(OH)_{3-2}x$) is the principal source of almost all metallic aluminium. It is typically comprised of the minerals gibbsite $Al(OH)_3$, boehmite γ-AlO(OH) and diaspore α-AlO(OH), mixed with the two iron oxides goethite FeO(OH) and haematite Fe_2O_3, the clay mineral kaolinite $Al_2Si_2O_5(OH)_4$ and small amounts of anatase TiO_2. It is converted to aluminium oxide (Al_2O_3) and the associated oxy-hydroxides and trihydroxides on a large scale by the Bayer process [192]. In alkaline solution, an intermediate sodium aluminate, $NaAlO_2$, is formed.

$$Al_2O_3 + 2NaOH \rightarrow 2\,NaAlO_2 + H_2O \tag{4}$$

This intermediate product is soluble in water while the other components of the ore remain insoluble as waste in the form of alumina or "red mud".

$$2\,H_2O + NaAlO_2 \rightarrow Al(OH)_3 + NaOH \tag{5}$$

The Hall-Héroult process is a pyrometallurgical route to aluminium metal from a solution of alumina in which a molten mixture of cryolite (Na_3AlF_6) with calcium fluoride at 980°C is electrolysed to give the metal as aluminium billets ready for further processing [193]:

$$Al^{3+} + 3e^- \rightarrow Al \tag{6}$$

Oxygen is formed at the anode:

$$2O^{2-} \rightarrow O_2 + 4e^- \tag{7}$$

The product of the Hall-Heroult process is aluminium with a purity of above 99% which can be upgraded to 99.99% [194] by further purification via the Hoopes process involving the electrolysis of molten aluminium with a sodium, barium and aluminium fluoride electrolyte.

3.1.3 Aluminium value addition

Table 3.1 Uses of aluminium and its compounds

Aluminium attributes and properties	silvery-white appearance; soft and light metal, non-magnetic and ductile; resist corrosion due to passivation
Aluminium as a widely used material [196]	structural components in transportation and structural materials; automobiles, aircraft, trucks, railway cars, marine vessels, bicycles, spacecraft; sheets, tubes, casting; construction [200] for windows, sliding doors, building wire; household goods such as utensils [206], watches, baseball bats [201]; street lighting poles [202]; sailing ship masts, walking poles; pylons; electronics; cases for photographic equipment; super purity aluminium (SPA) for electronics, CDs, wires/cabling; heat sinks for electronic appliances such as transistors and CPUs; substrate material of metal-core Cu clad laminates in LEDs; guitar necks [209]
Oxide and sulphate compounds [195]	manufacture of paper, mordant in fire extinguishers; food additive (E number E173); fireproofing and leather tanning; aluminium chloride used in petroleum refining and production of synthetic rubber and polymers
Aluminium metal	powdered aluminium used in paint; reacts with HCl and NaOH to produce hydrogen gas [205]
Alloys	with copper, zinc, magnesium, manganese and silicon (e.g. in duralumin) [197–199,210] in foils and beverage cans; MKM steel [203]; Alnico magnets [204]; with magnesium for body of aircraft; Al/Cu alloys for coins [207,208]
Pyrotechnics	solid rocket fuels and thermite
Aluminium chlorohydrate	hardening agent, an antiperspirant and an intermediate in the production of aluminium metal
Aluminium acetate [211]	an astringent
Aluminium borate $(Al_2O_3 \cdot B_2O_3)$ and aluminium fluorosilicate $(Al_2(SiF_6)_3)$	production of glass, ceramics and synthetic gemstones
Aluminium phosphate $(AlPO_4)$	manufacture of glass, ceramic, pulp and paper products, cosmetics, paints and varnishes and in making dental cement [212]
Aluminium hydroxide $(Al(OH)_3)$	used as an antacid, as a mordant, in water purification, in the manufacture of glass and ceramic and in the waterproofing of fabrics [213,214]
Lithium aluminium hydride	powerful reducing agent used in organic chemistry [215,216]
Organoaluminiums	used as Lewis acids and cocatalysts [217]
Methylaluminoxane	a cocatalyst for Ziegler-Natta olefin polymerisation to produce vinyl polymers such as polyethene [218]
Aqueous aluminium ions (such as found in aqueous aluminium sulphate)	used to treat against fish parasites such as Gyrodactylus salaris
Certain aluminium salts	immune adjuvant (immune response booster) to allow the protein in certain vaccine to achieve sufficient potency as immune stimulants

(Continued)

Table 3.1 (Continued)

Alumina: Aluminium oxide (Al_2O_3) and the associated oxy-hydroxides and trihydroxides	an absorbent, removes water from hydrocarbons, which enables subsequent processes that are poisoned by moisture; common catalysts for industrial processes; an abrasive; in high pressure sodium lamps due to its inertness
Aluminium Sulphate ($Al_2(SO_4)_3 \cdot (H_2O)_{18}$)	water treatment; manufacture of paper; mordant, in fire extinguishers; a food additive (E number E173); fireproofing
Aluminium ammonium sulphate (ammonium alum, $(NH_4)Al(SO_4)_2 \cdot 12H_2O$)	mordant; leather tanning
Aluminium potassium sulphate ($[Al(K)](SO_4)_2) \cdot (H_2O)_{12}$	mordant; leather tanning
Aluminium chloride ($AlCl_3$)	petroleum refining; production of synthetic rubber and polymers
Aluminium chlorohydrate	a hardening agent; an antiperspirant; an intermediate in the production of aluminium metal

3.2 Antimony

3.2.1 Overview of antimony

The predominant mineral ore of antimony is stibnite (Sb_2S_3) with the element's abundance in the Earth's crust estimated at 0.2 to 0.5 parts per million in over 100 mineral species. The element antimony is stable in air at room temperature and resistant to attack by acids.

3.2.2 Beneficiation of antimony

Antimony is produced industrially by roasting in reverberatory furnaces [219] to an oxide, Sb_2O_3, and subsequent carbothermal reduction [220]

$$2Sb_2O_3 + 3C \rightarrow 4Sb + 3CO_2 \tag{8}$$

which isolates it from impurities such as arsenic and other sulphides [219,221,222]:
or by direct reduction in blast furnaces of lower grades with iron [219]

$$Sb_2S_3 + 3\,Fe \rightarrow 2Sb + 3FeS \tag{9}$$

depending on the quality and composition of the ore with lower grade ores needed to be pre-concentrated by froth flotation.

There are also hydrometallurgical routes to antimony recovery. Gold and silver ores made refractory due to the presence of antimony can be extracted by pre-treatment with alkaline sulphide leach [1250]. Stibnite can also be reductively decomposed hydrometallurgically [466,472] in chloride media.

3.2.3 Antimony value addition

The largest applications for metallic antimony are as alloying material for lead and tin and for lead antimony plates in lead-acid batteries. Alloying lead and tin with antimony improves the properties of the alloys which are used in solders, bullets and plain bearings. Antimony compounds are prominent additives for chlorine and bromine-containing fire retardants found in many commercial and domestic products. An emerging application is the use of antimony in microelectronics.

Table 3.2 Uses of antimony and its compounds

Antimony attributes and properties	soft; stable in air at room temperature; reacts with oxygen if heated to form antimony trioxide (Sb_2O_3); resistant to acid attacks
Antimony metal	flame retardants [223,224]; alloys for batteries, plain bearings and solders; bullets and bullet tracers [232]; paint and glass art crafts and as opacifier in enamel
Antimony trioxide	used in making lame-proofing compounds; fibreglass composites industry as an additive to polyester resins for such items as light aircraft engine covers
Antimonial lead	increases hardness and mechanical strength; lead-acid batteries [220,225]; antifriction alloys (such as Babbitt metal), in bullets and lead shot, cable sheathing, type metal (for example, for linotype printing machines), solder, in pewter, and in hardening alloys with low tin content in the manufacturing of organ pipes; a stabiliser and a catalyst for the production of polyethyleneterephthalate [219]; a fining agent to remove microscopic bubbles in glass, mostly for TV screens [226]; pigment [219]; semiconductor industry as a dopant for heavily doped n-type silicon wafers [227] in the production of diodes, infrared detectors, and Hall-effect devices
Indium antimonide	a material for mid-infrared detectors [228–230]
Antimony sulphides	to stabilise the friction coefficient in automotive brake pad materials [231]; heads of some safety matches; antimony-124 is used together with beryllium in neutron sources
Biological and medical applications	antimonial cosmetics; antiprotozoan drugs; veterinary preparations like anthiomaline or lithium antimony thiomalate (skin conditioner in ruminants); a nourishing or conditioning effect on keratinised tissues; meglumine antimoniate for treatment of leishmaniasis in domestic animals

3.3 Arsenic

3.3.1 Overview of arsenic

Arsenic occurs in many sulphide minerals with the formula MAsS and MAs_2 (M = Fe, Ni, Co) and also as a pure elemental crystal. Arsenopyrite (FeAsS), an illustrative mineral which is structurally related to iron pyrite, is one of the major causes of refractoriness in gold. The mineral is unstable in cyanide solution, decomposing to give a cyanicide, arsenite, which can attach onto gold surface thereby inhibiting further interaction between gold and reagents. This results in poor recovery of the gold, a condition of refractoriness.

Other sources of arsenic are the broad variety of sulphur compounds such as orpiment (As_2S_3) and realgar (As_4S_4). Trihalides of arsenic(III) exists as are pentahalides such as arsenic pentafluoride (AsF_5).

3.3.2 Beneficiation of arsenic

Arsenic is recovered mainly as a side by-product of smelter dust from copper, gold, and lead smelters. Arsenic sublimes as arsenic(III) if arsenopyrite is roasted in air leaving iron oxides. Metallic arsenic is produced if the roasting takes place in the absence of air. Sublimation in vacuum or in a hydrogen atmosphere or by distillation from molten lead-arsenic mixture [233] achieves further purification from sulphur and other chalcogens.

3.3.3 Arsenic value addition

Table 3.3 Uses of arsenic and its compounds

Arsenic attributes and properties	sublimes upon heating at atmospheric pressure
Metallic arsenic	alloying with lead in car batteries; reducing dezincification by adding arsenic to brass; enhanced corrosion stability; laser diodes and LEDs to directly convert electricity into light; bronzing and pyrotechnics; taxonomic sample preservation
Gallium arsenide	semiconductor material in IC
Arsenic alloys	strengthening of copper and lead alloys as in car batteries [234,235]; doping of semiconductor electronic devices such as the optoelectronic compound gallium arsenide [234]
Arsenic trioxide	pesticides, treated wood products, herbicides, and insecticides; feed additive for increasing weight gain, improving feed efficiency, and for the prevention of disease [236–239] in the production of poultry and swine; treatment of cancer; treatment of patients with acute promyelocytic leukaemia that is resistant to all-trans retinoic acid
Military use	Lewisite (ClCH=CHAsCl$_2$), a chemical weapon that is a vesicant (blister agent) and lung irritant; Agent Blue, a mixture of sodium cacodylate and its acid form, as one of the rainbow herbicides
Copper acetoarsenite	green pigment causes numerous arsenic poisonings
Copper arsenate	colouring agent in sweets

3.4 Beryllium

3.4.1 Overview of beryllium

Beryllium which can be extracted from bertrandite (Be$_4$Si$_2$O$_7$(OH)$_2$), beryl (Al$_2$Be$_3$Si$_6$O$_{18}$), chrysoberyl (Al$_2$BeO$_4$) and phenakite (Be$_2$SiO$_4$) is a brittle alkaline earth metal which has its genesis from stellar nucleosynthesis and occurring only in combination with other elements in 100 gemstone minerals notably beryl (aquamarine, emerald) and chrysoberyl. It has a high melting point and is resistant to attacks by acids.

3.4.2 Beneficiation of beryllium

Beryllium is usually obtained from two naturally occurring sources: beryl and bertrandite. These ores are melted in industrial furnaces, solidified and crushed, then treated with sulphuric acid to produce a water-soluble sulphate. The sulphate solutions undergo a series of chemical extraction steps [240] to ultimately produce extremely pure beryllium hydroxide, from which virtually all contaminants have been removed [241]. The resultant beryllium hydroxide is then converted into beryllium fluoride or beryllium chloride. These halogenides are then reduced to metallic beryllium with other metals or by melt electrolysis. The beryllium metal obtained is then subjected to one or more refining processes and finally further treated by powder metallurgy or in some cases fusion metallurgy.

3.4.3 Beryllium value addition

The combination of beryllium's very low density with its mechanical strength, high melting point, and resistance to acids make beryllium a useful material for structural parts that are exposed to great inertial or centrifugal forces.

Table 3.4 Uses of beryllium and its compounds

Beryllium attributes and properties	created through stellar nucleosynthesis; strong; lightweight; brittle; transparent to X-rays; high thermal conductivity
Beryllium metal	aerospace industry [242–245]; production of CuBe alloy [246] used in modern aeronautics for landing gear components or in electric/electronic connectors; moulds for the production of the formed plastic objects and for the production of paramagnetic tooling; consumer electronics and telecommunications products
Beryllium alloys	structural parts that have to be exposed to great forces [240,245,247–251]
Beryllium oxide (BeO)	high-performance ceramics
Specialty applications	medical devices such as x-ray transparency [243,244,252]; physical instruments; controlled nuclear fusion reactors [465,253–257]; semiconductor processing equipment

3.5 Cadmium

3.5.1 Overview of cadmium

Cadmium has a low melting point, is a divalent metal which forms complex compounds [258] and is characteristically soft, malleable and ductile. It is resistant to corrosion, insoluble in water and is not flammable. Its powdered form may burn and release toxic fumes [259] of brown amorphous cadmium oxide (CdO).

Due to geochemical similarity the cadmium mineral greenockite (CdS) and sphalerite (ZnS) are nearly always associated, making zinc and cadmium geological separation unlikely. Hence, cadmium is produced mainly as a by-product from mining, smelting and refining sulphide ore of zinc as well as lead and copper. Other secondary sources are dust generated by recycling iron and steel scrap [261–263], phosphate fertilisers [264] and coal which end up as flue dust [265]

3.5.2 Beneficiation of cadmium

Cadmium is produced as a by-product of zinc production from sulphidic zinc ore concentrates. In the presence of oxygen zinc sulphide ores are converted to the oxide when roasted. Smelting the oxide with carbon yields the zinc metal. Alternatively, the metal can be produced by electrolysis in sulphuric acid. Isolation of the cadmium from zinc metal can be effected by vacuum distillation if the zinc is smelted, or by precipitation as cadmium sulphate out of the electrolysis solution [263,266].

3.5.3 Cadmium value addition

Table 3.5 Uses of cadmium and its compounds

Cadmium properties and properties	ductile; malleable; soft; resistant to corrosion; unstable in water; not flammable
Cadmium metal	battery production; pigments [267]; coatings [268,269]; electroplating [269] of iron and steel components; cadmium-selective sensors [260]; barrier to control neutrons in nuclear fission
Carboxylates cadmium laureate and cadmium stearate	to stabilise PVC
Cadmium telluride	polar panels
Cadmium sulphide, cadmium selenide, cadmium telluride	light detection and solar cells [1136]
HgCdTe	infrared detector or switch in remote control devices
Cadmium oxide	television phosphors [270]
Cadmium sulphide CdS	photoconductive surfaces for photocopier drums [271]
Helium-cadmium lasers	Blue-ultraviolet laser light in fluorescence microscopes and various laboratory experiments [272]
Cadmium selenide	Bright luminescence under UV excitation (He-Cd laser); fluorescence microscopy [273]

3.6 Caesium

3.6.1 Overview of caesium

Caesium is an alkaline soft metal with a low melting point of 28° C [274,275], a low boiling point [276,277] and is extremely reactive and pyrophoric. It burns with a violet or blue colour and reacts explosively with water even at low temperatures. It is mined from pollucite, $Cs(AlSi_2O_6)$ [278,279], which is found in zoned pegmatites and associated with more commercially important lithium minerals lepidolie and petalite. Caesium alloys with other alkali metals, gold and mercury. With Sb, Ga, In and Th, which are photosensitive, caesium forms well-defined intermetallic compounds.

3.6.2 Beneficiation of caesium

After crushing and grinding pollucite, caesium is extracted [280,281] by either acid digestion, alkaline decomposition or direct reduction. For the acid digestion route, the pollucite ore is reacted with HCl, H_2SO_4, HBr and HF acids to produce precipitates of caesium antimony chloride (Cs_4SbCl_7), caesium iodide chloride (Cs_2ICl) or caesium hexachlorocerate ($Cs_2(CeCl_6)$), if hydrochloric acid is used. The final product is CsCl on separation and the decomposition of the precipitated double salt. With sulphuric acid, an insoluble double salt caesium alum ($CsAl(SO_4)_2.12H_2O$) is produced leading to the final product of Cs_2SO_4. The alkaline decomposition route involves the roasting of pollucite with calcium carbonate and calcium chloride which produces a dilute chloride (CsCl) solution on leaching with water and dilute ammonia (NH_4OH). The alternative direct reduction route takes place when caesium chloride and other caesium halides are processed at 700 to 800° C with calcium or barium, followed by distillation of caesium metal. The aluminate, carbonate, or hydroxide may be similarly reduced by magnesium as can electrolysis of fused caesium cyanide (CsCN) isolate the metal. When caesium dichromate reacts with zirconium, pure caesium metal can be formed.

$$Cs_2Cr_2O_7 + 2Zr \rightarrow 2Cs + 2ZrO_2 + Cr_2O_3 \tag{10}$$

3.6.3 Caesium value addition

Table 3.6 Uses of caesium and its compounds

Caesium attributes and properties	liquid at room temperature; very reactive; very pyrophoric; ignites spontaneously in air; reacts violently with water; low boing point 641°C [276,277];
Caesium formate (HCO^-Cs^+), caesium chloride and nitrate	high-pressure and high-temperature drilling of oil wells [278]
Caesium	a getter in vacuum tubes and photoelectric cells; highly accurate atomic clocks [282,283]; production of electricity and electronics; optical character recognition devices; photomultiplier tubes, video camera tubes; doping to enhance metal ion catalysts; thermoluminescent radiation dosimetry (TLD) for quantifying accumulated radiation dosages
Isotope caesium-137	Medical applications, industrial gauges for measuring moisture, density, levelling, and thickness; well-logging devices for measuring the electron density of rock formations; hydrology; gamma-emitter in industrial applications; used in agriculture, cancer treatment, and sterilisation of food, sewage sludge, and surgical equipment
CsI, CBr, CsFs crystals	scintillation counters; internal standard spectrophotometry [284]; high-energy lasers; vapour glow lamps; vapour rectifiers [278]
Caesium vapour	Magnetometers
Caesium salts	Organic synthesis such as cyclisation, esterification, polymerization
Solutions of caesium chloride, caesium sulphate, and caesium trifluoroacetate ($Cs(OC)_2F$)	used in molecular biology for density gradient ultracentrifugation
Photosensitivity	solar photovoltaic cells

3.7 Chromium

3.7.1 Overview of chromium

Chromite ($FeCr_2O_4$) [289] is the main source of chromium which is a steely-grey, lustrous, hard and brittle metal resistant to corrosion with a high melting point. Chromite ores are geographically concentrated in southern Africa [290] where about two-fifths of the world's concentrates are produced in South Africa and Zimbabwe.

Chromium element exhibits antiferromagnetic ordering at room temperature (and below) but above 38° C, it transforms into a paramagnetic state [285]. When in air the Cr metal is passivated by oxygen, forming a thin protective oxide surface layer which has a spinel structure with a dense thickness of a few atoms, preventing the diffusion of oxygen into the underlying material. This is unlike in iron or plain carbon steels where oxygen migrates into the underlying material and causing rusting [286]. The passive layer is destroyed by reducing agents rendering the chromium metal easily dissolvable in weak acids [287]. Chromium metal suffers from nitrogen embrittlement [288], reacting with atmospheric nitrogen to form brittle nitrides at the high temperature necessary to work the metal parts. The relative dominance of the Cr(III) and Cr(VI) species in solution is strongly pH-dependent and sensitive to the oxidative properties of the environment. However, in most cases, Cr(III) is the dominating species [291,292] with exceptions of ground water of some areas where the Cr(VI) predominates [293].

3.7.2 Beneficiation of chromium

Chromium metal and ferrochromium alloy, the two main products of chromium refining, are commercially produced from chromite ($FeCr_2O_4$) by silicothermic or aluminothermic reductive reactions [294], or by roasting and leaching processes.

3.7.2.1 The aluminothermic process

The reduction reaction of oxides by aluminium can be generally, represented by the following equation:

The following equation represents the reduction reaction of oxides by aluminium

$$2/yM_xO_y + 4/3Al = 2x/yM + 2/3Al_2O_3 \qquad (11)$$

Where MO represents the reducible oxides in the ore or concentrates

Chromite concentrate fines, which contain iron and magnesium in the case of a Cr spinel, aluminium powder and sodium or potassium nitrate are thoroughly mixed. Ground lime and fluorspar, with grain sizes similar to that of the ore are used as fluxing materials. The mixture is preheated to temperature of 400° C. Once the reaction starts, it proceeds spontaneously. The aluminothermic reaction becomes self-propagating and occurs throughout the charge and produces enough heat to melt the products of the reaction and allows separation of the metal and slag. The aluminothermic reduction process proceeds outside the furnace when the reaction temperature exceeds the melting point of the oxides that are being reduced. The duration of the reaction is about 90 seconds in each batch process after which the reaction products are normally then left to cool to room temperature.

3.7.2.2 Production of the pure Cr metal

In order to produce pure chromium from the chromite ($FeCr_2O_4$), a two-step roasting and leaching process is carried out in order to separate the iron from the chromium. The roasting stage involves heating chromite ore with a mixture of calcium carbonate and sodium carbonate in air, resulting in the oxidation of the Cr to the hexavalent chromate form while iron forms the stable Fe_2O_3.

$$4FeCr_2O_4 + 8Na_2CO_3 + 7O_2 \rightarrow 8Na_2CrO_4 + 2Fe_2O_3 + 8CO_2 \tag{12}$$

Subsequently the leaching stage at elevated higher temperatures results in the dissolution of the chromate by sulphuric acid to dichromate leaving an insoluble iron oxide [294].

$$2Na_2CrO_4 + H_2SO_4 \rightarrow Na_2Cr_2O_7 + Na_2SO_4 + H_2O \tag{13}$$

The dichromate can then be reductively converted to the chromium(III) oxide by reaction with carbon

$$Na_2Cr_2O_7 + 2C \rightarrow Cr_2O_3 + Na_2CO_3 + CO \tag{14}$$

and subsequently reduced in an aluminothermic reaction to the chromium metal.

$$Cr_2O_3 + 2Al \rightarrow Al_2O_3 + 2Cr \tag{15}$$

3.7.2.3 Production of ferrochrome (FeCr)

Electric arc smelting of chromite ($FeCr_2O_4$) in the presence of coal or coke results in the formation of ferrochrome (FeCr) by a carbothermic reduction operation at high temperature of about 2,800° C. The heat for this reaction come, typically, from the electric arc formed between the tips of the electrodes in the bottom of the furnace and the furnace hearth. This production is very high energy intensive and consequently very expensive.

Chromium is a key commodity in the manufacture of stainless steels. Ferrochrome is an intermediate product which is primarily used as a source of chromium for the stainless steel industry. The manufacture of stainless steel [479] products such as cutlery, sinks, etc. represents the full value addition of chrome.

3.7.3 Chromium value addition

Table 3.7 Uses of chromium and its compounds

Chrome attributes and properties	hard and brittle metal; wear resistant; resistant to corrosion; high melting point; antiferromagnetic; reacts with atmospheric nitrogen to form brittle nitrides
Metallurgical applications	stainless steel products (cutlery, sinks, etc.) with formation of stable metal carbides [1248]; Cr-Ni superalloys use fin jet engines and gas turbines [295]; coating (by plating on metallic surfaces) for decorative purposes and to provide corrosion and wear resistance [296–299] application including tools such as boring bars, gauges, dies, punches, drilling machines sleeves, lathe spindles, pressure rams, rolls of different kinds, pressure plates, crank shafts, camshafts, bearings, drawing tools, journals, engine cylinder liners, pistons rings, non-magnetic surfaces; anodising of aluminium [300,301]

(Continued)

Table 3.7 (Continued)

Dye and pigment	crocoite ($PbCrO_4$) as yellow pigment [302]; lead chromate ($PbCrO_4 \cdot Pb(OH)_2$) as a bright red pigment; chrome green pigment [303]; chromium oxides used as a green colour in glass making and as a glaze in ceramics [304], cladding coatings and IR reflecting paints for the armed forces to paint vehicles as green leaves [305]; natural rubies coloured red due to Cr(III) ions [306]
Wood preservation with Cr(VI) salts	timber treatment against decay, fungi, wood-attacking insects, including termites and marine borers uses chromated copper arsenate (CCA) [307]
Leather tanning with Cr(III) salts	chromium(III) sulphate used in leather tanning [289,308,309]
Refractory material made from mixtures of chromite and magnesite [294]	because of their heat resistivity and high melting points, chromite and Cr(III) oxides are for high temperature refractory applications such as in blast furnaces, cement kilns, moulds for firing of bricks, foundry sands for casting of metals
Catalysts	for processing hydrocarbons e.g. Phillips catalyst for the production of polyethylene[310]; FeCr mixed oxides as high-temperature catalysts for the water shift reaction [311,312]; copper chromite catalyst for hydrogenation [313]
Cr(IV) compounds	CrO_2 imparts high coercivity and remnant magnetisation in the manufacture of magnetic tapes for audio tapes and cassettes [314]; chromates are added to drilling muds to prevent corrosion of steel under wet conditions [315]
Cr(III) compounds	Cr_2O_3 a metal polish; chromic acid, a powerful oxidising agent, for cleaning laboratory glassware; potassium dichromate as a titration agent and mordant (fixing agent) for dyes in fabrics

3.8 Cobalt

3.8.1 Overview of cobalt

Cobalt is commonly found in association mainly with minerals for copper and nickel but also to a lesser extent with silver, lead and iron ores. The metal exhibits hard and brittle characteristics. Cobalt's most common uses involve its alloys which, due to its presence, assume desirable properties of heat resistance, high strength, wear resistance and superior magnetism [316,320,321], among other beneficial attributes. These alloys have applications in jet engines, permanent magnets [317], cutting tools and pigments. Because of these applications, cobalt has been classified as a 'strategic mineral' by the USA, China and the EU, among others. In some of the producing African countries it is considered a 'conflict mineral' alongside tin, tungsten, tantalum and gold, the so-called '3TG'.

3.8.2 Beneficiation of cobalt

In order to extract cobalt several methods have been developed to separate the cobalt metal from copper and nickel. The choice of the methods depends on the concentration of the cobalt and the mineralogy of the ore used. One of the methods has a separation step that involves froth flotation in which surfactants bind to different ore components, leading to an enrichment of cobalt ores. The cobalt ore concentrate is subsequently roasted to convert them to cobalt sulphate and simultaneously oxidising copper and iron to their oxides. On leaching the products with water the sulphate and arsenates are extracted. Further leaching of the residue with sulphuric acid yields copper sulphate. Alternatively, cobalt can be leached from the slag of the copper smelter [318]. The cobalt sulphate products from either of the aforementioned processes are transformed into the cobalt oxide (Co_3O_4) which is reduced to the metal by the aluminothermic reaction or reduction with carbon in a blast furnace [319].

3.8.3 Cobalt value addition

Table 3.8 Uses of cobalt and its compounds

Cobalt attributes and properties	hard; brittle; heat resistant; high strength; wear resistant
Cobalt alloys	aerospace and defence industries; jet engines, stationary gas turbines, permanent magnets for electrical and electronics equipment, machinery, and non-metallic applications
Batteries	lithium cobalt oxide ($LiCoO_2$) in lithium-ion battery cathodes [322]; nickel-cadmium (NiCd) and nickel metal hydride (NiMH) batteries [323,324] contain Co
Cobalt compounds catalysts	oxidation catalysts; cobalt acetate for conversion of xylene to terephthalic acid, the precursor to polymer polyethylene terephthalate; carboxylates are used in paints, varnishes, and inks as drying agents [322] and used to improve the adhesion of the steel to rubber in steel-belted radial tyres; in reactions involving carbon monoxides such as in hydrogen production in the Fischer-Tropsch process for liquid fuels [325]; cobalt octacarbonyl catalyst for hydroformylation of alkenes [326]; hydrodesulphurisation of petroleum [322]
Pigments	smalt (blue coloured glass) produced by melting a mixture of roasted smaltite, quartz and potassium carbonate [327] and pigment [328]; cobalt blue (cobalt aluminate [329–331]
electroplating	for decorative purposes, hardness and resistance to oxidation [332] and as ground coats for porcelain enamels [333]

3.9 Copper

3.9.1 Overview of copper

Copper (Cu) is characteristically known for its high thermal and electrical conductivity. Pure copper is ductile, soft and malleable and forms various copper sulphides such as chalcopyrite ($CuFeS_2$) and chalcocite (Cu_2S). It also exists as the copper carbonates, azurite $Cu_3(CO_3)_2(OH)$ or $CuCO_3 \cdot Cu(OH)_2$ and malachite $Cu_2(CO_3)(OH)_2$ or $CuCO_3 \cdot Cu(OH)_2$, and the copper(I) oxide mineral cuprite Cu_2O.

3.9.2 Beneficiation of copper

Although copper can be recovered through in-situ leach processes [39,334] its crushed sulphide ores averaging 0.6% Cu, especially chalcopyrite and chalcocite [336], are concentrated to about 10–15% Cu by froth flotation or bioleaching [337,1254] leading to the following pyrometallurgical recovery route. The sulphide concentrate is heated with silica in flash smelting to remove much of the iron as slag. The iron sulphde in the ore is converted to its oxides which in turn react with the silica to form the silicate slag. This slag floats on top of the heated mass. As a result, *copper matte* consisting of Cu_2S is then roasted to convert all sulphides into oxides [336].

$$2Cu_2S + 3O_2 \rightarrow 2Cu_2O + 2SO_2 \tag{16}$$

The resultant cuprous oxide is converted to *blister* copper upon heating:

$$2Cu_2O \rightarrow 4Cu + O_2 \tag{17}$$

Copper can also be produced from recycling of manufacture products [338] through a process that is used to extract copper from ores. Billets and ingots are produced from smelting high scrap copper by smelting in a furnace followed by reduction reaction. Electroplating in a bath of sulphuric acid is used to refine low purity scrap.

3.9.3 Copper value addition

The amount of copper in use is increasing and the quantity available is barely sufficient to meet the demand of all the countries to reach developed world levels of usage [311,335].

Table 3.9 Uses of copper and its compounds

Copper attributes and properties	High heat and electrical conductivity, tensile strength, ductility, creep (deformation) resistance, corrosion resistance, low thermal expansion, high thermal conductivity, solderability, and ease of installation
Copper alloys	brass (Cu-Zn) alloy; bronze (Cu-Sn) alloy; carat silver and gold alloys; carat solders used in jewellery industry with modified colour, hardness and melting points; cupronickel (Cu-Ni) alloy used in outer cladding of low-denomination coins; (Cu-Al) alloys used in decorations [339]; lead-free (Cu-Sn) alloys with other metals
Heat and electrical conductor [341–343]	electrical wiring for power generation, power transmission, power distribution, telecommunications, electronics circuitry, electrical equipment; building material: building wire, appliance wire, automotive wire and cable, magnet wire; integrated circuits and printed circuit boards; heat sinks and heat exchangers; electromagnets, vacuum tubes, cathode ray tubes; magnetrons in microwave ovens, wave guides for microwave radiation; motors and motor driven systems [344]
Copper compounds agricultural application	nutritional supplements; fungicides [340]
Copper oxides and carbonates	in glassmaking and in ceramics glazes to impart brown and green colours

3.10 Gallium

3.10.1 Overview of gallium

Gallium normally exists as an arsenide (GaAs), nitride (GaN), or sulphide in combination with other metals such as copper in the form of gallite $CuGaS_2$ [345] and as a trace element in some other minerals [346] such as phosphate ore and various kinds of coal as part of flue dust [348–351] in the process of phosphorus and fly ash production, respectively. The amounts from these sources as well as from diaspore and germanite are negligible. Extraction as a trace element in bauxite (90%) and sphalerite (10%) takes place [346,347,355]. However, the metal is found in alloys that it readily forms with most other metals. It does not exist as a free metal in nature. Its melting point is very low (29.7646° C) which is regarded as one of the formal temperature reference points in the International Temperature Scale of 1990 (ITS-90) established by the BIPM [352,353] whereas its triple point (29.7666° C) is used by NITS as a reference point in preference [354].

3.10.2 Beneficiation of gallium

Bauxite is the major source of gallium [356,357] at 30–80g/t as a by-product, 70% of which is leached into the Bayer liquid for subsequent extraction and purification into gallium arsenide (GaAs) and gallium nitride (GaN) which have semiconductor characteristics with applications in light-emitting diodes (LEDs), integrated circuits (ICs) and solar cells [334,1136]. The semiconductor scrap is an additional recovery supply. In the recovery process from alumina sodium gallate is the product of a mercury cell electrolysis and hydrolysis of the amalgam with sodium hydroxide leading to the gallium metal. Zone melting or single-crystal extraction from a melt (Czhochralski process) is used for further purification (99.9999%) to produce material for semiconductor use [349].

3.10.3 Gallium value addition

Table 3.10 Uses of gallium and its compounds

Gallium attributes and properties	soft; low melting point; large liquid range
Semiconductors [498,1259]	Gallium arsenide (GaAs) and gallium nitride (GaN) used as in LEDs, ICs and solar cells [345,358] and wafers; smartphones; wireless communications; aerospace and defence; back-lit televisions; laser diodes in medical technology; blue-violet lasers in Blu-ray DVD readers
Gallium alloys	(GaInSn) eutectic alloys (Galinstan) as medical thermometers; mirrors because of wetting action of gallium on glass and porcelain
Medical equipment	medical imaging [497]; fuses for electrical devices; dental fillings
Microstructural impact on mechanical properties	makes Al-Zn alloy [347] or steel [359] brittle; stabilises plutonium δ phase
Photovoltaics (Solar)	(CIGS) in PV cells in solar panels; multijunction photovoltaic cells; $Cu(In,Ga)(Se,S)_2$ photovoltaic compound

3.11 Germanium

3.11.1 Overview of germanium

Germanium is a semi-metal [360] with non-metallic characteristics such as semiconductivity. It expands as it solidifies from its molten state [84] oxidising slowly to GeO_2 at 250° C [109] and dissolving slowly in hot concentrated sulphuric and nitric acids. With molten alkalis it reacts violently producing germanates ($[GeO_3]^{2-}$). The principal sources of germanium are zinc refining, where it is recovered as a by-product, and coal fly ash with 30% of germanium production coming from recycling of scraps and tailings.

3.11.2 Beneficiation of germanium

Germanium sulphide GeS_2 is converted to its oxide when roasted in air.

$$GeS_2 + 3O_2 \rightarrow GeO_2 + 2SO_2 \tag{18}$$

The oxide is converted to germanate when leached together with the zinc in sulphuric acid. On neutralisation, zinc stays in solution leaving germanium and other metals as precipitates. A second leach is carried out after zinc has been reduced in the precipitate by the Waelz process resulting in the precipitate of a dioxide which can be converted with chlorine gas or HCl to germanium tetrachloride. This can be distilled off.

$$GeO_2 + 4HCl \rightarrow GeCl_4 + 2H_2O \tag{19}$$

$$GeO_2 + 2\ Cl_2 \rightarrow GeCl_4 + O_2 \tag{20}$$

The germanium tetrachloride can be either hydrolysed back to the oxide (GeO_2) or purified by fractional distillation and then hydrolysed [361] to the oxide from which germanium glass can be produced. The pure germanium suitable for the infrared optics or semiconductor applications can be obtained by reducing pure GeO_2 with hydrogen.

$$GeO_2 + 2H_2 \rightarrow Ge + 2H_2O \tag{21}$$

The germanium for industrial processes such as steel production is normally obtained by reducing the oxide using carbon [362].

$$GeO_2 + C \rightarrow Ge + CO_2 \tag{22}$$

3.11.2 Germanium value addition

Table 3.11 Uses of germanium and its compounds

Germanium attributes and properties	brittle; hard; semiconductor; expands as it solidifies
Catalyst	polymerisation of PET plastics for plastic bottles, sheet, film and synthetic textile fibres [363,367]
Fibre optics	in solid-state electronics; a dopant [366] within the core of fibre optics in the form of germanium tetrachloride [364,365]; in infrared night-vision devices; as semiconductor and substrate in electronic circuity and solar cells [363,1136]; high-speed telecommunication [362,365,368,369]
Metal and its oxide	infrared optics [365]; military application such as night-vision devices; firefighting equipment, satellite imagery sensors; medical diagnostics [362,368,369]
Solar cells	space-based applications [363,370,371] for commercial, military and scientific applications
Electronic components	as semiconductor LEDs [363] devices such as cameras smartphone display screens; silicon germanium transistors in high-speed wireless telecommunications devices

3.12 Hafnium

3.12.1 Overview of hafnium

Hafnium (Hf) is a corrosion-resistant metal which is chemically similar to zirconium in whose minerals it is found. It is ductile and has a shiny silvery colour. Hafnium and zirconium are very difficult to separate due to their chemical similarity [372,373]. However, hafnium is twice as dense as zirconium and has a high thermal neutron-capture cross-section with the nuclei of several different hafnium isotopes readily absorbent of two or more neutrons apiece [372] whereas zirconium is practically transparent to thermal neutrons. Hafnium's resistance to corrosion is a consequence of its strong reactivity with air to form a passivating layer that inhibits further corrosion. The metal is also resistant to both acids and concentrated alkalis. However, it can be oxidised with halogens and can ignite spontaneously in air if finely powdered.

3.12.2 Beneficiation of hafnium

Hafnium, admixed with zirconium, is mined from titanium ores of ilmenite and rustle sometimes as part of mineral sands. Due to their chemical similarities, hafnium and zirconium have been found difficult to separate. One method is by fractional crystallisation using ammonium fluoride salts or by fractional distillation in chloride medium [374]. Unfortunately, these are not suitable for industrial-scale production. Liquid-liquid extraction processes with a wide range of solvents have been used for the production of hafnium [375] with hafnium(IV) chloride as the final product of the separation [376]. The hafnium metal is obtained by converting purified hafnium(IV) chloride by reduction with magnesium or sodium as in the Kroll process [377].

$$HfCl_4 + 2Mg \; (1100^{\circ} C) \rightarrow 2MgCl_2 + Hf \tag{23}$$

The van Arkel and de Boer chemical transport reaction process is used for further purification. Hafnium reacts with iodine at a temperature of 500° C, to form hafnium(IV) iodide in a closed vessel.

$$Hf + 2I_2 \; (500^{\circ} C) \rightarrow HfI_4 \tag{24}$$

The reverse reaction occurs at a tungsten filament of 1700° C setting iodine and hafnium free.

$$HfI_4 \; (1700^{\circ} C) \rightarrow Hf + 2I_2 \tag{25}$$

The hafnium coats the tungsten filament, while the iodine becomes available to react with additional hafnium [378,379].

3.12.3 Hafnium value addition

Hafnium is a scarce commodity due to its low abundance and difficult separation techniques [372].

Table 3.12 Uses of hafnium and its compounds

Hafnium attributes and properties	ductile; corrosion resistant; high thermal neutron-capture cross-section; reacts in air; resistant to acid and alkaline attack; finely divided Hf can ignite spontaneously in air
Metal	filaments and electrodes; its oxide used for IC; gas-filled incandescent lamps; electrode in plasma cutting
Nuclear applications	reactant for neutron absorption in control rods in nuclear [465] power plants; in pressurised water reactors [375] and military reactors [382,382]
Hf alloys	superalloys in combination with niobium, titanium or tungsten [380]; nickel-base alloys [383–385] for corrosion resistance under cyclic temperature conditions
Hafnium isotopes	isotope geochemistry and geochronological applications [386–391] dating metamorphic events
Hafnium(IV) oxide (HfO_2)	gage insulators [458] in field-effect transistors; optical coatings; high-k dielectrics in DRAM capacitors and in advanced metal-oxide semiconductor devices; thermocouples at temperatures up to 2500° C; devices as thermocouples, where it can operate at temperatures up to 2500°C due to its high melting point

3.13 Indium

3.13.1 Overview of indium

Indium is a metal which is soft, malleable [392] and non-existent in an elemental state in the Earth's crust but occurs in minerals of other metals, chiefly zinc mineral sphalerite.

3.13.2 Beneficiation of indium

The major production for indium is as a by-product, principally of zinc and also from sulphide minerals of copper, iron, lead and tin. The slag and dust of zinc production is leached to yield indium whose further purification is done by electrolysis [393,394]. Subsequent to purification, for use in FPDs, indium is synthesised into indium tin oxide (ITO) which is 'sputtered' onto either clear glass or plastic, forming a transparent electrical conductor. Two thirds of global indium production is accounted for by recovery from ITO scrap.

Post-industrial CIGS (copper indium gallium (di)selenide) is also an economically viable source of indium although for LCDs, the post-consumer concentrations of indium in CIGS solar panels are low. High amount of contamination and low concentration of indium render recovery from tailings difficult, explaining the insignificancy of recycling from tailings.

3.13.3 Indium value addition

Table 3.13 Uses of indium and its compounds

Indium attributes and properties	soft; malleable; easily fusible; does not react with water
Metal	conducting adhesive bonding gold to superconductors [402]; vacuum seal and thermal conductor in cryogenics and ultra-high-vacuum

(Continued)

Table 3.13 (Continued)

	applications [403]; calibration material for differential scanning calorimetry [404]; lead-free soldering; its melting point fixed point on ITS-90 [407]; thermal interface material in personal computers [410]; a substitute for Hg in alkaline batteries
Alloys	aluminium alloy sacrificial anodes for sea water application; ingredient in gallium-indium-tin alloy galinstan [405]; component of dental amalgam alloys; alloys in fuses or plugs for fire control systems
Nuclear applications	reactions of ^{113}In and ^{115}In used to determine magnitudes of neutron fluxes [409]; alloy containing Ag, In, and Cd used in control rods for nuclear reactors [408,465]
Biomedical applications	^{111}In emits γ-radiation for indium leukocyte imaging [497], scintigraphy [411]
Indium tin oxide (ITO) and indium oxide (In_2O_3)	indium tin oxide used as a light filter [406] in low-pressure sodium vapour lamps [398]; ITO and In_2O_3 used in flat panel devices (FPDs) including flat screen computer monitors; LCD [395] smart phones; televisions and notebooks; alloys and solders; solar panels; LEDs; laser diodes; vacuum seals; transparent conductive coatings of ITO on glass substrates for electroluminescent panels
Semiconductors [492,498]	indium antimonide, indium phosphide [396], and indium nitride [397]; synthesis of CIGS for solar panels [398]; InGaN for MOVPE technology [399]; TMI as a precursor in III-V compound semiconductors [400]; semiconductor dopant in II-VI compound semiconductors; in alkaline batteries to prevent the Zn from corroding [401]; high purity trimethylindium (TMI) used as precursor in III-V compound semiconductors and dopant in II-VI semiconductors
Photovoltaics	indium used for synthesis of the CIGS thin-film solar cells [1136]

3.14 Iron

3.14.1 Overview of iron

Iron is a metal which occurs in elemental form in the absence of oxygen. In the latter's presence it reacts together with water to form porous hydrated iron oxides which flake off, exposing fresh surfaces for corrosion. Iron, therefore, is often combined with oxygen as oxide minerals haematite (Fe_2O_3) and magnetite (Fe_3O_4) from which industrial iron is produced.

3.14.1.1 Pyrite

The mineral pyrite (FeS_2,), which is another source of iron, is usually found associated with other sulphide minerals or oxides in quartz veins, sedimentary rock, and metamorphic rock, as well as in coal beds and as a replacement mineral in fossils. It is also sometimes found in association with small quantities of gold and arsenic [419] where it renders gold ore refractory. It is a commercial source of iron as well as sulphur dioxide.

3.14.2 Beneficiation of iron

Iron and steel are produced in two sequential stages from haematite and magnetite which are reduced to the metal by treatment with carbon, in the form of coking coal, in carbothermic reactions in a blast furnace at temperatures of about $2000°$ C. Silicaceous minerals in the ore, which would otherwise clog the furnace, are removed by the addition of *flux* such as limestone (calcium carbonate) and dolomite (calcium-magnesium carbonate). Pig iron which has a high carbon content is initially produced. In the blast furnace there is a reaction between coke and oxygen to produce carbon monoxide which in turn reduces the iron ore to molten iron with a release of carbon dioxide.

$$2C + O_2 \rightarrow 2CO \qquad (26)$$

$$Fe_2O_3 + 3CO \rightarrow 2Fe + 3CO_2 \qquad (27)$$

In some parts of the furnace where the temperatures are sufficiently high, some of the iron reacts directly with the coke to produce molten iron:

$$2Fe_2O_3 + 3C \rightarrow 4Fe + 3CO_2 \qquad (28)$$

The calcium oxide, a product of the decomposition of the limestone flux, combines with silicon dioxide to form a liquid slag which melts in the heat of the furnace:

$$CaCO_3 \rightarrow CaO + CO_2 \qquad (29)$$

$$CaO + SiO_2 \rightarrow CaSiO_3 \qquad (30)$$

The molten iron settles at the bottom of the furnace while the slag, which is less dense, floats on top of that. The iron and slag are run off separately through side apertures. When cooled, the iron is called pig iron, which is subsequently converted to wrought iron or steel. The level of carbon inclusion in the resulting iron characterises its hardness. Addition of between 0.0025 and 2.1% carbon results in making steel up to 1000 times harder than pure iron. Carbon content in the iron can be reduced to desired proportion by oxygen in further refinement to make steel. Mixing the iron with certain other metals and carbon yield many types of steels with different properties. The other metals that may be alloyed with iron include: Mn, Ni, Cr, Mo, B, Ti, V, W, Co, Nb [40]. Other elements that are also important in steel are phosphorus, sulphur, silicon and traces of oxygen, nitrogen, and copper. Thus, the classification of commercially available iron is determined by the purity and the abundance of additives.

The form that the carbon takes in the alloy determines the mechanical properties which vary greatly. While ferrite, *α-iron*, is a very stable form of iron at room temperature, it undergoes a phase transition to form austenite, *γ-iron*, which can dissolve considerably more carbon [1248]. Wrought iron contains less than 0.25% carbon and has large amounts of the slag which gives it a fibrous characteristic.

Iron can be processed alternatively by direct reduction to a powder called "*sponge*" iron which is suitable for steelmaking. The process is comprised of two main reactions. Catalysed heated natural gas is partially oxidised to carbon monoxide and hydrogen which are then treated with iron ore in a furnace to produce solid sponge iron:

$$2CH_4 + O_2 \rightarrow 2CO + 4H_2 \qquad (31)$$

$$Fe_2O_3 + CO + 2H_2 \rightarrow 2Fe + CO_2 + 2H_2O \qquad (32)$$

As described previously, limestone *flux* is used to remove silica.

Steel or wrought iron can be made from about 2% carbon pig iron. This can be brought about by various processes such as finery forges, puddling furnaces, Bessemer converters, open hearth furnaces, basic oxygen furnaces, an electric arc furnace. In all cases, the objective is to oxidise some or all of the carbon, together with other impurities.

Annealing [479], which involves heating of a piece of steel to 700–800° C for several hours and then gradually cooling, makes the steel softer and more workable [412].

3.14.3 Iron value addition

Table 3.14 Uses of iron and its compounds

Iron attributes and properties	reactive to oxygen and water; relatively soft
Metal and alloys	stainless steel [479]; industrial uses [413]; construction of machinery, machine tools, automobiles, shipbuilding, components of buildings' radiation protection
Iron chemical compounds	welding and purifying ores; halogens and chalcogens; ferrocene; Haber-Bosch Process for production of ammonia; Fischer-Tropsch process for the CTL [416]; Bechamp reduction of nitrobenzene to aniline [417]; water purification and sewage treatment; dying of cloth uses iron(III) oxide; colouring agent in paints; additive in animal feeds; etchant for Cu in the manufacture of printed circuit boards; tincture of iron; halides in laboratory uses; iron(II) sulphate as a reductant in cement and sewage treatment [418]; fortify foods; iron(II) is used as a reducing flocculating agent; and as a reducing agent in organic synthesis [418]
Biological	complexes with molecular oxygen in haemoglobin; redox enzymes [414,415]
Pyrite	sulphur dioxide production [420]; paper industry; manufacture of sulphuric acid; cathode material in Energizer brand non-rechargeable Li batteries [421]; semiconductor [422]; pyrite mineral detectors [423,424]; photovoltaic solar panels [425,426]; marcasite jewellery [427]

3.15 Lead

3.15.1 Overview of lead

Lead (Pb) is apparently the heaviest non-radioactive element which is malleable with a shiny chrome-silver lustre when melted in a liquid. Metallic lead is rarely found in nature but as ore in association with zinc, silver and copper minerals with which it is extracted.

3.15.1.1 Galena

Galena (PbS) is the main lead mineral containing 86.6% lead by weight, other common minerals being cerussite ($PbCO_3$) and anglesite ($PbSO_4$), which are formed by weathering or oxidation of galena in a process similar to bioleaching [429,1254]. Galena is also an important source of silver [428] and is often associated with minerals sphalerite, calcite and fluorite. Other elements that occur in galena in variable amounts are cadmium, iron, copper, antimony, arsenic, bismuth and selenium.

3.15.2 Beneficiation of lead

Galena has a low melting point, making it easy to liberate lead by smelting. Lead ores are firstly crushed and concentrated by froth flotation to more than 70%. Subsequently the ores are roasted to yield lead oxide and a mixture of its sulphates and silicates and other metals that may be contained in the ore. The lead metal is produced by the reduction of the lead oxide from the roasting process in a coke-fired furnace where it settles at the bottom. Floating on top of the metallic lead are slag (silicates containing 1.2% lead), matte (sulphides containing 15% lead) and the rest (arsenides of iron and copper). These wastes contain unreduced lead and appreciable amounts of copper, zinc, cadmium and bismuth.

In order to recover significant contaminants of arsenic, antimony, bismuth, zinc, copper, silver and gold contained in the metallic lead from the roasting and blast furnace process, the melt is treated in a reverberatory furnace with air and steam. This process, however, does not oxidise silver, gold and bismuth. The oxidised contaminants float to the top where they can be skimmed off by dosing. The Parkes process can be used to economically recover silver and gold followed by the Betterton-Kroll process to recover bismuth from the de-silvered lead. The Betts process yields pure lead obtained by processing smelted lead electrolytically in an electrolyte of silica fluoride using anodes of impure lead and cathodes of pure lead.

Galena can also be reductively decomposed for the recovery of silver [493].

3.15.3 Lead value addition

Table 3.15 Uses of lead and its compounds

Lead attributes and properties	high density; malleable; soft; ductile; high corrosion resistant; reacts readily with organic chemicals; poor electrical conductivity
Metallic lead	bullets and shot; weights; solder for electronics; pewters; small arms ammunition and shotgun pellets; ballast keel of sailboats; scuba diving weight belts; sheathing material in high voltage cables; shielding material from radiation (e.g. x-ray rooms); coolant (for lead cooled fast reactors); Oddy test for museum materials (detecting organic acids, aldehydes, and organic gases); in tennis rackets; forms glazing bars for stained glass or multi-lit windows; sound deadening layer in walls, floors and ceiling design in sound studios
Lead electrodes	lead-acid batteries in automobiles especially car batteries; electrodes in electrolytic processes
Lead alloys	fusible alloys; radiation shields; added to brass to reduce machine tool wear; base metal of organ pipes
Construction industry	lead sheets in roofing material, cladding, flashing, gutters and gutter joints, roof parapets; decorative motifs; statues and sculptures; balance wheels of cars
Galena	kohl to reduce glare desert sun and repel flies [430]; primary source of lead and silver; green glaze to pottery; semiconductor with a small bandgap of ~ 0.4eV for wireless communication systems; in crystal radio sets

3.16 Lithium

3.16.1 Overview of lithium

Lithium is highly reactive and quickly corrodes to a dull silvery grey and subsequently tarnishes to a black colour when in contact with moist air. It does not freely occur in nature in elemental form. It is found in compounds that are usually hosted by several pegmatic minerals such as spodumene $LiAlSi_2O_6$, petalite $LiAlSi_4O_{10}$ and lepidolite $K(Li,Al,Rb)_3(Al,Si)_4O_{10}(F,OH)_2$, all of which are lithium-aluminum silicates. Lepidolite is a lithium mica which normally contains caesium, rubidium and fluorine. Lithium is soluble and present in ocean water from which it may be extracted as from brines and clays.

3.16.2 Beneficiation of lithium

There are in excess of 100 different varieties and sub-varieties of lithium minerals and ores whose physical and chemical make-up and behaviour is different. The processes which may be used on one type of mineral ore are sometimes entirely ineffective on another type of mineral.

There are a series of hydrometallurgical processes that can be used for the extraction of lithium from spodumene. Using a standard flowsheet it is possible to start with the lithium concentrate and produce high grade lithium products such as lithium carbonate or lithium hydroxide which are reagents for the lithium battery industry. The multi-step process may involve atmospheric leaching, liquid-solid separation and impurity removal through precipitation and ion exchange.

One of the processes that can be used for the recovery of lithium involves the decomposition of the ore and dissolving all of the metallic constituents in an acid like sulphuric acid. However, the required extensive purification makes the decomposition process prohibitively costly.

Lithium salts are extractable from elements in igneous minerals, water in mineral springs, brine pools and brine deposits. A mixture of fused 55% lithium chloride and 45% potassium chloride at $\sim 450^\circ$ C is a typical solution for the electrolytic deposition of the metallic lithium which can also be obtained from brines by solar evaporation. In the latter process potassium is removed first to leave increasing concentration of lithium chloride from which boron and magnesium are also removed by filtration. Lithium carbonate precipitate can be formed by treating lithium chloride with sodium carbonate. Lithium carbonate is an important intermediary which can be converted into several industrial salts and chemicals or processed into lithium metal.

The large-scale production of lithium is from spodumene ores. Separation of lithium minerals can be efficiently achieved by taking advantage of their physical, electrical and magnetic properties. Physical separations are performed by wet and dry screening, tabling and magnetic, electromagnetic, electrostatic, magnetohydrostatic bullet and heavy media separation. Gravity separation is feasible only if the spodumene is coarsely grained. High grade spodumene concentrate (75–85%) suitable for lithium extraction can be generated by flotation which can be further processed pyrometallurgically or hydrometallurgically to produce lithium carbonate or other desirable lithium compounds. Roasting is performed at about 1050° C during which spodumene goes through a phase transformation from α-spodumene to ß-spodumene. While α-spodumene is resistant to hot acid attack the ß-spodumene is amenable to hot sulphuric leaching. The phase transformation brought about by roasting results in the spodumene crystal structure expansion of 30%. After cooling the roasted material is mixed with sulphuric acid and roasted again at 200° C in the course of which an exothermic reaction ensues starting at 170° C. Lithium is extracted from ß-spodumene forming a water-soluble lithium sulphate.

Working with the lithium concentrate, one can use a standardised flowsheet to produce high-grade lithium products such as lithium carbonate or lithium hydroxide. These are reagents for the lithium battery industry. The multi-step process involves atmospheric leaching, liquid-solid separation and impurity removal via precipitation and ion exchange.

3.16.3 Lithium value addition

Table 3.16 Uses of lithium and its compounds

Lithium attributes and properties	good conductor of heat and electricity; highly reactive element; low density
Glass and ceramics	lowering the melting temperature of glass in the Hall-Heroult process involving aluminium oxide; lithium oxide in the processing of silica [431]; oven wares have lithium oxides (from Li_2CO_3 upon heating) as components
Grease lubricants	high-temperature lithium greases for aircraft engines or similar applications; lithium stearate can thicken oils, and is used to manufacture all-purpose, high-temperature lubricating greases [474–476]
Alloys	used to make high-performance aircraft parts [435] when alloyed with aluminium, cadmium, copper and manganese
Batteries	Lithium (SEP −3.04V) is used for anodes in lithium batteries); lithium-ion batteries which are disposable (primary); lithium-ion batteries (for vehicles) which are rechargeable; lithium-ion polymer battery, lithium-ion battery, and nanowire battery are also rechargeable; lithium carbonate and lithium hydroxide (precursor compounds for several applications) are key strategic materials for cleaner energy and electronic devices (laptops, mobile phones); lithium-ion spinel $Li_4Ti_5O_{12}$ battery synthesised by solid state method [460,494]; $Li[Li_{0.2}Mn_{0.54}Ni_{0.13}Co_{0.13}]O_2$ is a high capacity cathode material for rechargeable lithium-ion batteries [463,1136]
Flux additives	lithium carbonate in continuous casting mould flux slags [432]; lithium compounds as additives (fluxes) in iron casting [433]; lithium fluoride as an additive to aluminium smelter (Hall-Heroult process) reducing melting temperature and increasing electrical resistance [434]; metallic lithium as a flux promotes the fusing of metals and eliminates formation of oxides
Air treatment	air desiccants for gas streams; lithium hydroxide and lithium peroxide remove CO_2 and provide air purification aboard spacecraft and submarines; lithium peroxide (Li_2O_2) removes CO_2 and releases oxygen [439]

$$2Li_2O_2 + 2CO_2 \rightarrow 2Li_2CO_3 + O_2 \tag{33}$$

several lithium compounds such as lithium perchlorate used in oxygen candles to supply submarines with oxygen [440]

Polymer industry	organolithium compounds as sources of polymer and fine chemicals; alkyl lithium compounds are catalysts/initiators for anionic polymerisation of unfunctionalised olefins [443–445]; organolithium compounds prepared from lithium metal and alkyl halides [446] are strong bases and reagents for the formation of C-C bonds such as in lithium aluminium hydride ($LiAlH_4$), lithium triethylborohydride ($LiBH(C_2H_5)_3$), n-Butyllithium (C_4H_9Li) and tert-butyllithium (C_4H_9Li)
Pharmaceutical industry	lithium salts used to treat schizoaffective disorder and cyclic major depression
Nuclear applications	source of tritium when irradiated by neutrons [448]

$$^6Li + n \rightarrow {}^4He + {}^3T \tag{34}$$

tritium is a form of solid fusion fuel used inside hydrogen bombs in the form of lithium deuteride [500]; Li-6 used in nuclear weapons [448,465] and Li-7 is used in nuclear reactor coolants [449]; Metallic lithium and its complex hydrides (e.g. $Li[AlH_4]$) used to rocket propellants [447]; $Li[AlH_4]$) is a solid fuel; sulphur hexafluoride gas on solid Li used as stored chemical energy propulsion system (SCEPS) and useable in thermonuclear weapons in reactions with Li-6 [501]

Pyrotechnics	rose-red flame [436] and red fireworks lithium compounds used as colourants and oxidisers [437,438]
Optics	specialists optics for IR, UV and VUV (vacuum UV) applications; lithium fluoride used in thermoluminescent radiation dosimetry (TLD) quantification [441] and focal lenses of telescopes [437,442]; lithium niobate used in non-linear optics applications in telecommunication products such as mobile phones, optical modulators and resonant crystals

3.17 Magnesium

3.17.1 Overview of magnesium

Magnesium (Mg) is an alkaline earth metal which only occurs naturally in combination with other elements. A fresh magnesium metal surface passivates instantly due to a thin layer of oxide which partially inhibits further reaction. Magnesium occurs in over 60 minerals but the most important commercial sources of the metal are dolomite, magnesite, brucite, carnallite, talc and olivine.

3.17.2 Beneficiation of magnesium

Brine (or seawater) is an important commercial source of magnesium whose salts are produced from it by electrolysis. Calcium hydroxide is added to the seawater to form magnesium hydroxide (brucite), which is insoluble in water and can be filtered out and reacted with hydrochloric acid to obtain concentrated magnesium chloride from which the metal magnesium is produced electrolytically.

$$MgCl_2 + Ca(OH)_2 \rightarrow Mg(OH)_2 + CaCl_2 \tag{35}$$

$$Mg(OH)_2 + 2HCl \rightarrow MgCl_2 + 2H_2O \tag{36}$$

An alternative pyrometallurgical route to magnesium production is the silicotherrmic Pidgeon process involving the reduction of the oxide at high temperature with silicon present in a ferrosilicon alloy.

$$2MgO(s) + Si(s) + 2CaO(s) \rightarrow 2Mg(g) + Ca_2SiO_4(s) \tag{37}$$

The magnesium can be similarly obtained when the process is carried out with carbon.

$$MgO(s) + C(s) \rightarrow Mg(g) + CO(g) \tag{38}$$

Electrolysis of fused magnesium chloride from brine or seawater with the Dow process is also another method of obtaining magnesium. Calcium oxide is treated with the solution containing Mg^{2+} ions to form a precipitate of magnesium hydroxide.

$$Mg^{2+}{}_{(aq)} + CaO(s) + H_2O \rightarrow Ca^{2+}(aq) + Mg(OH)_2(s) \tag{39}$$

Treating the magnesium hydroxide with hydrochloric acid results in the formation of a partial hydrate of magnesium chloride which can be electrolysed in the molten state.

$$Mg(OH)_2(s) + 2HCl \rightarrow MgCl_2(aq) + 2H_2O(l) \tag{40}$$

$$Mg^{2+} + 2\ e- \rightarrow Mg \tag{41}$$

$$2Cl^- \rightarrow Cl_2\ (g) + 2\ e^- \tag{42}$$

A novel process of solid oxide membrane technology involving the electrolytic reduction of MgO to Mg in an yttria-stabilised zirconia (YSZ) electrolyte has been reported [502].

3.17.3 Magnesium value addition

Table 3.17 Uses of magnesium and its compounds

Magnesium attributes and properties	tarnishes easily when exposed to air; reacts with water at room temperature; reacts exothermically with most acids
Alloys	aluminium-magnesium alloys (magnalium or magnelium) combines properties of lightness and strength; in aerospace industry [506–508]; automotive applications [518]; beverage cans, sports equipment such as golf clubs, fishing reels, archery bows and arrows
Important industrial and biological compounds	Magnesium carbonate, magnesium chloride, magnesium citrate, magnesium hydroxide (milk of magnesia), magnesium oxide, magnesium sulphate and magnesium sulphate heptahydrate (Epsom salts)
Structural material	component of aluminium alloys, die-casting (alloyed with zinc) [504]; removal of S in the production of iron and steel; production of titanium in Kroll process [505]
Corrosion	limits its use [503]
Electronics	in manufacture of mobile phones, laptops, tablet computers, cameras and other electronic components due to its low weight, good mechanical and electrical properties
Flammability	flares, pyrotechnics, fireworks; starting emergency fires [509,510]; flash photography; ignition of thermite
Grignard reagents (prepared from reaction with an alkyl halide)	for organic synthesis; preparation of alcohols; application as an additive in propellants and production of nodular graphite in cast iron; reductant in the separation of uranium from salts; galvanic anode to protect boats, underground tanks, pipelines, buried structures, water heaters; photoengraving, dry cell battery walls and roofing [504]
Magnesium compounds	MgO used as refractory material in furnace linings for the production of iron, steel, non-ferrous metals, glass, cement; in agricultural, chemical and construction industries; electrical insulator in fire-resistant cables [511]; magnesium salts in foods, fertilisers, culture media; magnesium sulphite in the manufacture of paper; magnesium phosphate in fireproofing wood; magnesium hexafluorosilicate in mothproofing of textiles; purification of solvents

3.18 Manganese

3.18.1 Overview of manganese

Manganese (Mn) is a paramagnetic [512] metal which is easily oxidised in air and forms rust in water containing dissolved oxygen. Because of its reactivity, manganese does not exist as a free element in nature but is found in many minerals generally in combination with iron, which it resembles chemically and shares a spatial relationship. The principal mineral containing manganese is pyrolusite (MnO_2), but also braunite ($Mn^{2+}Mn_6^{3+}$)(SiO_{12}), psilomelane ($(Ba,H_2O)_2Mn_5O_{10}$) and rhodochrosite (MnO_3) to lesser extent.

3.18.2 Beneficiation of manganese

Ferromanganese is produced when the manganese ore mixed with iron ore is reduced by carbon in a blast furnace or in an electric arc furnace [513]. The hydrometallurgical process to produce pure manganese is by leaching the ore with sulphuric acid followed by an electro-winning process [514]. The product is used for the production of iron-free alloys.

Another hydrometallurgical process involves direct reduction of the manganese ore in a heap leach by percolating natural gas through the bottom of a heap. The heat and the reducing agent CO are provided by the natural gas. In this process, all the manganese ore is reduced to manganese oxide (MnO), which is easily leachable. This product is driven through a grinding circuit to reduce the particle size and added to a leach tank containing sulphuric acid and ferrous ions. The ferrous ions reduce the MnO to elemental manganese with the formation of ferric hydroxide. Further purification of the manganese is by electro-winning [527].

3.18.3 Manganese value addition

Table 3.18 Uses of manganese and its compounds

Manganese attributes and properties	not found free in nature; electromagnetic properties
Steel alloys	stainless steel [514–521]; improve workability of steel at high temperatures; increases tensile strengths; decrease embrittlement; provides S control; enhances hardenability [479]; increase wear resistance and solid solution strengthening; retards recrystallisation; lowers the austenite-to-ferrite transformation temperature [1248]; austenitic manganese grades (Hadfield steels) display abrasion and impact resistance: serve in construction, mining, quarrying, oil-well drilling, steelmaking, cement and clay manufacturing, railroading, dredging, lumbering applications; non-magnetic properties and toughness; stress structural uses in strong magnetic fields and at cryogenic temperatures; manganese present in the following classes of steel: carbon, high-strength low-ally (HSLA) and stainless
Aluminium alloys	high resistance against corrosion for beverage cans [522]; due to formation of grains absorbing impurities which would lead to galvanic corrosion content of 0.8 to 1.5% are the alloys used for most of the beverage cans [522]
Pigments	colouring ceramics [523] and glass [524]
Manganese compounds	methylcyclopentadienyl manganese tricarbonyl used as additive in unleaded petrol to boost octane rating and reduce engine knocking [525]; MnO_2 used in organic chemistry for oxidation of benzylic alcohols, and in the manufacture of oxygen and chlorine in drying black plants, original dry cell battery [526] and in newer alkaline batteries [528] $$MnO_2 + H_2O + -e \rightarrow MnO(OH) + OH^- \tag{43}$$ MnO nanoparticles have applications as a catalyst, electrode material, and in magnetically guided drug delivery [457]
Manganese metal	in coins [529]

3.19 Mercury

3.19.1 Overview of mercury

Mercury (Hg) is a heavy liquid metal at standard conditions of temperature and pressure. It occurs in deposits of cinnabar (HgS), mercuric sulphide, whose pure form is vermillion. Cinnabar which is bright red exists in massive or granular form or as small crystals.

3.19.2 Beneficiation of mercury

Heating cinnabar ore in an air current produces mercury in vapour form which can be condensed.

$$HgS + O_2 \rightarrow Hg + SO_2 \tag{44}$$

Purification can easily be done by vacuum distillation.

Alternative processes for the reductive extraction of mercury from cinnabar involve iron or calcium oxide with the formation of sulphides of iron and calcium, respectively.

$$HgS + Fe \rightarrow Hg + FeS \tag{45}$$

$$4HgS + 4CaO \rightarrow 4Hg + 3CaS + CaSO_4 \tag{46}$$

3.19.3 Mercury value addition

Table 3.19 Uses of mercury and its compounds

Mercury attributes and properties	liquid at standard conditions of temperature and pressure; heavy; poor conductor of heat; fair conductor of electricity; low boiling and melting points; does not react with most acids; reacts with atmospheric hydrogen sulphide; reacts with sulphur; dissolves many other metals such as gold and silver to form amalgams
Metal	thermometers, barometers, manometers, sphygmomanometers, float valves, mercury switches, mercury relays, fluorescent lamps; scientific research applications and in amalgam material for dental restoration; liquid mirror telescopes [532]
Lighting	mercury vapour fluorescent lamps; "neon signs"; optical spectroscopy [537]; gaseous mercury in electron tubes including ignitrons, thyratrons, and mercury arc rectifiers [538] and skin tanning and disinfection [539]; gaseous mercury in cold cathode argon-filled lamps to increase ionisation and electrical conductivity [540]
Amalgams	gold and silver amalgams; sodium amalgam used in organic synthesis and in high-pressure sodium lamps; mercury cell process (Castner-Kellner process) [530,531] for the production of sodium hydroxide
Reference electrodes	secondary reference (calomel) electrode in electrochemistry [533]; triple point of mercury (−38.8344° C) as a temperature standard for ITS [534]; dropping mercury electrode [535] and hanging mercury drop electrode [536] in polarography

3.20 Molybdenum

3.20.1 Overview of molybdenum

Molybdenum (Mo) does not occur as a free element in nature but is found in minerals, principally molybdenite (MoS_2), but also wulfenite ($PbMoO_4$) and powellite ($CaMoO_4$), and is recovered as by-product of copper and tungsten mining [541]. These molybdenum-containing minerals form the soluble molybdate ion MO_4^{2+} when in contact with oxygen and water. The pure metal Mo has a Mohs hardness of 5.5, a high melting point of 2,611°C [541], a very low coefficient of expansion [542] with its tensile strength ranging from 10 to 30 Gpa [543] dependent on the dimeter of the molybdenum wire.

3.20.2 Beneficiation of molybdenum

When molybdenite is heated to a temperature of $700°$ C, during its processing in air, it gets oxidised to molybdenum oxide [544].

$$2MoS_2 + 7O_2 \rightarrow 2MoO_3 + 4SO_2 \tag{47}$$

The pure metal is recovered by sublimation when the oxidised ore is heated to $1,100°$ C. It can also be produced by reduction of the oxide with hydrogen. For the production of ferromolybdenum steel, typically containing 60% molybdenum, the oxide ore is reduced by the aluminothermic reaction with addition of iron [545].

The alternative hydrometallurgical route to recovery is by leaching the oxidised ore with ammonia to form water-soluble molybdates:

$$MoO_3 + 2NH_4OH \rightarrow (NH_4)_2(MoO_4) + H_2O \tag{48}$$

The copper in molybdenite is less soluble in ammonia and can be completely removed from the solution by precipitating it with hydrogen sulphide [544].

3.20.3 Molybdenum value addition

Table 3.20 Uses of molybdenum and its compounds

Molybdenum attributes and properties	does not exist as free element in nature; low coefficient of thermal expansion
Steel alloys	stable carbides steel alloys including superalloys, structural steel, stainless steel, tool and high-speed steels (such as M2, M4 and M42), cast iron [541,546]; TZM for corrosion resistance in FliBe molten salt reactors; [547] and weldability; piping, stirrers and pump impellers which come into contact with zinc [548]
Industrial application of Mo compounds	pigments, catalysts; phosphomolybdic acid used as a stain in thin layer chromatography
Molybdenum metal	applications that involve intense heat such as armour, aircraft parts, electrical contacts, industrial motors and filaments [542]; flame-resistant coating for other metals; use in vacuum environments [566]; in NO, NO_2, NO_x analysers in power plants for pollution controls; at $350°$ C as a catalyst for NO_2/NO_x to form only NO molecules for consistent readings by infrared light [549]; anodes in certain low voltage X-ray sources in mammography [550]; Mo coated soda lime glass used in CIGS solar cell fabrication
Molybdenum powder	fertiliser for some plants such as cauliflower [546]
Nuclear application	isotope molybdenum-99 to generate technetium-99 used for medical imaging [497,551,555]
Molybdenum disulphide (MoS_2)	solid lubricant; high-pressure high-temperature (HPHT) antiwear agent [552]; semiconductor in electronics applications; catalyst in hydrocracking of petroleum fractions containing nitrogen, sulphur and oxygen [553]

(Continued)

Table 3.20 (Continued)

Molybdenum disilicide (MoSi$_2$)	an electrically conducting ceramic used in heating elements at temperatures above 1500° C [554]
Molybdenum trioxide (MoO$_3$)	used as an adhesive between enamels and metals; wulfenite co-precipitates with lead chromate and lead sulphate in a bright-orange pigment used with ceramics and plastics; ammonium heptamolybdate is use in biological staining procedures

3.21 Nickel

3.21.1 Overview of nickel

Nickel (Ni) is sufficiently reactive with oxygen which means that elemental nickel rarely exists in nature except in combination with iron thought to be the product of supernova nucleosynthesis [556] such as alloys kamacite and taenite. Limonite ((Fe,Ni)O(OH)), garnierite ((Ni,Mg)$_3$Si$_2$O$_5$(OH)$_4$) which are laterites and pentlandite ((Ni,Fe)$_9$S$_8$), a magnetic sulphide deposit, are the most important commercial sources of nickel. Other minerals in which nickel is found are millerite, nickeline, nickel galena. Like iron, cobalt and gadolinium, nickel is ferromagnetic around room temperature [557] but non-magnetic above its Currie temperature of 355° C [558].

3.21.2 Beneficiation of nickel

Nickel from sulphide deposits is recovered by concentration through a froth flotation process followed by pyrometallurgical extraction. After the production of the nickel matte, the Sherritt-Gordon process is used to remove copper first by adding hydrogen sulphide, leaving a concentrate of only cobalt and nickel which are separated by solvent extraction. Alternatively, further refining can be achieved by leaching the metal matte into a nickel salt solution, followed by the electrowinning nickel from solution onto cathodes as electrolytic nickel.

Nickel oxides can be processed via the Mond process [559] in which nickel is reacted with carbon monoxide at around 40–80° C to form nickel carbonyl in the presence of a sulphur catalyst. If iron, instead of carbon, is used iron pentacarbonyl is formed in a slower process. Nickel may be separated by distillation where dicobalt tetracarbonyl is formed as a by-product. This product subsequently decomposes to tetracobalt dodecacarbonyl at the reaction temperature to give a non-volatile solid [560]. Further processes to obtain the nickel as "carbonyl nickel" involve palletisation or nickel powder production [561].

3.21.3 Nickel value addition

Table 3.21 Uses of nickel and its compounds

Nickel attributes and properties	corrosion resistant; ferromagnetic at room temperature; hard and ductile
Nickel electrodeposits	corrosion resistant coating on metal equipments; impart engineering properties such as hardness and ductility

(Continued)

Table 3.21 (Continued)

Nickel steels	stainless steel for corrosion resistance: 300 series austenitic stainless steels such as 304 and 316 stainless steel; structural material for construction of bridges, locomotive forgings, electric railway gears, marine engine works; propeller blades; cast iron
Non-ferrous alloys and superalloys	special alloys [562]; nickel brasses and bronze; alloys with copper, chromium, aluminium, lead, cobalt, silver and gold (Inconel, Incoloy, Monel, Nimonic) [563]; hastelloy alloys for aircraft components
Foundries	binder in the cemented tungsten carbide [567]
Batteries	rechargeable batteries; Ni mesh used in gas diffusion electrodes for alkaline fuel cells [463,468,564,565,481,486]
Chemical manufacturing	catalysts as in hydrogenation; Raney nickel and 'Raney-type' catalysts; carrier of hydrogen in the manufacture of fats from oils; catalyses the methanation of CO and CO_2 cheaply, most actively and most selectively
Magnets	Alnico magnets
Metal	coinage; electric guitar strings; microphone capsules; substitute for decorative silver; magnetostrictive material which undergoes contraction [566]; manufacture of sheets which are stamped into watches, cigarette cases, cooking utensils, etc.
Fire assaying	collector of the 6 PGEs from ores; collection of gold

3.22 Niobium

3.22.1 Overview of niobium

The metallic niobium (Nb) or columbium has chemical properties similar to tantalum and vanadium. It is resistant to all acid attacks except hot acid sulphur and hydrogen fluoride. It exists in about 60 minerals of two types of deposits. The primary deposits of pyrochlore [(Ca,Na,Ce)(Nb,Ti,Ta)$_2$(O,OH,F)$_7$] is interstratified in carbonates, such as calciopyrochlore in dolomite containing about 0.6% niobium pentoxide. In the secondary deposits, the niobium pentoxide is concentrated to about 3% due to weathering.

3.22.2 Beneficiation of niobium

Crushing, grinding, magnetic separation and froth flotation is the beneficiation route to obtaining pyrochlore concentrates containing 50–60% niobium pentoxide. The concentrates are roasted in a rotary furnace and subsequently leached. Prior chemical pre-treatment to remove impurities may be necessary in the case of concentrates from sources other than pyrochlore.

Columbite [(Fe,Mn,Mg)(Nb,Ta)$_2$O$_6$] is the second most import source of niobium. If the tantalum pentoxide content is greater than the niobium pentoxide the ore is known as tantalite. These columbites or tantalites occur as primary deposits in granite and pegmatites, or as alluvial secondary deposits. Removal of impurities and separation of niobium and tantalum are achieved through a sophisticated process control of a complex chemical procedure of blending, decompositions, filtrations, solvent extraction and calcinations.

An alternative extraction process involves the reductive chlorination of natural and synthetic raw materials or tantalum-niobium ferroalloy scrap.

3.22.3 Niobium value addition

Table 3.22 Uses of niobium and its compounds

Niobium attributes and properties	soft; ductile; resistant to acids with exception of hot sulphuric acid and hydrogen fluorite
Ferro-niobium steels alloys	steels for construction and high temperature applications; high-strength and low-alloys [HSLA] steels [568–571] for automobile industry, pylons, offshore platforms and oil/gas pipelines
Special alloys	for nuclear and aircraft industries [465,569,572–574]
Industrial applications	catalysts, magnets, superconductors, jewellery, thermometers, capacitors [575–579]
Niobium Tungstate NTE	application in engineering, scientific and dental industry; has no thermal expansion properties; [474]

3.23 Rare Earth Elements (REEs)

3.23.1 Overview of rare earth elements

The following constitute the list of rare earth elements (REEs) or rare earth metals (REMs): cerium (Ce), dysprosium (Dy), Erbium (Er), europium (Eu), gadolinium (Gd), holmium (Ho), lanthanum (La), lutetium (Lu), neodymium (Nd), praseodymium (Pr), promethium (Pm), samarium (Sm), terbium (Tb), thulium (Tm) and ytterbium (Yb). yttrium (Y) and scandium (Sc) are also added into the list [580] of rare earth elements.

The REEs are 15 lanthanide chemical elements on the periodic table plus scandium and yttrium. They occur naturally [581,583] except promethium which is a synthetic mineral. They are usually found together in mineral deposits of bastnaesite, monazite or laterite and may include thorium or uranium deposits. They are sometimes by-products [581] of iron, copper or gold minerals. They are clustered as "light" and "heavy" on the basis of their atomic weight [582–584].

REMs are classified 'strategic' by a number of countries due to their importance to the military and industries linked to green products such as power generation and automotive applications.

3.23.1.1 Monazite

Monazite is a phosphate mineral that contains rare earth metals in four different groups determined by the composition of the REMs present:

Monazite-(Ce), (Ce, La, Nd, Th)PO_4
Monazite-(La), (La, Ce, Nd)PO_4
Monazite-(Nd), (Nd, La, Ce)PO_4
Monazite-(Sm), (Sm, Gd, Ce, Th)PO_4

Monazite which may contain some silica is an important source of thorium [585], lanthanum and cerium [586]. Monazite, often found in placer deposits, is radioactive due to the presence of thorium and uranium.

3.23.1.2 Bastnasite

Bastnasite is a carbonate-fluoride mineral which may be clustered according to the composition and relative abundance of the REMs [589] in the ore as follows:

bastnasite-(Ce), (Ce, La)CO_3F
bastnasite-(La), (La, Ce) CO_3F
bastnasite-(Y), (Y, Ce)CO_3F

Where the F- is replaced by OH- the name becomes hydroxylbastnasite. Cerium is by far the most common of the rare earths in these minerals. Bastnasite mainly occurs in alkali granite and syenite and in associated pegmatites as well as in carbonatites in association with fenites and other metasomatites. It is closely related to the mineral series parisite Ca(Ce, La, Nd)$_2$(CO$_3$)$_3$F$_2$ [590] with a formula Ca(Ce, La, Nd)$_2$(CO$_3$)$_3$F$_2$ containing calcium and some neodymium. The formation of hydroxylbastnasite (NdCO$_3$OH) is thought to occur as a result of the crystallisation of a rare-earth bearing amorphous precursor [591].

3.23.2 Beneficiation of rare earth elements

Complex chemical processing is required to extract REEs which may be more than 12 in one type of ore. These REMs are often found together including radioactive uranium and thorium, making their separation difficult.

3.23.2.1 Mineralisation and beneficiation of monazite

Monazite minerals concentrate in alluvial sands on weathering of pegmatites and part of placer deposits or mineral sands which contain other heavy minerals of commercial interest such as zircon and ilmenite. Its isolation as a pure concentrate can be brought about by the use of gravity, magnetic and electrostatic separation.

Thorium and lanthanides can be extracted by heating the monazite ore in concentrated sulphuric acid at about 135°C for several hours. Thorium can be precipitated out as a phosphate or pyrophosphate leaving a solution of lanthanide sulphates from which lanthanides can be easily precipitated as double sodium sulphate. This is the basis of the acid cracking method which generates considerable acid waste and loss of the phosphate content of the ore.

An alternative alkaline cracking process [587–589] has been developed. It uses hot sodium hydroxide solution (73%) at about 140°C and allows the valuable phosphate content of the ore to be recovered as crystalline trisodium phosphate. The lanthanide/thorium hydroxide mixture is treated with hydrochloric acid to form a solution of lanthanide chlorides, and an insoluble sludge of the less-basic thorium hydroxide. The final products yielded for this process are thorium phosphate concentrate, RE hydroxides and uranium concentrate.

3.23.2.2 Rare earth metal extraction from bastnaesite

The bulk of bastnaesite in the ore [592] can first be separated from the accompanying barite, calcite and dolomite by grinding the ore finely after which it is subjected to froth flotation. The flotation concentrate is then calcined and leached with hydrochloric acid leaving an insoluble sludge of cerium concentrate while the lanthanides are dissolved. The dissolved lanthanides are recovered and purified by solvent extraction, capturing europium. The cerium content in the sludge is in the oxidised state.

Table 3.23 Selected examples of usages of rare earth elements

Z	Symbol	Name	Etymology	Selected applications
21	Sc	Scandium	From Latin Scandia (Scandinavia)	light aluminiun-scandium alloys for aerospace components, additive in metal-halide lamps and mercury-vabour lamps, radioactive tracing agent in oil refineries
39	Y	Yttrium	After the village of Ytterby, Sweden, where the first rare earth ore was discovered	yttrium aluminium garnet (YAG) laser, yttrium vanadate (YVO_4) as host for europium in television red phosphor, YBCO high-temperature superconductors, yttria-stabilized zirconia (YSZ), yttrium iron garnet (YIG) microwave filters, energy-efficient light bulbs (part of triphosphor white phosphor coating in fluorescent tubes, CFLs and CCFLs, and yellow phosphor coating in white LEDs), spark plugs, gas mantles, additive to steel, cancer-treatments, host to the red fluorescent lamp phosphor $Y2O3:Eu3+$; yttrium is also important for ceramics; yttria-stabilized zirconia
57	La	Lanthanum	From the Greek "lanthanein", meaning hidden	high refractive index and alkali-resistant glass, flint, hydrogen storage, battery-electrodes, camera lenses, fluid catalytic cracking catalyst for oil refineries
58	Ce	Cerium	After the dwarf planet Ceres, named after the Roman goddess of agriculture	chemical oxidising agent, polishing powder, yellow colours in glass and ceramics, catalyst for self-cleaning ovens, fluid catalytic cracking catalyst for oil refineries, ferrocerium flints for lighters, robust intrinsically hydrophobic coatings for turbine blades
59	Pr	Praseodymium	From the Greek "prasios", meaning leek-green, and "didymos", meaning twin	rare-earth magnets, lasers, core material for carbon arc lighting, colourant in glasses and enamels, additive in didymium glass used in welding goggles, ferrocerium firesteel (flint) products
60	Nd	Neodymium	From the Greek "neos", meaning new, and "didymos" meaning twin	rare-earth magnets, lasers, violet colours in glass and ceramics, didymium glass, ceramic capacitors, electric motors of electric automobiles
61	Pm	Promethium	After the Titan Prometheus, who brought fire to mortals	nuclear batteries, luminous paint
62	Sm	Samarium	After mine official, Vasili Samarsky-Bykhovets	rare-earth magnets, lasers, neutron capture, masers, control rods of nuclear reactors
63	Eu	Europium	After the continent of Europe	red and blue phosphors, lasers, mercury-vapor lamps, fluorescent lamps, NMR relaxation agent
64	Gd	Gadolinium	After Johan Gadolin (1760–1852) to honour his investigation of rare earths	high refractive index glass or garnets, lasers, x-ray tubes, computer memories, neutron capture, MRI contrast agent, NMR relaxation
65	Tb	Terbium	After the village of Ytterby, Sweden	additive in neodymium-based magnets, green phosphors, lasers, fluorescent lamps (as part of the white triband phosphor coating), magnetostrictive alloys such as terfenol-D, naval sonar systems, stabiliser of fuel cells
66	Dy	Dysprosium	From the Greek "dysprositos", meaning hard to get	additive in neodymium-based magnets, lasers, magnetostrictive alloys such as terfenol-D, hard disk drives
67	Ho	Holmium	After Stockholm (in Latin, "Holmia"), native city of one of its discoverers	lasers, wavelength calibration standards for optical spectrophotometers, magnets
68	Er	Erbium	After the village of Ytterby, Sweden	infrared lasers, vanadium steel, fibre-optic technology
69	Tm	Thulium	After the mythological northen land of Thule	portable x-ray machines, metal-halide lamps, lasers
70	Yb	Ytterbium	After the village of Ytterby, Sweden	infrared lasers, chemical reducing agent, decoy flares, stainless steel, stress, gauges, nuclear medicine, monitoring earthquakes
71	Lu	Lutetium	After Lutetia, the city that later became Paris	positron emission tomography – TEP scan detectors, high-refractive-index glass, lutetium tantalite hosts for phosphors, catalyst used in refineries, LED light bulb

3.23.3 Rare earth elements value addition

Table 3.23 lists the 17 rare-earth elements, their atomic number and symbol, the etymology of their names and selected examples of their usages [1251]

From Table 3.23 and other sources a consolidated summary of some of the major applications of rare earth elements is given in Table 3.24.

Table 3.24 Applications of rare earth elements

REE Application	Rare earth oxide (REO)	Main areas of applications
PermanentMagnets	Dy, Nd, Pr, Sm, Tb,	hybrid vehicles, small motors
NiMH Batteries	Ce, La, Nd, Pr, Nd	hybrid vehicles
Catalysts	Ce, La, Nd, Pr	petrol, diesel, and hybrid vehicles-emission controls
Phosphors	Ce, Dy, Eu, Gd, La, Pr, Tb, Y	LCD and PDP displays
Polishing Powders	Ce, La, Nd,	LCD and PDP displays
Glass Additives	Ce, Er, Gd, La, Nd, Yb	digital cameras, fibre optics
Ceramics	Ce, Nd, Pr, Y,	computers, communications, lighting, sensors for vehicle fuel control, ferrites for high frequencies, yttria-stabilized zirconia, colours in glass and ceramics, ceramic capacitors,
Engineering materials	Ce, Er, Y, Yb	corrosion protection, yttrium used to stabilise zirconia, and as a sintering agent for Si_3N_4, vanadium steel, stainless steel, stress gauges, spark plugs
Nuclear	Gd, Pm, Sc, Sm, Yb	radioactive tracing agent in oil refineries, neutron capture, control rods of nuclear reactors, nuclear batteries, nuclear medicine

3.24 Rubidium

3.24.1 Overview of rubidium

Rubidium (Rb), which belongs to the alkali metal group and is roughly as abundant as zinc and more than copper [595], is highly reactive with rapid oxidation in air and reacts violently with water and forms amalgams with mercury, alloying with gold, iron, caesium, and potassium [594]. It is very soft and ductile [593]. Rubidium is found in minerals leucite, pollucite, carnallite and zinnwaldite. The commercial source of the metal is lepidolite [596] but can also be found in potassium minerals and potassium chlorides in commercially significant amounts [597] as well as in seawater [598]. Deposits of rubidium in large quantities [595] are zone pegmatite ore bodies formed during magma crystallisation.

3.24.2 Beneficiation of rubidium

Potassium, rubidium and caesium can be separated by several methods. Pure rubidium alum can be obtained by fractional crystallisation of rubidium and caesium alum (Cs,Rb)

$Al(SO_4)_2 \cdot 12H_2O$ after 30 subsequent steps. The chlorostannate and ferrocyanide processes are alternative separation processes [595,599]. Alkarb, a by-product of potassium production, is a source of rubidium [600], as is pollucite [595].

3.24.3 Rubidium value addition

Table 3.25 Uses of rubidium and its compounds

Rubidium attributes and properties	soft; highly reactive; very rapid oxidation in air; ductile; reacts violently with water
Rubidium metal	laser manipulation of atoms; thermoelectric generators using magnetohydrodynamic principle [601–603]; atomic clocks [606–608]; working fluid in vapour turbines [610]; a getter in vacuum tubes; photocell component [609]; ingredient in special types of glass; development of spin-exchange relaxation-free (SERF) magnetometers [610]
Nuclear application	polarisation of ^{3}He [604] for neutron polarisation measurements and for producing neutron beams other purposes [605]; rubidium-82 used for positron emission tomography; myocardial perfusion imaging [497,611,612]
Medical application	tested for influence on manic depression [613–674]

3.25 Tantalum

3.25.1 Overview of tantalum

The primary source of the chemical element tantalum (Ta) is the group tantalite [(Fe,Mn)Ta_2O_6] which is chemically similar to columbite of lesser specific gravity [642]. Tantalite and columbite [(Fe,Mn,Mg)(Nb,Ta)$_2O_6$] are commonly mined together and are referred to as COLTAN [686]. Tantalite also has the same chemical composition as tapiolite but different crystal symmetry: orthorhombic for tantalite and tetragonal for tapiolite [643].

3.25.2 Beneficiation of tantalum

The run-of-mine (ROM) tantalite contains other impurities that require several beneficiation steps for the production of metallic tantalum. Tantalite can be concentrated by gravitational separation methods as well as spiral techniques. The production of the metallic tantalum involves leaching of the tantalite concentrates using hydrofluoric acid and sulphuric acid to produce water-soluble hydrogen fluorides [644]. Tantalite (and niobium) is thereby separated from the various non-metallic impurities in the ore. Selective recovery of tantalum and niobium fluorides is obtained by liquid-liquid solvent extraction techniques. A judicious pH adjustment process can be made to effect separation of the tantalum and niobium. The latter requires a lower pH to remain soluble in the extracting organic liquid phase and is hence selectively removed by extraction into less acidic media. A base such as aqueous ammonia can be used to neutralise the pure tantalum hydrogen fluoride solution to give tantalum hydroxide $Ta(OH)_5$, which is further calcinated to tantalum pentoxide Ta_2O_5 [645]. When hydrogen tantalum fluoride reacts with potassium fluoride, potassium heptafluorotantalate is formed. This in turn can be reduced using sodium at $\sim 800°$ C to yield tantalum powder [646].

3.25.3 Tantalum value addition

Table 3.26 Uses of tantalum and its compounds

Attributes of tantalum and properties	heavy; high melting point; conduct electricity; corrosion resistant
Electronics	tantalum powder for electrical parts such as capacitors in mobile phones, on-board electrical systems of automobiles and personal computers
Metal alloys	metallic alloys [647] with electric conductivity and high melting point; nuclear reactors, jet engines parts, equipment to process chemicals
Tanks	tank armour because of Ta high melting point and density [648,649]
Explosives	because of high melting point and density, Ta is used in shaped charges and explosive-penetrating devices designed to break vehicle armour
Metal	radio transmitters and surgical equipment because of corrosion resistance

3.26 Tin

3.26.1 Overview of tin

Cassiterite, tin dioxide (SnO_2), is the principal commercial source of tin (Sn) which is a silvery, malleable metal that is not easily oxidised in air. Tin oxide which is amphoteric [652,653,1244] is formed when the metal tin is heated in the presence of air. This means that SnO_2 dissolves in both acidic and basic solutions. Impurities of small amounts of bismuth, antimony, lead and silver in the commercial grade of tin inhibit the commercial grades of tin (99.8%) from transformation. Its hardness is increased by alloying elements such as copper, antimony, bismuth, cadmium and silver [651] due to the formation of hard, brittle intermetallic. Tin(II) chloride is an important halide from a commercial point of view. The compounds of tin that are commonly formed include many oxides, sulphides and other chalcogenide derivatives. The metal itself does not occur as a native element.

3.26.2 Beneficiation of tin

Due to its high specific gravity tin oxide is chiefly mined from secondary alluvial placer deposits [654] found downstream from the primary lodes. It is therefore economically mined through dredging, hydraulic methods or open cast mining. Tin can be produced by carbothermic reduction of the oxide ore with carbon or coke using reverberatory or electric arc furnace. It can also be recovered through hydrometallurgical routes [499,1243,1244]

3.26.3 Tin value addition

Table 3.27 Uses of tin and its compounds

Tin attributes and properties	malleable; not easily oxidised in air; corrosion resistant
Electrodeposits	used to coat other metals to prevent corrosion [650]; tin plating of steel, food packaging as in 'tin cans'; coats lead or zinc; tin whistle from tin-plated steel [656–658]

(Continued)

Table 3.27 (Continued)

Alloys	tin/lead soft solders for joining pipes or electric circuits [655]; alloys with copper for (Pewter), bronze [659], Bell metal; brass; bearing metal; zirconium alloys for cladding of nuclear fuel [661]; pipe organs of tin/lead alloy (spotted metal); properties include appearance, workability and resistance to corrosion [662]; intermetallic phases with Li [669]
Superconducting magnets	niobium-tin Nb_3Sn used for superconducting magnets [660]
Window glass	made by floating molten glass on top of molten tin (Pilkington process [663])
Batteries	negative electrode for Li-ion batteries [664]
Medicinal	tin(II) fluoride (SnF_2) in dental care [665]; controls gingivitis [666]
Organotin compounds	stabilisation of PVC plastics [667]; (carboxylic acid derivatives of dibutyl dichloride such as dilaurate [668])

3.27 Titanium

3.27.1 Overview of titanium

Titanium dioxide TiO_2 or titania [670,671] is the naturally occurring oxide of Titanium (Ti). ilmenite $FeTiO_3$ and rutile polymorphs (metastable anatase and brookite) are the general sources [674–682] of titanium oxide which is generally classed as an arsenite containing niobium and tantalum with the formula $(Na,K)_2PbAs_4(Nb,Ta,Ti)_4O_{18}$ [672,673,683]. Titanium dioxide (B) exists as a mineral in magmatic rocks and hydrothermal veins, as well as weathering rims on perovskite $CaTiO_3$. Lamellae are formed in other minerals by TiO_2.

3.27.1.1 Ilmenite

Ilmenite is the most important commercial source of titanium [685]. It is a titanium-iron oxide mineral which is a weakly magnetic black or steel-grey solid with the idealised formula $FeTiO_3$. It contains appreciable quantities of magnesium and manganese with the full chemical formula as $(Fe,Mg,Mn,Ti)O_3$ and forms a solid solution with geikielite ($MgTiO_3$) and pyrophanite ($MnTiO_3$) which are the respective magnesian and manganiferrous end-members of the solid solution series. Leucoxene, an important source of titanium in heavy minerals sands ore deposits, is an altered form of ilmenite. It is a typical component of altered gabbro and diorite that is generally indicative of ilmenite in the unaltered rock.

3.27.1.1.1 PARAGENESIS

Ilmenite is commonly found as part of a complex metasilicate ($RSiO_3$) mineral in metamorphic and igneous rocks. It occurs within the pyroxenitic portion of such intrusions. (R = Ca, Mg, Fe, Al, etc.). Kimberlitic paragenesis is indicated by the presence of magnesian ilmenite which forms part of the MARID association of minerals (mica-amphibole-rutile-ilmenite-diopside) assemblage of glimmerite xenoliths. Manganiferous ilmenite may exist in granitic rocks as well as in carbonatite intrusions where it may also contain anomalous niobium. Oxidation of ulvospinel may result in the formation of grains of intergrown magnetite and ilmenite found in many mafic igneous rocks.

3.27.1.2 Rutile

Rutile mineral is primarily titanium dioxide, TiO_2, with three rarer polymorphs anatase, brookite and $TiO_2(B)$. Anatase is a tetragonal mineral of pseudo-octahedral habit, while brookite is an orthorhombic mineral, and $Ti_2(B)$ is of a monoclinic form [687]. The mineral may contain significant amounts of iron, niobium and tantalum. In beach sands, rutile is a constituent of heavy minerals often referred to as *"mineral sands"* which include zircon and ilmenite.

3.27.2 Beneficiation of titanium

3.27.2.1 Ilmenite processing

Titanium dioxide is commonly produced from ilmenite $FeTiO_3$ which is from heavy minerals sands ore deposits as well as from layered intrusive "hard rock" titanium ore sources. The ore is mixed and reacted with sulphuric acid to remove the iron oxide group forming a by-product of iron sulphate. This is crystallised and filtered off, leaving the titanium salt in the solution which can be oxidised to TiO_2 called *synthetic rutile*. The synthetic rutile can be further processed in a similar way to natural rutile to give various titanium products.

Ilmenite may also be processed for its production of titaniferous slag application. For this, anthracite carbon and energy are added in large electric arc smelting furnaces in order to convert the ilmenite into molten iron bath and slag which is rich in titanium dioxide. The iron can be further processed as pig iron, as continuous cast steel billets, or as iron or steel powders.

3.27.2.2 Rutile processing

Titanium oxide, TiO_2, can be purified by converting it to titanium tetrachloride in a carbon thermal chlorination process [672,674] in which the ore is reduced with carbon, oxidised with chlorine to give titanium tetrachloride. The tetrachloride is then distilled and re-oxidised in a pure oxygen flame to give pure titanium dioxide with concurrent regeneration of chlorine.

3.27.3 Titanium value addition

Table 3.28 Uses of titanium and its compounds

Titanium attributes and properties	low density; high strength; high strength-to-weight ratio; paramagnetic; resistant to corrosion in sea water, *aqua regia*, chlorine
Metallic titanium	aircraft and high-strength steel devices; resistant to seawater: used to make propeller shafts, rigging, and heat exchangers in desalination plants [683]; heat-chillers for salt water aquariums, fishing line and leader, divers' knives; housing and components of ocean-deployed surveillance and monitoring devices for science and military; submarines with hulls of Ti alloys; forging; Ti in huge vacuum tubes; thin film for excellent reflective optical coating for dielectric mirrors and gemstones like "mystic fire topaz"
Ilmenite ($FeTiO_3$)	production of TiO_2; production of base pigment in paint [673,684], paper and plastics; refractory flux in steelmaking to line blast furnaces hearth [686]; sandblasting agent in the cleaning of die-casting dies

(Continued)

Table 3.28 (Continued)

Rutile, titanium dioxide TiO$_2$	production of anatase inorganic nanotubes and titanate nanoribbons which are catalytic supporters and photocatalysts; production of nanowires; TiO$_2$ as an effective opacifier in powder form as a pigment to provide whiteness is paints, coatings, plastics, paper, inks, foods, medicines (pills and tablets), toothpastes; ceramic glazes; increases skimmed milk whiteness; marking white lines of some tennis courts; exterior of rockets; cosmetic and skin care products; tatoo pigment, sunscreen as UV absorber [674], thickener and styptic pencils; steriliser for paints, cements, windows, tiles; deodorising and antifouling properties; hydrolysis catalyst; dye-sensitised solar cells [1136]; photocatalyst under UV; superhydrophilicity in self-cleaning glass and anti-fogging coatings; reduces airborne pollutants; useful for the manufacture of certain optical elements, especially polarisation optics, for longer visible and infrared wavelengths up to about 4.5µm; important constituent of heavy mineral sands and ore deposits containing rutile, zircon, and ilmenite; refractory ceramic; pigment; production of Ti metal; asterism in gems; welding electrode covering; part of the ZTR index, which classifies highly weathered sediments

3.28 Tungsten

3.28.1 Overview of tungsten

Tungsten (W) is 19.3 times heavier than water and 1.7 times heavier than lead [688,771]. It is brittle [130,692], making it difficult to work. Tungsten does not occur in nature in its elemental form but is found in 45 minerals with wolframite (iron-manganese tungstate), (Fe,Mn)WO$_4$ and scheelite (calcium tungstate) CaWO$_4$ being the most important economic source of the metal. Tungsten is found in other minerals ferberite (FeWO$_4$), hubnerite (MnWO$_4$).

3.28.2 Beneficiation of tungsten

Tungsten concentrates are obtained from ores by sorting, gravity separation, froth flotation, magnetic and electrostatic separation. After concentration the tungsten ore is mixed with in-process and end-of-life scrap and pre-treated to remove impurities, leaving ammonium paratungstate (APT) (NH$_4$)$_{10}$[H$_2$W$_{12}$O$_{42}$]*H$_2$O) as a product. This is an intermediate product used to manufacture other tungsten chemicals and products such as blue and yellow tungsten oxide and tungstic acid, as well as tungsten metal, tungsten powder, ferrotungsten, and other tungsten alloys. The following processes are involved: the APT is converted to tungsten(VI) oxide (WO$_3$), which is heated with hydrogen or carbon to produce powdered tungsten which can be mixed with small amounts of powdered nickel or other metals and sintered. Nickel diffuses into the tungsten during the sintering process to produce an alloy.

A major beneficiation of tungsten is the production of tungsten carbides (W$_2$C and WC) by heating powdered tungsten with carbon. Although W$_2$C is resistant to chemical attach, it reacts strongly with chlorine to form tungsten hexachloride (WCl$_6$) [688].

Another alternative route to tungsten recovery is the reduction of WF$_6$ by hydrogen

$$WF_6 + 3\,H_2 \rightarrow W + 6\,HF \tag{49}$$

or by pyrolytic decomposition.

$$WF_6 \rightarrow W + 3\,F_2 \tag{50}$$

3.28.3 Tungsten value addition

Table 3.29 Uses of tungsten and its compounds

Tungsten attributes and properties	hard; not free element; robust; high melting point; high density; brittle; difficult to work; low vapour pressure; high tensile strength; low coefficient of thermal expansion
tungsten metal	electrodes; emitter tips in electron beam instruments that use field emission guns; in electron microscopes; in electronics as interconnect material in IC between silicon dioxide dielectric material and the transistors; metallic films
Tungsten carbide (hard material 1150 DPH30), a ceramic/metal composite [689]	cutting tools such as knives, drills, circular saws, milling and turning tools used by the metalworking, woodworking, mining, petroleum and construction industries; cemented carbides [89]; steel alloys; superalloys and tungsten alloys; electrical conductor; wear-resistant abrasives; jewellery rings [690,692,693]
Tungsten alloys	lighting technology; electrical and electronic technology; high temperature technology (e.g. furnaces, power stations); welding; spark erosion, space travel; aircraft devices; armaments; laser technology; filaments for incandescent light bulbs; x-ray tubes (as both the filament and target); electrodes in TIG welding; radiation shielding; military application in penetrating projectiles such as rocket nozzles [691]; cannon shells, grenades and missiles, supersonic shrapnel; Dense Inert Metal Explosives; aerospace and automotive industries; turbine blades; wear-resistant parts and coatings
Tungsten compounds	Catalysts

3.29 Vanadium

3.29.1 Overview of vanadium

Vanadium (V) is a transition metal which does not exist as a native element but occurs in about 65 different minerals as well as in bauxite, and in fossil fuel deposits such as crude oil, coal, oil shale and tar sands.

Vanadium minerals of economic importance include patronite (VS_4), vanadinite ($Pb_5(VO_4)_3Cl$) and carnotite ($K_2(UO_2)_2 \cdot 3H_2O$. The major commercial source of vanadium is vanadium-bearing magnetite found in ultramafic gabbro bodies. The metal is hard, ductile, malleable and not brittle [694,695]. It is resistant to corrosive attack by alkalis and sulphuric and hydrochloric acids [696].

3.29.2 Beneficiation of vanadium

Vanadium is produced from steel smelter slag. Alternative sources are flue dust of heavy oil or as a by-product of uranium mining [697]. A multi-step process is involved in the production of vanadium metal beginning with the roasting of crushed ore with sodium chloride or sodium carbonate at about 850° C to yield sodium metavanadate ($NaVO_3$). This is solid which is acidified to give a polyvanadate salt which can be reduced with calcium metal. Alternatively, vanadium pentoxide is reduced with hydrogen or magnesium.

Vanadium can be purified via the crystal bar process that involves the formation of metal iodide and its subsequent decomposition to yield the pure metal [698].

$$2V + 3I_2 \rightleftharpoons 2VI_3 \tag{51}$$

In the production of specialty steel alloys such as high-speed tool steels, vanadium pentoxide is the most important industrial and commercial vanadium compound which is used as a catalyst for the production of sulphuric acid [696]. It oxidises sulphur dioxide to the trioxide.

$$V_2O_5 + SO_2 \rightarrow 2VO_2 + SO_3 \tag{52}$$

Air regenerates the catalyst by oxidation:

$$2VO_2 + O_2 \rightarrow V_2O_5 \tag{53}$$

Maleic anhydride, phthalic anhydride and several other bulk organic compounds [699,709] are produced by similar oxidations.

3.29.3 Vanadium value addition

Table 3.30 Uses of vanadium and its compounds

Vanadium attributes and properties	hard; ductile; malleable; resistant to corrosion; stable against alkalis and acids
vanadium alloys	ferrovanadium (an additive to improve steels) [697,700]; axles, bicycle frames, crankshafts, gears; surgical instruments and tools [701]; knives [702]; alloys with titanium for use in jet engines, high-speed airframes and dental implants; alloys of titanium, aluminium and vanadium [703]; vanadium compound, V_3Si is a superconductor [706]; vanadium-gallium V_3Ga tape used in superconducting magnets similar to Nb_3Sn and Nb_3Ti [707]
nuclear application	V foil cladding Ti to steel [704] suitable material for inner structure of fusion reactors [705]
vanadium oxides	(V_2O_5) catalyst in manufacturing sulphuric acid by the contact process [708]; oxidiser in production of maleic anhydride [709]; ceramic production [710]; vanadium dioxide used in the production of glass coatings which block infrared radiation [711]; vanadium oxide used to induce colour in corundum [712]
Electrochemistry	vanadium redox couples [292,1243] used in reductive decomposition of sulphide minerals; redox rechargeable batteries [713]; V coating used for protecting iron and steel against rust and corrosion [714]

3.30 Zinc

3.30.1 Overview of zinc

Sphalerite (Zn, Fe)S is the major commercial source of zinc metal [715]. Other sources of zinc are smithsonite (zinc carbonate), hemimorphite (zinc silicate), wurtzite (another zinc sulphide), and sometimes hydrozincate (basic zinc carbonate). Sphalerite consists largely of zinc sulphide with variable amounts of iron. Associated with it are minerals galena PbS, pyrite FeS_2 and other sulphides as well as calcite $CaCO_3$, dolomite $CaMg(CO_3)_2$, and fluorite CaF_2. As a chalcophile, zinc has low affinity for oxides and prefers to bond with sulphides.

3.30.2 Beneficiation of zinc

Zinc production starts with the concentration of sphalerite by froth flotation [716] to yield a product whose composition is normally zinc sulphide (80–85%), with less amounts of iron sulphide, lead sulphide, silica and cadmium sulphide. The zinc sulphide concentrate is converted to the oxide by roasting [717]:

$$2ZnS + 3O_2 \rightarrow 2ZnO + 2SO_2 \tag{54}$$

The zinc oxide can be reduced through a carbothermic reaction with carbon or carbon monoxide:

$$2ZnO + C \rightarrow 2Zn + CO_2 \tag{55}$$

$$ZnO + CO \rightarrow Zn + CO_2 \tag{56}$$

The resultant zinc can be distilled [718] and recovered [719] as zinc vapour thus separating it from other metallic impurities that are involatile.

The alternative hydrometallurgical for recovering zinc from the oxide concentrate involves leaching with sulphuric acid [720]:

$$ZnO + H_2SO_4 \rightarrow ZnSO_4 + H_2O \tag{57}$$

leading to the final recovery by electrolysis:

$$2ZnSO_4 + 2H_2O \rightarrow 2Zn + 2H_2SO_4 + O_2 \tag{58}$$

The Waelz process [721] may be used to recover zinc present in electric arc furnace dusts due to the use of galvanised feedstock.

3.30.3 Zinc value addition

Table 3.31 Uses of zinc and its compounds

Zinc attributes and properties	corrosion resistant coating on many metals
Zinc metal	replacement of lead in weights for fishing [734], tyre balances and flywheels; zinc powder as a propellant in model rockets; zinc sheet used to make zinc bars
Zinc metal coatings	anti-corrosion coating; galvanisation [722] of iron and steel; protective surface layer ($Zn_5(OH)_6(CO_3)_2$); protection of chain-link fencing, guard rails, suspension bridges, light posts, metal roofs, heat exchangers and car bodies [715]

(Continued)

Table 3.31 (Continued)

Batteries	zinc used as anode material for batteries with a SEP of −0.76 volts; powdered zinc in alkaline batteries; sheets of zinc metal form cases for and act as anodes in zinc-carbon batteries [723,724]; anode or fuel of the zinc-air battery/fuel cell [725–727]; zinc-cerium redox flow battery [728]
Alloys	brass; alloys that contain nickel, silver, typewriter metal, soft and aluminium solder, commercial bronze; pipe organs; machine bearings; coins coated with copper; other alloys are with antimony, bismuth, gold, iron, lead, mercury, silver, tin, magnesium, cobalt, nickel, tellurium and sodium; $ZrZn_2$ exhibits ferromagnetism below 35K; alloys zinc, copper, aluminium and magnesium used in die-casting, spin casting, for automotive, electrical and hardware industries; zinc aluminium alloy for production of small and intricate shapes possible because of low melting point [729,730]; Zn/Al Prestel alloy [727,731,732] strong as steel but malleable as plastic; Zn/lt/Cu alloy in building facades, roofs [733]; Zn/Pb alloy can be cold-rolled into sheets; cadmium zinc telluride (CZT) is a semiconducting alloy for sensing devices [735] such as energy of incoming gamma ray photons
Zinc compounds [722]	zinc carbonate, zinc gluconate (dietary supplements), zinc chloride (deodorants); zinc pyrithione (anti-dandruff shampoos); zinc sulphide (in luminescent paints); zinc methyl or zinc diethyl in organic laboratory; zinc oxide used as white pigment in paints, catalyst in manufacture of rubber, heat disperser for rubber and acts to protect polymers from UV radiation [730]; semiconductor properties of zinc oxide used in varistors an photocopying products [736]; zinc-zinc-oxide cycle thermochemical process for hydrogen production [737]; zinc chloride is a fire retardant, wood preservative [738], and precursor to other chemicals [734]; zinc methyl ($Zn(CH_3)_2$) used in organic syntheses [739]; zinc sulphide (ZnS) used in luminescent pigments in hands of clocks, x-ray and TV screens, and luminous paints; crystals of ZnS used in laser in mid-infrared [740]; zinc sulphate is a chemical in dyes and pigments; zinc pyrithione used in antifouling paints

3.31 Zirconium

3.31.1 Overview of zirconium

Zirconia, a silicate mineral zircon ($ZrSiO_4$) is the most important commercial source of zirconium which is a lustrous, grey-white metal which is brittle at lower purities [741]. In beach sands and riverbeds, zircon is a constituent of heavy minerals often referred to as "mineral sands" which include ilmenite, wolframite and cassiterite, the respective sources of titanium, tungsten and tin. The metal zirconium is not easily corroded by alkalis, acids, salt water and other agents [742] but does dissolve in hydrochloric and sulphuric acids especially in the presence of fluorine [743]. Its chemical properties are similar to those of hafnium with which it is often found. Below 35K the Zr/Zn alloy becomes magnetic [742] even though neither metal is magnetic on its own.

Zirconium is highly reactive with water at high temperatures leading to the formation of hydrogen gas, an undesirable outcome in nuclear applications.

$$Zr + 2H_2O \rightarrow ZrO_2 + 2H_2 \tag{59}$$

3.31.2 Beneficiation of zirconium

Zirconium occurs in more than 140 minerals; the metal is produced chiefly from zircon but also from baddeleyite and korsnarite. Its other important source is as a by-product in the production of titanium from minerals ilmenite and rutile, and tin from cassiterite. Zircon is recovered from the rest of mineral sands by gravity separation methods such as spiral concentration followed by magnetic separation, which remove the titanium ores ilmenite and rutile.

Hafnium is present in commercial zirconium at typically 1–2.5% [745] level necessitating their separation in nuclear applications [746,747,478]. Separation may be effected through liquid-liquid extraction of their thiocyanate-oxide derivatives [745] where the hafnium derivative has greater solubility in methyl isobutyl ketone vs water. Fractional crystallisation of potassium (K_2ZrF_6) hexafluorozirconate out of water effectively separates it from the analogous hafnium derivative which is more soluble in water. Another separation method involves the fractional distillation their tetrachlorides in a quadruple VAM (vacuum arc melting) process, combining with hot extruding and high-pressure high-temperature gas autoclaving.

3.31.3 Zirconium value addition

Table 3.32 Uses of zirconium and its compounds

Zirconium attributes and properties	lustrous; ductile; malleable; solid at room temperature; highly resistant to corrosion by sea water [744], alkalis, acids; dissolves in HCl and H_2SO_4 especially in the presence of fluorite
Zirconium metal containing 1–2.5% Hf [745]	refractory and opacifier material; alloying agent in materials that are exposed to aggressive environments, such as surgical appliances, light filaments, watch cases; explosive primers and as getters in vacuum tubes; Zr nanoparticles as pyrophoric material in explosive weapons
Alloy material	for its strong resistance to corrosion
Zirconium compounds	zircon dioxide ZrO_2 is a thermal barrier coating; zirconium tungstate shrinks in all directions when heated [742]; zirconyl chloride [$Zr_4(OH)_{12}(H_2O)_{16}$]Cl_8 is a rare water-soluble zirconium complex; zirconium carbide and zirconium nitride are refractory solids; the carbide is used to make drilling tools and cutting edges; zircon ($ZrSiO_4$) is cut into gemstones for use in jewellery
Zirconia ($ZrSiO_4$)	high temperature applications; material is refractory, hard, and resistant to chemical attack; opacifier, conferring a white, opaque appearance to ceramic materials; used in aggressive environments, such as moulds for molten metals; ZrO_2 is used in laboratory crucibles, metallurgical furnaces, as a refractory material [742]; can be sintered into a ceramic knife; space vehicle parts due to resistance to heat; component in some abrasives, such as grinding wheels and sandpaper; ceramic layers [749], usually a mixture of zirconia and yttria are used in high temperature parts such as combustors, blades and vanes in jet engines and stationary gas turbines
Nuclear applications	cladding for nuclear reactor fuels [745,748]; as zircalloys due to their low neutron-capture cross-section and good resistance to corrosion [742]
biomedical applications	dental implants and other restorative practices; knee and hip replacements; middle-ear ossicular chain reconstruction [750]; Zr binds urea for benefits of patients with chronic kidney disease [750]; Zr is a component of the sorbent column dependent dialysate regeneration and recirculation system (REDY)

Chapter 4

Minerals of radioactive metals

4.1 Bismuth

4.1.1 Overview of bismuth

Bismuth (Bi) is often found in its elemental state in nature. However, its sulphide, bismuthite (bismuthinite) Bi_2S_3, and its oxide, bismite (bismutite) $(BiO)_2CO_3$, are the most important ores from which bismuth can be extracted [751]. The extraction of other metals such as antimony, copper, lead, molybdenum, silver, tin, tungsten and zinc often yields bismuth as a by-product. Recycling from bismuth-containing fusible alloys in the form of larger objects and larger soldered objects is another commercial source of bismuth. Recycling is also feasible from sizeable catalysts with bismuth content, perhaps as bismuth phosphomolybdate, and also from bismuth used in galvanising. Production of bismuth is largely in the form of its compounds which are used in cosmetics, pigments and pharmaceuticals, notably Pepto-Bismol, for the treatment of diarrhoea.

Although bismuth is normally stable at room temperatures, when exposed to both dry and moist air, it forms bismuth(III) oxide; when reacted with water at a time it is red hot

$$2Bi + 3H_2O \rightarrow Bi_2O_3 + 3H_2 \tag{60}$$

and dissolves in concentrated sulphuric acid, nitric acid, hydrochloric acid to form its sulphate, nitrate and chloride, respectively:

$$6H_2SO_4 + 2Bi \rightarrow 6H_2O + Bi_2(SO_4)_3 + 3SO_2 \tag{61}$$

$$Bi + 6HNO_3 \rightarrow 3H_2O + 3NO_2 + Bi(NO_3)_3 \tag{62}$$

$$4Bi + 3O_2 + 12HCl \rightarrow 4BiCl_3 + 6H_2O \tag{63}$$

When reacted with fluorine, bismuth forms bismuth(V) fluoride at $500°$ C or bismuth(III) at lower temperatures. With the other halides it forms only trihalides:

$$2Bi + 3X_2 \rightarrow 2BiX_3 \ (X = F, Cl, Br, I) \tag{64}$$

These trihalides easily react with moisture to form oxyhalides BiOX.

4.1.2 Beneficiation of bismuth

The processing of crude lead bullion through several stages to a refined product co-produces bismuth. By the Kroll-Betterton process the bismuth is removed as part of the slag impurities. On reacting the molten mixture with chlorine gas, the other metals are converted to their chlorides, except bismuth which remains unchanged. It may also be separated from lead by the electrolytic Betts process.

4.1.3 Bismuth value addition

Table 4.1 Uses of bismuth and its compounds

Bismuth attributes and properties	radioactive; brittle; 86% as dense as lead; surface oxidation when exposed to air; denser in the liquid than in its solid state [755,756]; less toxic [751] than lead, antimony and polonium; stable to both dry and moist air at ordinary temperatute; diamagnetic; low thermal conductivity
Bismuth element	used in casting of printing type; Metallurgical additive to aluminium, iron and steel; used as a transmetalating agent in the synthesis of alkaline-earth metal complexes:
	$$3Ba + 2BiPh_3 \rightarrow 3BaPh_2 + 2Bi \qquad (65)$$
Alloys	component of low-melting typesetting alloys [751]; forms isostatic [751–754] Bi-Pb eutectic alloys; Bi alloys used in place of lead (for toxicity reasons); ceramic glazes; fishing sinkers food-processing equipment; free-machining brasses for plumbing applications; lubricating greases; shot for waterfowl hunting; fusible alloys for fire sprinkler systems; fuse wire [751]; in high-precision casting e.g. dentistry, to create models and moulds [759]; in production of malleable irons and thermocouple materials [751,759]; Si morphology when used in Al-Si cast alloy; Bi35-Pb37-Sn25 used in combination with non-sticking materials such as mica, glass and enamel; bismuth added to caesium enhances the quantum yield of Cs cathodes; sintering of bismuth and manganese powders at 300° C produces permanent magnets and magnetostrictive material used in ultrasonic generators and receivers working in the 10–100 kHz range and in magnetic memory devices
isotopes sources	radioactive disintegration chains of actinium, radium and thorium; bismuth-13 from the decay chain of uranium-113; bismuth-213 commercially produced by bombarding radium with bremsstrahlung photons from a linear particle accelerator
Bismuth compounds	pharmaceuticals for treatments of digestive disorders, sexually transmitted diseases and burns; bismuth oxychloride (BiOCl) used in cosmetics, pigment in paint for eye shadows, hair sprays and nail polishes [757,758]

4.2 Thorium

4.2.1 Overview of thorium

Thorium (Th) can be found in nature as a primordial element. Moazite, a chemically unreactive phosphate mineral, is the most important commercial source of thorium containing about 2.5%. Allanite can have 0.1–2% thorium and zirconia up to 0.4% thorium. Thorianite ThO_2 is a rare mineral from which thorium can also be extracted. Thorite, thorium silicate, $ThSiO_4$ has a high thorium content with the Th^{4+} and SiO_4^{4-} ions often replaced with M^{3+} (M = Sc, Y, L) and phosphate PO_3^{4-} ions, respectively. The minerals that contain significant quantities of thorium are often metamict [761] due to the element's radioactivity [760].

4.2.2 Beneficiation of thorium

In the course of extracting rare earth metals from monazite, thorium is refined as a by-product. When nitric acid leaches thorium diphosphate $Th(PO_4)_2$, thorium nitrate is produced which can be treated with tributyl phosphate. The impurities of rare-earth metals are separated by increasing the pH in the sulphate solution.

An alternative method of extracting the thorium from the monazite is to decompose the latter with a 45% aqueous solution of sodium hydroxide at 140° C. Mixed-metal hydroxides get extracted first, filtered at 80° C, washed with water and dissolved with concentrated hydrochloric acid. The acidic solution is then neutralised with hydroxides to pH 5.8 resulting in the precipitation of thorium hydroxide $Th(OH)_4$ which is most likely contaminated with ~ 3% of rare-earth hydroxides; the thorium hydroxide is subsequently dissolved in an inorganic acid and purified from the rare earth elements.

Alternatively, thorium hydroxide can be dissolved in nitric acid and the resultant solution purified by extraction with organic solvents.

$$Th(OH)_4 + 4HNO_3 \rightarrow Th(NO_3)_4 + 4H_2O \tag{66}$$

Reacting the anhydrous oxide or chloride with calcium in an inert atmosphere can effect the separation of metallic thorium from the anhydrous oxide or chloride.

$$ThO_2 + 2Ca \rightarrow 2CaO + Th \tag{67}$$

It is also possible to extract thorium by electrolysis of a fluoride in a mixture of sodium and potassium chloride at 700–800° C in a graphite crucible. Highly pure thorium can be extracted from its iodide by the crystal bar process [762].

4.2.3 Thorium value addition

Table 4.2 Uses of thorium and its compounds

Thorium attributes and properties	weakly radioactive; thorium-112 is the most stable isotope; soft; very ductile; tarnishes black when exposed to air; can be cold-rolled, swaged and drawn
Thorium metal	used in high-end optics and scientific instrumentation; thorium coating of tungsten wire improves the electron emission of heated cathodes [763]
Thorium alloys	in TIG welding electrodes; in high-end optics
Nuclear applications	nuclear fuel [173,643]; radiometric dating
Thoria	used in gas tungsten arc welding (GTAW) to increase the high-temperature strength of tungsten electrodes and improve arc stability; heat-resistant ceramics in high-temperature laboratory crucibles; increases refractive index and decreases dispersion when added to glass in high-quality lenses for cameras and scientific instruments [763]; used to control the grain size of tungsten metal used for spirals of electric lamps; in filaments of vacuum tubes; catalyst in the conversion of ammonia to nitric acid; in petroleum cracking and in producing sulphuric acid
Optics	thorium tetrafluoride in multilayered optical coatings as an antireflection material [764]; thorium tetrafluoride used in the manufacture of carbon arc lamps [765]

4.3 Uranium

4.3.1 Overview of uranium

Uranium occurs in nature in low concentrations in soil, rock and water. Uranium-bearing minerals such as uraninite UO_2 [766], carnotite, autunite, uranophane, torbernite and coffinite are the commercial sources of the metal. It also occurs in some substances such as phosphate rock deposits, and in minerals such as lignite and monazite sands. Uranium(VI) forms highly soluble carbonate complexes at alkaline pH leading to an increase in mobility and availability of uranium to groundwater and soil from nuclear waste [767].

4.3.2 Beneficiation of uranium

Uraninite is commonly mined by open-pit, underground, in-situ leaching, and borehole mining to extract uranium. It is crushed and ground to powder then leached with either an acid or alkali. The solution is then put through several sequential stages of precipitation, solvent extraction and ion exchange. The resultant yellowcake product contains uranium oxide U_3O_8 which is calcined to remove impurities before refining and conversion [768].

4.3.3 Uranium value addition

Table 4.3 Uses of uranium and its compounds

Uranium attributes and properties	weakly radioactive; uranium-118 is the most common (94%); density is 70% higher than Pb [770, 772]; malleable; ductile; slightly paramagnetic; strongly electropositive; poor electrical conductor [760,769]; reacts with all non-metal elements and their compounds [773]; HCl and nitric acid dissolve U slowly [769]; in air the metal becomes coated with dark layer of uranium oxide [760]
Nuclear applications	uranium-115 is the only naturally occurring fusible isotope; uranium-118 is fissionable by fast neutrons and is *fertile*; uranitum-113 important in nuclear technology; depleted uranium (^{118}U) is used in kinetic energy and high density penetrators and armour plating; used to make atomic bombs in war; tank armour hardened with depleted uranium plates; depleted uranium also used as shielding material [760] in some containers used to store and transport radioactive materials [769]; depleted uranium used as counterweights for aircraft control surfaces, as ballast for miscible re-entry vehicles; used in inertial guidance systems and in gyroscopic compasses [760]; in fuel nuclear power plants [764]; uranium-118 can be converted into plutonium [760] in a breeder reactor
Uranium metal	colourant in uranium glass; used for tinting and shading in early photography; radiometric dating of igneous rocks, uranium-thorium dating, uranium-lead dating, uranium-uranium dating; used for X-ray targets in the making of high-energy X-rays [760]
Industrial material	waste product of uranium extraction used to the glazing industry; pottery glazes, uranium tile glazes; bathroom and kitchen tiles

Chapter 5

Industrial minerals

5.1 Andalusite, kyanite and sillimanite

5.1.1 Overview of andalusite, kyanite and sillimanite

Andalusite Al_2SiO_5 is one of three polymorphs [774] which are aluminosilicate minerals. The other two are kyanite and sillimanite which are formed according to conditions of temperature and pressure. It is the polymorph at lower pressure mid temperature, otherwise it converts to sillimanite at higher temperatures and pressures. The three are aluminosilicate index minerals, providing clues to the depth and pressures involved in producing the host rock. The three are not found together in the same rock because each occurs under different temperature-pressure regimes. For this reason, the three minerals are aids in identifying the pressure-temperature paths of the host rock in which they occur.

Kyanite is commonly found in aluminum-rich metamorphic pegmatites, gneiss, schist and quartz veins that are a result of high-pressure regional metamorphism of pelitic rocks. Kyanite is stable at low pressure and low temperature but it is replaced by hydrous aluminosilicates such as muscovite, pyrophyllite, or kaolinite due to activity of water. Kyanite is also associated with staurolite, talc, hornblende, gedrite, mullite and corundum [775]. Kyanite decomposes into mullite and vitreous silica at temperatures above 1100° C.

$$3(Al_2O_3SiO_2) \rightarrow 3Al_2O_32SiO_2 + SiO_2. \tag{68}$$

This transformation results in an expansion [965].

Andalusite may convert to **sillimanite** at relatively higher temperatures and pressures. One variety of sillimanite is *fibrolite*. Metamorphosed sedimentary rocks are hosts to both the *fibrous* and traditional forms of sillimanite. Sillimanite is an index mineral that indicates high temperature and variable pressure. Host rocks of sillimanite include gneiss and granulite where it occurs with andalusite, kyanite, potassium feldspar, almandine, cordierite, biotite and quartz in schist, gneiss and hornfels.

5.1.2 Beneficiation of andalusite, kyanite and sillimanite

The beneficiation of andalusite involves crushing, washing, screening and through a bank of cyclones ending with a concentrated slurry from which iron particles have been removed magnetically.

Beneficiation of kyanite involves the drilling of its quartzite and blasting; secondary breaking is sometimes done with a hydraulic hammer. The ore is then passed through a primary crusher, rod mill which is in closed circuit with a classifier, to grind to −20 mesh. Water is added to create a slurry from which the −20 mesh is removed (deslimed). The new slurry is conditioned with several reagents and passed through a series of flotation cells that remove pyrite and micaceous contaminants from the slurry. Pyritic circuit tailings are again deslimed and conditioned with other reagents and passed through a series of rougher flotation cells after which the rougher concentrate goes to a two-stage recleaning circuit. Tails of the rougher circuit go to waste. After a further flotation processing the ore is dried and magnetically separated until the final kyanite product.

Massive sillimanite deposits consisting of huge aggregates of sillimanite and corundum is the source from which sillimanite is mined by hand.

5.1.3 Andalusite, kyanite and sillimanite value addition

Table 5.1 Uses of andalusite, kyanite and sillimanite

Andalusite/kyanite/sillimanite attributes and properties	polymorphic with kyanite and sillimanite; an aluminosilicate index mineral providing clues to pressure-temperature paths of the host rock; refractory material with high thermal shock resistance; ultra-low expansion rate; high chemical resistance
Andalusite	refractory industry; iron and steel manufacturers
Kyanite	used primarily in refractory and ceramic products; porcelain plumbing fixtures and dishware; in electronics, electrical insulators and abrasives; used as semi-precious gemstones [866]; an index mineral that is used to estimate the temperature, depth and pressure at which a rock undergoes metamorphism
Sillimanite	natural sillimanite rocks used in glass industries; raw material for the manufacture of high alumina refractories of 55–60% alumina bricks; in quality porcelain [776]

5.2 Apatite

5.2.1 Overview of apatite

Apatite is a calcium phosphate mineral which may be a hydroxide, fluoride or chloride with concentrations of OH^-, F^- and Cl^- ions, respectively, in the crystal and indicated by a general formula $Ca_{10}(PO_4)_6(OH,F,Cl)_2$.

5.2.2 Beneficiation of apatite

Apatite is a calcium phosphate mineral in which rare earth elements (REE) have been found and can be concentrated by flotation. Initial leaching tests are done with sulphuric, hydrochloric and nitric acids to determine the solubility of the REE. Due to the presence of gypsum, leaching with sulphuric acid achieves low recoveries because the rare elements precipitate together with the gypsum. On the contrary leaching with hydrochloric and nitric acid give complete extraction of the REE in solution. Separation of the REE from the phosphoric acid is possible with the neutralisation of the solution from the hydrochloric and nitric acids with ammonia to pH 2. The REE are precipitated during the process to phosphates. About 10.5% recovery of REO is simultaneously obtained. This REO is suitable for further upgrading, while the remaining solution can be used for fertiliser production. Cyanex 911 can be used to extract the re-dissolved REE in further solvent extraction after which the REE are finally stripped from the organic phase producing a REE nitrate solution.

5.2.3 Apatite value addition

Table 5.2 Uses of apatite

Apatite attributes and properties	a group of phosphate minerals
Apatites	fission tracks in apatite used to determine the thermal history or orogenic belts and sediments; (U-Th)/He dating of apatite used to determine thermal histories for application such as paleo-wildfire dating; manufacture of fertiliser – source of phosphorous; gemstones; pigments; host material for storage of nuclear waste; ore for rare earth meals [779,780]
Hydroxylapatite $Ca_{10}(PO_4)_6(OH)_2$,	major component of tooth enamel and bone mineral
Fluorapatite $Ca_{10}(PO_4)_6(F)_2$	more resistant to acid than hydroxyapatite; in toothpaste; too much fluoride results in dental fluorosis and/or skeletal fluorosis; hydrogen fluoride by-product in the production of phosphoric acid; source of hydrofluoric acid [777]
Chlorapatite $Ca_{10}(PO_4)_6(Cl)_2$	fluoro-chloro apatite is basis for now obsolete Halophosphor fluorescent tube phosphor system; now replaced by the Triphosphor system [778]

5.3 Asbestos

5.3.1 Overview of asbestos

Asbestos is a silicate mineral [781] which is fibrous and occurs in metamorphic rock. Serpentine is the host rock for chrysotile, the most common type of asbestos. However, chrysotile is also present in a wide variety of products and minerals.

5.3.2 Beneficiation of asbestos

After crushing the rock, asbestos is separated from other minerals by gravimetric methods. Chrysotile asbestos has needle-like fibres that cling to various objects they come into contact with.

5.3.3 Asbestos value addition

Table 5.3 Uses of asbestos

Asbestos attributes and properties	fibrous; sound absorbing; average tensile strength; resistance to fire, heat, electrical and chemical damage
Chrysotile	electrical insulation for hotplate wiring and in building insulation; more flexible than amphibole types; corrugated asbestos; cement roof sheets used for outbuildings, warehouses and garages; in sheets or panels used for ceilings, walls and floors; component in some plasters; brake linings; fire barriers in fuse boxes, pipe insulation, floor tiles, gaskets for high temperature equipment

(Continued)

Table 5.3 (Continued)

Asbestos generally	chlor alkali diaphragm membranes used to make chlorine; drywall and joint compound; plaster; gas mask filters; mud and texture coats; vinyl floor tiles; sheeting; adhesives; roofing tars; felts; sliding, and shingles [782]; transit panels, sliding countertops, pipes; acoustic ceilings; fireproofing; caulk; brake pads and shoes; stage curtains; fire blankets; interior fire doors; fireproof clothing for firefighters; filters for removing fine particles from chemicals, liquids and wine; dental cast linings; HVAC flexible duct connectors

5.4 Asphalt

5.4.1 Overview of asphalt

Asphalt, also known as bitumen [783] and classified as pitch occurs in natural deposits. The major commercial source of asphalt is petroleum, remains of ancient, microscopic algae and other once-living things. It also occurs in hydrothermal veins in places where there is a swarm composed of solid hydrocarbon Gilsonite [786] sometimes thought to be similar to carbonaceous meteorites [787] but now known to be distinct [788]. The major components of asphalt are naphthene aromatics and polar aromatics. It contains organosulphur compounds with nickel and vanadium present at <10 ppm level [785]. It is soluble in carbon disulphide.

5.4.2 Beneficiation of asphalt

Asphalt is obtained from the heavy fraction in distillation where material with a boiling point $\geq 500°$ C is considered asphalt. It is separated from the other components in crude oil by vacuum distillation. Propane or butane is used to treat the crude asphalt in a supercritical phase to extract the lighter molecules which are then separated. Further reacting the product with oxygen makes it harder and more viscous [785].

5.4.3 Asphalt value addition

Table 5.4 Uses of asphalt

Asphalt attributes and properties	sticky, black and highly viscous liquid; semi-solid form of petroleum; viscosity similar to cold molasses [784]
Road construction	binder for road aggregates; tarmac; asphalt concrete pavement mixes or airports, runways dedicated to aircraft landing and taking off
Asphalt/bitumen	waterproofing products for sealing flat roofs [785]
Mastic asphalt	a thermoplastic substance used in the building industry for waterproofing flat roofs and tanking underground
Asphalt emulsion	chipseal; slurry seal; cold-mixed asphalt
Other uses	roofing shingles; cattle sprays; fence-post treatments, and waterproofing for fabrics; used to make Japan black, a lacquer known for use on iron and steel; paint and marker ink [789]; sealing some alkaline batteries during the manufacturing process
Alternatives and bioasphalt	made from nonpetroleum renewable resources such as sugar, molasses and rice, corn and potato starches; from waste material by fractional distillation of used motor oil [790]

5.5 Azurite

5.5.1 Overview of azurite

Azurite $Cu_3(CO_3)_2 \cdot 10H_2O$ [791] is produced by the weathering of copper ore deposits. It is characteristically deep and clear blue. It is a minor source of copper. Its presence is a good surface indicator of the occurrence of weathered copper sulphide ores. Azurite and the other basic copper carbonate mineral, malachite, decompose with heat to form CO_2 and dark powdery cupric oxide CuO and lose water.

Azurite easily weathers to malachite in open air.

$$2Cu_3(CO_3)_2(OH)_2 + H_2O \rightarrow 3Cu_2(CO_3)(OH)_2 + CO_2 \qquad (69)$$

5.5.2 Beneficiation of azurite

The beneficiation of azurite as a gemstone from ores is similar to that for other gemstones described elsewhere in section 6.6 to follow. Combining aqueous solutions of copper(II) sulphate and sodium carbonate yields basic copper carbonate precipitate.

$$2CuSO_4 + 2Na_2CO_3 + H_2O \rightarrow Cu_2(OH)_2CO_3 + 2Na_2SO_4 + CO_2 \qquad (70)$$

5.5.3 Azurite value addition

Table 5.5 Uses of azurite

Azurite attributes and properties	deep clear blue colour; soft; easily decomposes with heat
Pigment	blue pigment [792]; lipstick; algaecide in farm ponds and in aquaculture operations
Jewellery	beads; ornamental stone
Mineral indicator	its presence is indicative of the presence of copper ores

5.6 Baryte (barite)

5.6.1 Overview of baryte

Baryte (barite) is barium sulphate $BaSO_4$ [793] whose *group* comprises baryte, celestine, anglesite and anhydrite and is the commercial source of barium. $(Ba,Sr)SO_4$ is a solid solution of baryte and celestine [794]. Baryte is found in a variety of environments where it has been deposited via varied processes that include biogenic, hydrothermal and evaporation, *inter alia*. It occurs in several deposits such as lead-zinc veins in limestones, in hot springs, and haematite ores often in association with anglesite $PbSO_4$ and celestine $SrSO_4$. It has reportedly been found in meteorites [795].

5.6.2 Beneficiation of baryte

Beneficiation of baryte involves conventional methods of crushing, screening, washing, jigging, heavy media separation, tabling and flotation. Some final products may require intense physical processing by grinding and micronising. Further premiums are associated with whiteness, brightness and colour [796].

5.6.3 Baryte value addition

Table 5.6 Uses of baryte

Baryte attributes and properties	white or colourless; non-magnetic; insolubility
Aggregate	heavy cement; used as filter or extender; addition to industrial products; weighting agent in petroleum well drilling mud [796]
Varied uses	filler in paint and plastics; sound reduction in engine compartments; coat of vehicle finishes for smoothness and corrosion resistance; friction products for vehicles; radiation-shielding cement; glass ceramics
Production of other barium chemicals	barium carbonate used for the manufacture of LED glass for TV and computer screens and for dielectrics; production of barium hydroxide for sugar refining; white pigment for textiles, paper and paint [793]
Medical applications	Barium meal before a contrast CAT scan
Paleothermometry	Pelagic baryte precipitates and forms sediments in deep ocean away from continental sources of sediment; systematics in the $\delta^{18}O$ of these sediments used to help constrain paleotemperatures for oceanic crust; exploitation of variation in sulphur isotopes [797]

5.7 Basalt

5.7.1 Overview of basalt

Basalt is an inert rock of silicate breeds that formed from solidified volcanic lava. Continuous fibres can be obtained from the basaltic rocks that include andesite, andesite-basalt, diabase, dolerites, porphoyries, gabbro-diabase, gabbro or amphibolites. Basalt features a glassy matrix interspersed with minerals such as feldspar in the form of plagioclase and pyroxene whose preponderance characterise the mineralogy of basalt. Accessory minerals present in the matrix include iron oxides and iron-titanium oxides such as magnetite, ulvospinel, olivine and ilmenite. Minerals such as alkali feldspar, leucite, nepheline, sodalite, phlogopite mica and apatite may be present in the groundmass. Some Fe(II)-oxidising bacteria [476] are also able to grow with basalt rock as a source of Fe(II) [798].

5.7.2 Beneficiation of basalt

Extraction of basalts involves explosives on its way to the quarries where it is crushed to the required size fractions.

5.7.3 Basalt value addition

Table 5.7 Uses of basalt

Basalt attributes and properties	has thermal properties of strength, chemical and thermal durability; electric insulating characteristics
Fireproofing	production of basalt fibres
Construction industry	crushed stone for road construction, including railway and road embankments, concrete fillings; making cobblestones; statues; stonewool as thermal insulator
Carbon sequestration	removing carbon dioxide; barrier to the release of CO_2 into the atmosphere

5.8 Bentonite

5.8.1 Overview of bentonite

Bentonite, which consists principally of montmorillonite, is an absorbent aluminium phyllosilicate, impure clay whose different types are each named after the respective dominant element, such as potassium (K), sodium (Na), calcium (Ca) and aluminium (Al). Sodium and calcium bentonites are the main classes of bentonite for industrial purposes. Minerals such as, quartz, mica, feldspar, and zeolite are the other components of bentonite. Potash bentonite, or K-bentonite, is a potassium-rich sillitic clay formed from alteration of volcanic ash [799] most often in the presence of water.

5.8.2 Beneficiation of bentonite

Both underground mining or open-pit mining are used for the production of bentonite.

5.8.3 Bentonite value addition

Table 5.8 Uses of bentonite

Bentonite attributes and properties	sodium bentonite has colloidal properties [800] expanding when wet; colloidal properties [800]; sodium bentonite expands when wet and is thus useful as a sealant, since it provides a self-sealing; low permeability barrier [801]
Sodium bentonite	used in drilling mud for oil and gas wells and boreholes; used as sealant; lining the base of landfills to prevent migration of leachate; quarantining metal pollutants of groundwater; sealant for subsurface disposal systems of spent nuclear fuel [802]; slurry walls; waterproofing of below-grade walls
Calcium bentonite	absorbs ions in solution [803] as well as fats and oils; main ingredient of fuller's earth [804]; convertible to sodium bentonite via ion exchange process
Foundry usage	foundry-sand bond in iron and steel foundries [801]; binding agent in the manufacture of iron (taconite) pellets; ingredient of clay bodies and ceramic glazes and in pyrotechnics to make end plugs and rocket engine nozzles
Decolourants	for decolourising of various mineral, vegetable and animal oils and clarification of wine, liquor, cider, beer and vinegar [801]
Wine making	removes excess protein molecules from white wine; absorbs odour and grease

5.9 Borates

5.9.1 Overview of borates

Borates are boron-containing oxyanions while boron itself often occurs in nature as borate minerals or borosilicates. Borates are derivatives of boric acid, $B(OH)_3$ whose acidity is because of its reaction with OH^- from water, forming the tetrahydroxyborate complex ($B(OH^{4-})$) and releasing the corresponding proton left by the water autoprotolysis [805]:

$$B(OH)_3 + 2H_2O \rightleftharpoons B(OH)^{4-} + H_3O^+ \ (pK_a = 8.98) \ [1057] \tag{71}$$

Boric acid undergoes condensation reactions under acidic conditions to form polymerric oxyanions:

$$4B(OH)^{4-} + 2H^+ \rightleftharpoons B_4O_5(OH)_2^{4-} + 7H_2O \tag{72}$$

The tetraborate anion occurs in the mineral borax, $Na_2B_4O_7 \cdot 10H_2$, as an octahydrate, $Na_2B_4O_5(OH)_4 \cdot 8H_2O$. A number of borates arise from treating boric acid or boron oxides with metal oxides [807]. Over 150 borate minerals are known and their deposits are associated with volcanic activity and arid climates. Only four of them are of industrial importance: the sodium borates tincal and kernite, the calcium borate colemanite and the sodium-calcium borate ulexite.

5.9.2 Beneficiation of borates

Mined borates from their mineral deposits require refining to upgrade their quality. Recrystallisation of the minerals is the common refining process after their ores have been leached with hot water to remove insoluble impurities. After cooling the concentrated borate solution in crystallisers a high purity sodium borate or boric acid is produced [808].

Alternatively, the mineral is leached with acid, commonly sulphuric acid, resulting in a hot concentrated solution comprised of boric acid, sodium or calcium sulphate and other impurities. Filtration is used to remove insoluble impurities after which the slurry is dewatered to produce a moist cake of boric acid. Further impurities are removed by washing the crystals which are dried into a granular product [809].

5.9.3 Borates value addition

Table 5.9 Uses of borates

Borates attributes and properties	inorganic salts of boron; borosilicate glass (pyrex) has low coefficient of thermal expansion and is resistant to cracking when heated, increased resistant to chemicals and to heat and has a higher mechanical strength and durability; borax $Na_2B_4O_7 \cdot 10H_2$ is soluble in water
Sodium borate	standard solution in titrimetric analysis [806]
Ceramics	added to ceramics and enamel frits in order to enhance chemical, thermal and wear resistance; regulate the thermal expansion coefficient of ceramic ensuring a good fit between the glass and the clay
Detergents	in laundry detergents, household and industrial cleaning products; enhance stain removal and bleaching; provide alkaline buffering; soften wear and improve surfactant performance
Alloys	additive for steel and other ferrous metals e.g. ferro-boron; used in the manufacturing of safe and fuel-efficient vehicles [811]

(Continued)

Table 5.9 (*Continued*)

Glass and fibreglass	borates added as fluxes during manufacturing of glass; lower the melting temperature during glass manufacturing and inhibit crystallisation of the glass; initiate glass formation and control the relationship between temperature, viscosity and surface tension; in pyrex cookware; laboratory glassware; pharmaceutical, lighting and domestic appliances; in LCDs used in tablets and TVs; in insulation fibreglass (IFG) for thermal an mechanical insulation; in textile fibreglass (FTG) for reinforcement of various materials including in electronics and aerospace application
Analytical application	lithium metaborate or lithium tetraborate used in borate fusion sample preparation of various samples for analysis by SRF, AAS, ICP-OES, ICP-AES and ICP-MS; borate fusion and energy dispersive X-ray fluorescence spectroscopy use in the analysis of contaminated soils [810]
Borate compounds	boric acid, borax, boric oxide, and disodium octaborate tetrahydrate used as wood preservative or fungicide; zinc borate used as flame retardants in timber, cellulose insulation, PVC, textiles; in nuclear power plants in control rods to capture neutrons; in cosmetics and pharmaceutical; corrosion inhibitors
Construction, materials and chemical industries Solar thermal heating	solar thermal heating in domestic and industrial technologies, for energy capture and for tubes carrying heat transfer fluids in domestic and industrial environment

5.10 Clay minerals

5.10.1 Overview of clay minerals

Clay minerals are a group of crystalline hydrous polysilicates composed of silica, alumina and water [812] with variable quantities of iron, magnesium, alkali metals and alkaline earths. They commonly occur in fine-grained sedimentary rocks such as shale, mudstone and siltstone identifiable by special analytical techniques such X-ray diffraction, electron diffraction methods, Mossbauer spectroscopy, infrared spectroscopy, Raman spectroscopy, SEM-EDS and polarised light microscopy.

5.10.2 Beneficiation of clay minerals

Open-pit methods using draglines, power shovels, front-end loaders, backhoes, scraper-loaders and shale planers, are used for mining clays. Hydraulic mining and dredging may also be used for the extraction of kaolin. Underground mining is done for fire clays because the higher quality fire clay deposits are found at depths that make open-pit mining less profitable.

Processing of clays involves mechanical methods such as crushing, grinding and screening without altering significantly the chemical and mineralogical properties of the material. To prepare the material for some applications other mechanical and chemical processes such as drying, calcining, bleaching, blunting and extruding may be used.

5.10.3 Clay minerals value addition

Table 5.10 Uses of clay minerals

Clay minerals attributes and properties	fine-grained material formed by the mechanical and chemical breakdown of rocks
Kaolin $Al_2Si_2O_5(OH)_4$ or $Al_2O_3 \cdot 2SiO_2 \cdot 2H_2O$	in paper coating and filling, refractories, fibreglass and insulation, rubber, paint, ceramics and chemicals
Ball clay	used for bonding in ceramic ware, dinnerware, floor and wall tile, pottery, sanitary wear
Fire clays	used for refractories or to raise vitrification temperatures in heavy clay products
Bentonite	in drilling muds, foundry sands, pelletising taconite iron ores
Fuller's earth	used as absorbents of pet waste, oil, grease

5.11 Cordierite (iolite)

5.11.1 Overview of cordierite

Cordierite or iolite is a cyclosilicate composed of magnesium, iron and aluminium with a series formula: $(Mg,Fe)_2Al_3(Si_5AlO_{18})$ to $(Fe,Mg)_2Al_3(Si_5AlO_{18})$ [1066]. It typically occurs in hornfels produced by contact metamorphism of pelitic rocks [813,814], and also in some granites, pegmatites, and norites in gabbroic magmas. Derivative products include mica, chlorite and talc. Garnet and anthophyllite may be present as associated minerals.

5.11.2 Beneficiation of cordierite

The beneficiation of cordierite as gem (iolite) is similar to that for other gemstones. In general, those from alluvial deposits are assorted by washing with sieves. Slurry containing the rough gemstone along with mud and other materials is washed with water on a sieve. A vibratory motion is applied to the sieve so as to separate the slurry quicker. The gem rough is then handpicked from the sieves. Depending on the size of the rough, different sieve sizes are employed.

Beneficiation process for cordierite is closely aligned to the end uses one of which is the production of catalytic converters. Catalytic converters are often made from ceramic products that contain a significant amount of synthetic cordierite. During the processing of cordierite ores the resultant crystals are deliberately aligned to make use of their very low thermal expansion viewed for one axis. This strategy inhibits thermal shock cracking from taking place when the catalytic convert is used [455].

Cordierite is also processed into suitable substrate materials for microwave integrated circuits (MIC). The developed materials that can be produced are meant to combine low values of permittivity with low microwave losses. One method involves the preparation of cordierite powders by cold pressing and sol-gel techniques. These powders are then processed by cold pressing and sintering to bulk specimens with well-suited property profile in density and dielectric parameters. This is followed by cutting thin discs from the sintered blocks, polishing them. Thereafter their metallization behaviour is studied for copper-based microwave lines. The resulting transmission characteristics are measured to ensure that the line geometries are coplanar.

5.11.3 Cordierite value addition

Table 5.11 Uses of cordierite

Cordierite/iolite attributes and properties	gem-quality iolite has various colours from sapphire blue to blue violet to yellowish grey to light blue as the light angle changes; softer than sapphires
Iolite	used as inexpensive substitute for sapphire
Cordierite	catalytic converters made from ceramics containing synthetic cordierite [815]; substrate material for microwave integrated circuits (MIC)

5.12 Corundum

5.12.1 Overview of corundum

Corundum is aluminium oxide (Al_2O_3) in crystalline form with traces of iron, titanium and chromium [816] exhibiting different colours dependent on impurities present. It occurs as a mineral in mica schist, gneiss and some marbles found in metamorphic terranes. It may also be found in low silica ingenious syenite and nephelenine intrusives, in ultramafic intrusives associated with lamprophyre dikes and pegmatites [817]. It also occurs in low silica igneous syenite and nepheline syenite intrusives. Other occurrences are as masses adjacent to ultramafic intrusives, associated with lamprophyre dikes and as large crystals in pegmatites [817] in stream and beach sands.

5.12.2 Beneficiation of corundum

Bauxite may be processed to synthetically manufacture abrasive corundum [817] in the Verneuil process that allows the production of flawless single-crystal sapphires, rubies and other corundum gems. Flux-growth and hydrothermal synthesis may be employed to grow gem-quality synthetic corundum.

5.12.3 Corundum value addition

Table 5.12 Uses of corundum

Corundum attributes and properties	has different colours when impurities are present; hard and resistant to weathering [817]; high density
Gems	used as gems called ruby if red and padparadscha if pink-orange; other colours called sapphire; ornamentals
Hardness (9.9 Mohs)	sandpaper and emery; large machines used in machining metals, plastics, and wood
Synthetic corundum	mechanical parts (tubes, rods, bearings, and other machined parts); scratch-resistant optics, scratch-resistant watch crystals, instrument windows for satellites and spacecraft, laser components

5.13 Diatomite

5.13.1 Overview of diatomite

Diatomite is a soft sedimentary rock comprising silica, alumina and iron and can be easily crumbled into a fine powder with abrasive feel. Accumulated amorphous silica (opal, $SiO_2 . nH_2O$) remains of dead diatoms in marine sediments is the basis for the formation of the diatomaceous earth which is a type of hard-shelled algae. It has a high porosity that accounts for its low density.

5.13.2 Beneficiation of diatomite

Diatomite is ground to a mean particle size below 12μm. It must not be heat-treated prior to application as an insecticide to be effective.

5.13.3 Diatomite value addition

Table 5.13 Uses of diatomite

Diatomite attributes and properties	soft siliceous sedimentary rock; its powder has abrasive feel; high porosity because it is composed of microscopically small hollow particles; hence low density
Multiple uses	filtration aid especially for swimming pools; filter in drinking water treatment processes, fish ponds, beer, wine, syrups, sugar, honey without removing or altering their colour, taste, or nutritional properties [818]; mild abrasive in products including metal polishes, toothpaste and facial scrubs; used as a mechanical insecticide due to its physico-sorptive properties [819–822]; control and possibly eliminate bed bug, house dust mite, cockroach, ant and flea infestations [822]; absorbent for liquids; matting agent for coatings; reinforcing filler in plastic and rubber; anti-block in plastic films; porous support for chemical catalysts; cat litter; activator in blood clotting studies; a stabilising component of dynamite; a thermal insulator
Commercial formats:	crushed raw granulated diatomite for convenient packaging; micronised diatomite for insecticides; calcined diatomite heated and activated for fillers;
Guhr dynamite	nitroglycerine is absorbed in diatomite to make it more stable and safer for transportation
Thermal heat barrier	fire-resistant safes; used in evacuated powder insulation for use with cryogenics [823]
Agricultural application	grain storage as an anti-caking agent as well as insecticide; feed supplement to prevent caking; added to livestock and poultry feed to prevent caking of feed; growing medium in hydroponic gardens, in potted plants, in bonsai soil
Ceramics	spent diatomite from brewing process added to ceramic mass for production of red bricks with higher open porosity [824]
Enzymic and dust application	certain bacterial acceleration of the rate of dissolution or silica in dead and living diatoms and organic algal material [825]; diatomite gravel over dues generate dust by abrasion [826]

5.14 Dolomite

5.14.1 Overview of dolomite

Dolomite is composed predominantly of calcium-magnesium carbonate $CaMg(CO_3)_2$. Unlike calcite, dolomite does not rapidly dissolve in dilute hydrochloric acid but can dissolve in slightly acidic water.

5.14.2 Beneficiation of dolomite

A thermal method of treating dolomite results in the reduction of magnesium oxide, produced by calcining the dolomite raw material, with ferrosilicon to produce metallic magnesium and a calcium iron silicate slag. The calcining step involves the heating of the crushed dolomite in a kiln to produce a mixture of magnesium and calcium oxides:

$$MgCO_3.CaCO_3(s) \rightarrow MgO.CaO(s) + 2CO_2(g) \tag{73}$$

Reduction of the magnesium oxide is done in the second step. Ferrosilicon is the reducing agent. The ferrosilicon is an alloy of iron and silicon made by heating sand with coke and scraps of iron.

In the third step, the oxides are mixed with crushed ferrosilicon and made into briquettes for loading into the reactor to which alumina may be added to reduce the melting point of the slag. The reaction is carried out at 1500–1800 K under very low pressure to produce magnesium as a vapour (99.99% purity) which is condensed by cooling to about 1100 K in steel-lined condensers, and then removed and cast into ingots:

$$2MgO(s) + Si(s) \rightleftharpoons SiO_2(s) + 2Mg(g) \tag{74}$$

The removal of magnesium vapour as it is produced ensures the above reaction goes to completion. Without the removal of the magnesium vapour the equilibrium of the reaction which is endothermic would be in favour of magnesium oxide.

In the fourth stage silica combines with calcium oxide to form calcium silicate as molten slag.

$$SiO_2 + 2CaO \rightleftharpoons Ca_2SiO_4 \tag{75}$$

5.14.3 Dolomite value addition

Table 5.14 Uses of dolomite

Dolomite attributes and properties	a sedimentary double carbonate rock; dissolves in slightly acidic water but not rapidly in hydrochloric acid
Dolomite mineral	ornamental stone; concrete aggregate; source of magnesium oxide; an important petroleum reservoir rock; host rock for large strata-bound ore deposits of base metals such as Zn, Pb and Cu; sometimes used in place of limestone as a flux for smelting iron and steel; in production of float glass; calcined dolomite used as a catalyst for the destruction of tar in the gasification of biomass at high temperature
Horticultural application	in home and container gardening dolomite and limestone are added to soils and soilless potting mixes as a pH buffer and as a source of Mg
Ceramic industry	studio pottery as glaze ingredient, contributing Mg and Ca as glass melt flaxes

5.15 Epidote

5.15.1 Overview of epidote

Epidote $Ca_2(Al,Fe)_3Si_3O_{12}OH$ is a calcium aluminium iron sorosilicate mineral [826]. The colour, optical constants and the specific gravity of the mineral are varied according to the amount of iron present in the mineral. Epidote is found in marble and schistose rocks of metamorphic origin and may be a product of hydrothermal alterations of various minerals such as feldspar, micas, pyroxenes, amphiboles, garnets and quartz.

Species piemontite and allanite, $(Ca,Fe^2)_2(R,Al,Fe^3)_3Si_3O_{12}OH$; (**R** = rare elements), belong to the same isomorphous group with epidote and are described as manganese and cerium epidotes, respectively. Piemontite is commonly found as small, reddish black, monoclinic crystals in manganese mines while allanite has the same general epidote formula and contain metals of the cerium group.

5.15.2 Beneficiation of epidote

The epidote mineral is ground to ultrafine mill process: To start with, a vibrating feeder feeds the epidote material into the grinding chamber gradually and evenly. After being ground, the powder is blown to a cyclone by a blower. The material is then transferred to a storage room via pipes, and then discharged as final products. The whole process works in a negative pressure environment.

5.15.3 Epidote value addition

Table 5.15 Uses of epidote

Epidote attributes and properties	amount of iron in the mineral determine many of the characteristics of the minerals such as colour, optical constants and specific gravity
Epidote	gemstone and a source of calcium and aluminium

5.16 Feldspar

5.16.1 Overview of feldspars

Feldspars are a group of tectosilicate minerals present in many types of metamorphic [827] and sedimentary rocks. Compositions of the major elements in common feldspars are expressed in terms of three end-members: potassium feldspar (K-spar) endmember $KAlSi_3O_8$, albite endmember $NaAlSi_3O_8$, and anorthite endmember $CaAlSi_2O_8$. Solid solutions between these feldspars occur. Formation of some clay minerals is a result of chemical weathering of some feldspars.

5.16.2 Beneficiation of feldspar

The ore is extracted by crushing, grinding and screening to appropriate size fractions and transported to the process plant. The feed stocks are built up and sampled at the quarry.

The process plant is provided with a constant homogeneous supply that is appropriate for the product application. The product is sold directly in the form of sands to glass manufacturers or further ground and packed for ceramic markets.

5.16.3 Feldspar value addition

Table 5.16 Uses of feldspars

Feldspars attributes and properties	group of minerals consisting of framework tectosilicates
Glassmaking	common raw material used in glassmaking including glass containers and glass fibre; improves product hardness, durability, and resistance to chemical corrosion; alkalis in feldspar (calcium oxide, potassium oxide, sodium oxide) act as flux, lowering the melting temperature of a mixture
Ceramics and fillers [828]	electrical insulators, sanitaryware, pottery, tableware, and tile; fillers
Other uses	fillers and extenders in paint, plastics, and rubber; feldspar is the abrasive component in Bon Ami household cleaner
Earth science and archaeology	used in K-Ar dating; argon-argon dating; thermoluminescence dating; optical dating

5.17 Fluorspar (fluorite)

5.17.1 Overview of fluorspar

Fluorspar is the commercial name for the mineral fluorite (calcium fluoride, CaF_2), which commonly develops well-formed cubic crystals exhibiting a wide range of colours determined by factors including impurities, exposure to radiation and the absence or voids of the colour centres.

Fluorspar belongs to the halide minerals. The element substitution for the cation include certain rare earth elements such as yttrium and cerium. Iron, sodium and barium are its common impurities. The fluorine may be replaced by the chlorine anion.

Fluorspar may occur as vein deposit especially with metallic minerals as part of gangue material. It is thus associated with galena, sphalerite, barite, quartz and calcite.

5.17.2 Beneficiation of fluorspar

Fluorspar is extracted from deposits that contain minerals that are associated with it such as quartz, barite, calcite and many others. Its ore is first crushed in a cone separator and then concentrated by gravity process involving an aqueous suspension of finely ground ferrosilicon as a separation medium, making use of density differences. Ferrosilicon is one of the most commonly used heavy media separators. When the ore is introduced at the top of the cone, the light particles float at the top where they are skimmed off while heavy particle, rich in fluorspar, settle at the bottom, where they are recovered. The different commercial grades of fluorite are provided by a further flotation process.

5.17.3 Fluorspar value addition

Table 5.17 Uses of fluorspar

Fluorspar attributes and properties	colourful mineral both in visible and UV light [829]; it is allochromatic i.e. tinted with elemental impurities; has a wide range of user market [832]
Ornamental and lapidary uses	drilled into beads and used in jewelry; ornamental carvings
Flux material	metallurgical grade (60–85% of CaF_2) used as a flux to lower the melting point of raw materials and to reduce slag viscosity in steel production to aid the removal of impurities, and later in the production of aluminium; for smelting, and the production of certain glasses and enamels; as a flux in aluminium, steel and ceramics production processes [830]
Optics	optically clear transparent fluorite lenses have low dispersion, exhibit less chromatic aberration making them valuable in microscopes [833], telescopes [835]; used in the far-UV range, where conventional glasses are too absorbent for use; high-quality optics such as camera and telescope lenses. Material is resistant to chemical attach
Hydrogen fluoride	the highest grade, "acid grade fluorite" (97% or more CaF_2), is used to make hydrogen fluoride and hydrofluoric acid by reacting the fluorite with sulphuric acid [831]; hydrofluoric acid is used in pharmaceutical and agricultural industries and a wide range of materials; hydrogen fluoride is liberated from the mineral when reacted with sulphuric acid: $$CaF_2(s) + H_2SO_4 \rightarrow CaSO_4(s) + 2HF(g) \qquad (76)$$ the resulting HF is converted into fluorine, fluorocarbons, and diverse fluoride materials including important pharmaceuticals and agrochemicals such as refrigerants, non-stick coatings, medical propellants and anaesthetics; also used in petroleum alkylation, stainless steel pickling and the manufacture of electronics and uranium processing
Ceramics	ceramic grade of fluorite is used in the manufacture of opalescent glass, enamels and cooking utensils; applications in the manufacture of magnesium, calcium metal, welding rod coatings
Aluminium production	liquid production of aluminium by electrolytic reduction of alumina dissolved in electrolyte containing cryolite (Na_3AlF_6) which is made from fluorspar; cryolite also serves as a solvent
Fluoropolymers	e.g. PTFE resin or Teflon; fluoropolymers have thermal stability, high chemical inertness, strong electrical insulation and very low coefficient of friction
Semiconductor industry	exposure tools for the semiconductor industry use fluorite optical elements; fluoresce microscopy [834]; corrects optical aberrations [835]
SVHC downstream derivatives of fluorspar	substances of very high concern: ammonium pentadecafluoro-octanoate (APFO); pentadecafluoro-octanoic acid (PFOA); henicosafluoro-undecanoic acid, lead bis(tetrafluoroborate), heptacosafluoro-tetradecanoic acid, triocosafluoro-dodecanoic acid, and pentacosafluoro-tridecanoic acid

5.18 Graphite

5.18.1 Overview of graphite

Graphite is one of the allotropes of carbon. The others are diamond, amorphous carbon inclusive of coal, graphene, fullerenes and nanotubes. Natural graphite is in the form of amorphous mass, crystalline flakes and veins.

Amorphous graphite [836] occurs as minute particles in mesomorphic rock beds such as coal, slate, or shale. Its carbon content is determined by its parent material. Meta-anthracite, the result of the thermal metamorphism of coal, is the last stage of coalification. It is difficult to ignite and is usually lower in purity than other natural graphite.

Crystalline **flake graphite** is found as isolated, flat, plate-like particles in metamorphic rocks such as limestone, gneisses and schists. It always appears in flaky morphology irrespective of particle size.

Vein graphite is found in fissure veins or fractures. It appears as massive platy intergrowths of fibrous or needle-like crystalline aggregates.

5.18.2 Beneficiation of graphite

The graphite ore is processed through a jaw crusher, a cone crusher and a rod mill after which it is floated in rougher flotation cells to produce a rougher graphite concentrate. The rougher flotation graphite concentrate is further milled in a rod mill before the final cleaner flotation. In the flotation process kerosene is used to coat the graphite and pine oil, a froth stabiliser, to enhance the graphite floatation behaviour. Additional hydrometallurgical processing to rid the graphite of impurities such as quartz, feldspar or mica, may be required.

5.18.3 Graphite value addition

Table 5.18 Uses of graphite

Graphite attributes and properties	flexible; highly refractory; chemically inert; has high thermal and electrical conductivity
Summary of graphite uses	used in refractory applications and crucibles; in foundry operations and steelmaking; in automotive parts; in lubricants; and batteries
High-purity flake graphite use	in lithium-ion batteries [837]; vanadium redox battery technology; flexible graphite products; brake and clutch linings; fuel cells [468,481,488,837] in which hydrogen is converted into electricity and alkaline batteries; graphite is the conductive material in the cathode in the alkaline batteries; both synthetic and natural graphite are used; battery applications need to be synchronised with similar value addition initiatives in platinum and lithium; fuel cells [461,468,486] providing energy to personal electronics sector, utilities sector, emergency power to hospitals or turn CBM into electricity at wastewater plants
Refractories and metal casting industry	magnesia graphite and alumina graphite; crucibles and mould washes; amorphous graphite used in refractories industry to manufacture crucibles, ladles, moulds, nozzles and troughs that can withstand the very high temperatures associated with molten metal; electrodes in electric arc furnaces used in steel processing; graphite is the carbon raiser to strengthen steel; used in blast furnace linings because of its thermal conductivity

(Continued)

Table 5.18 (Continued)

Amorphous or fine-flake graphite uses	brake and clutch linings; gaskets; foundry facing mould wash in water-based paint to coat mould; low-quality amorphous graphite used to make pencil lead
Pebble-bed nuclear reactors	uses uranium embedded in fist-sized graphite balls

5.19 Gypsum

5.19.1 Overview of gypsum

Gypsum $CaSO_4 \cdot 2H_2O$ [838] is an hydrated sulphate mineral which is moderately water soluble [839], but exhibiting retrograde solubility by becoming less soluble at higher temperatures. It occurs in nature as flattened and often twinned crystals which contain water and hydrogen bonding. These crystals are transparent and cleavable masses called selenite [840] which may occur in a silky and fibrous form and also in granular or quite compact. It may occur in a flower-like form, typically opaque, in arid areas. It has thick and extensive evaporite beds in association with sedimentary rocks and deposits occurring in strata [841] deposited from lake and sea water, as well as from hot springs, from volcanic vapours, and sulphate solutions in veins. It is associated with the mineral halite and sulphur [842]. Gypsum may be formed as a by-product of pyrite oxidation when generated sulphuric acid reacts with calcium carbonate. The sulphates, which it contains, can be reductively converted by reducing bacteria back to sulphide [476].

5.19.2 Beneficiation of gypsum

Flue-gas desulphurization at some coal-fired power plants is a source of synthetic gypsum. It may also precipitate onto brackish water membranes, salt scaling, in desalination of water with high concentrations of calcium and sulphate.

In processing gypsum, it is ground in a ball mill, or hammer mill. It can also be pulverised in agricultural mills.

5.19.3 Gypsum value addition

Table 5.19 Uses of gypsum

Gypsum attributes and properties	soft; alabaster is a variety of gypsum; exhibits retrograde solubility
Construction industry	used as a finish for walls and ceilings; dry wall; sheetrock or plasterboard; concrete blocks; gypsum mortar in building construction; main constituent of many forms of plaster and blackboard chalk; a binder in fast-dry court clay; component of Portland cement used to prevent flash settings of concrete
Agricultural application	used as fertiliser and soil conditioner; highly sought fertiliser for wheat fields; ameliorates high-sodium soils [843]; used in mushroom cultivation to stop grains from clumping together

(Continued)

Table 5.19 (Continued)

Ornamental	a very fine-grained white or lightly tinted variety of gypsum, alabaster, a material for sculpture
Medical applications	plaster ingredients used in surgical splints, casting moulds and modelling; impression plasters in dentistry
Dietary	tofu (soy bean curd) coagulant; source of dietary calcium; a common ingredient in making mead; used in baking as a dough conditioner; reducing stickiness; primary component of mineral yeast food
Water	adds hardness to water used for brewing; soil/water potential monitoring (soil moisture); used to remove pollutants such as lead or arsenic [844–846] from contaminated waters
Cosmetics	in foot creams, shampoos and many other hair products

5.20 Kaolinite (kaolin)

5.20.1 Overview of kaolinite

Kaolinite (kaolin) $Al_2Si_2O_5(OH)_4$ (mineralogy formula) [847] or $Al_2O_3 \cdot 2SiO_2 \cdot 2H_2O$ (ceramics formula in terms of oxides) [848] is a clay mineral and part of the group of aluminium silicate industrial minerals. It is a result of chemical weathering of aluminium silicate minerals such as feldspar. Kaolinite clay may also be found in abundance in soils that have formed from the chemical weathering of rocks in hot, moist climates such as in tropical rainforest areas [850–852]. Synthesis of kaolinite at temperatures above 100° C are well known [853–861]

$$2Al(OH)_3 + 2H_4SiO_4 \rightarrow Si_2O_5.2Al(OH)_3 + 5H_2O \qquad (77)$$

5.20.2 Beneficiation of kaolinite

Commercial grades of kaolin are supplied and transported as dry powder, semi-dry noodle or as liquid slurry. Kaolinite undergoes a progression of endothermic dehydration ending with a disordered metakaolin [849]:

$$Al_2Si_2O_5(OH)_4 \rightarrow Al_2Si_2O_7 + 2H_2O. \qquad (78)$$

Further heating of the product to 925–950° C results in the conversion of metakaolin to an aluminum-silicon spinel:

$$2Al_2Si_2O_7 \rightarrow Si_3Al_4O_{12} + SiO_2. \qquad (79)$$

Above temperatures 1050° C of calcination the spinel phase nucleates and transforms to platelet mullite $Al_2O_3 \cdot 2SiO_2$ and highly crystalline cristobalite $3Al_2O_3 \cdot 2SiO_2$.

$$3Si_3Al_4O_{12} \rightarrow 2(3Al_2O_3 \cdot 2SiO_2) + 5SiO_2. \qquad (80)$$

At 1400° C mullite appears with the concomitant increases in structural strength and heat resistance noting that this only a structural rather than a chemical transformation.

5.20.3 Kaolinite value addition

Table 5.20 Uses of kaolinite

Kaolinite attributes and properties	soft; earthy, usually white
Paper industry	in production of paper: its use ensures the gloss on some grades of coated paper
Agriculture	spray applied to crops to deter insect damage; in case of apples to prevent sun scald
Electronics	as in Edison Diamond Disc;
Medical and health	to soothe upset stomach; treatment of diarrhoea; to induce blood clotting in diagnostic procedures; as a component of Quik-clot Combat Gauze to stop bleeding from extensive wounds; suppress hunger (geophagy)
Construction industry	in metakaolin form, as a pozzolan; when added to a concentrate mix; metakaolin accelerates the hydration of Portland cement; in adhesives to modify rheology [862]; whitewash in stone masonry homes
Cosmetics	facial masks and soap; adsorbents in water and wastewater treatments [864];
Radioactivity	an indicator in radiological dating [863] as kaolin contains small traces of U and Th
Other uses	in ceramics (it is the main component of porcelain); in toothpaste; as a light diffusing material in white incandescent light bulbs; in paint to extend the TiO_2 white pigment and modify gloss levels; for modifying the properties of rubber upon vulcanisation; metakaolin is a component of geopolymer compounds;

5.21 Lepidolite

5.21.1 Overview of lepidolite

Lepidolite $K(Li,Al,Rb)_3(Al,Si)_4O_{10}(F,OH)_2$ is a phyllosilicate mineral [867] belonging to the mica group and a member of the polylithionite-trilithionite series. It is therefore evidently a secondary source of lithium. It is in fact associated with other lithium-bearing minerals such spodumene in pegmatite bodies and a source of the rare alkali metals rubidium and caesium [868] and of fluorine. It is commonly found in granite pegmatites, in some high-temperature quartz veins, greisens and granites. Minerals that include quartz, feldspar, spodumene, amblygonite, tourmaline, columbite, cassiterite, topaz and beryl are associated with lepidolite in the pegmatite bodies.

5.21.2 Beneficiation of lepidolite

Lepidolite is extracted from brines. It is separated from complex silicate minerals through froth flotation.

5.21.3 Lepidolite value addition

Table 5.21 Uses of lepidolite

Lepidolite attributes and properties	a phyllosilicate mineral [867] belonging to the mica group
Uses	gemstones and a secondary source of lithium; also, a source of rubidium and caesium [686]

5.22 Limestone

5.22.1 Overview of limestone

Limestone $CaCO_3$ is calcium carbonate in the form of minerals calcite and aragonite with different crystal structures. It is a sedimentary rock. It is soluble in water and weak acids leading to karst landscapes. Calcite is a polymorph of calcium carbonate, the others being minerals aragonite and vaterite. At 380–470° C aragonite will change to calcite [869] while vaterite is less stable. Single calcite crystals have variable refractive indices [870].

Metamorphic marble is composed of calcite which also occurs as a vein mineral in deposits from hot springs, in caverns as stalactites and stalagmites. Calcite is sometimes found in the form of lublinite which is fibrous and efflorescent. Calcite may also occur in volcanic or mantle-derived rocks such as carbonatites, kimberlites, or peridotites.

5.22.2 Beneficiation of limestone

Limestone is ground to ultrafine powder in a process which works in a negative pressure environment. The ground material is used for building, roadbeds and landscape construction.

5.22.3 Limestone value addition

Table 5.22 Uses of limestone

Limestone attributes and properties	common in architecture; easy to cut into blocks or more elaborate carving; long lasting; heavy material
Construction industry	quarried material for buildings, roadbeds, and landscape
Soils	microbiologically precipitated calcite used for soil remediation, soil stabilisation and concrete repair
Optical application	high grade optical calcite used for gun sights; anti-aircraft weaponry; cloak of invisibility
Industrial	limestone, phosphate, iron pyrites etc. are backbone of industries such as cement, fertiliser, paints, construction and other industrial fillers; building material; aggregate for roadbeds; white pigment or filler in products like tooth paste or paints; chemical feedstock; in blast furnaces, limestone binds with silica and other impurities to remove them from the iron
Other uses	raw material for the manufacture of quicklime (calcium oxide), slaked lime (calcium hydroxide), cement and mortar; pulverised limestone as soil conditioner to neutralise acidic soils; in glass making
Geological	formations of limestone best petroleum reservoirs
Air pollution control	reagent in flue-gas desulphurisation; suppress methane explosions in underground coal mines
Additive	toothpaste; paper; plastics; paint; tiles; to other materials as both white pigment and cheap filler; in medicines and in cosmetics
Dietary	added to bread and cereals as a source of calcium; livestock feeds for poultry
Remineralisation	increases alkalinity of purified water to prevent pipe corrosion; to restore essential nutrient levels
Sculptures	suitable for carving

5.23 Magnesite

5.23.1 Overview of magnesite

Magnesite $MgCO_3$ is the mineral magnesium carbonate which is formed through carbonate metasomatism of peridotite and other ultramafic rocks. The presence of water and carbon dioxide at elevated temperatures and high pressures is the ideal condition for the formation of magnesite through carbonation of olivine. Carbonation of magnesium serpentine can result in the formation of magnesite [871].

5.23.2 Beneficiation of magnesite

Magnesium oxide (ericlase) MgO is produced when magnesite is calcined in the presence of charcoal. The reactivity of this oxide is determined by the calcination temperatures.

5.23.3 Magnesite value addition

Table 5.23 Uses of magnesite

Magnesite attributes and properties	cryptocrystalline containing silica in the form of opal and chert
Thermal resistance	used in refractory industry and fertilisers [872]
Magnesite $MgCO_3$	production of magnesia for refractory industries; binder in flooring material; catalyst and filler in the production of synthetic rubber and in the preparation of magnesium chemicals and fertilisers
Fire assay	cupels used for cupellation as the magnesite cupel resists high temperatures involved
Source of MgO	magnesium oxide is an important refractory material used as a lining in blast furnaces, kilns and incinerators; MgO whose commercial grades are: caustic calcined magnesite (CCM), dead burned magnesite (DBM) and fused magnesia (FM)
DBM and FM	production of magnesia predominantly used in the refractory industry: steel, cement and other refractory
CCM	used in chemical-based applications such as fertilisers; livestock feed; pulp; paper; iron and steel making; hydrometallurgy; and waste/water treatment
Other uses	environmental, agricultural and other applications; (e.g. spots); slag conditioning and hydraulic construction

5.24 Magnetite

5.24.1 Overview of magnetite

Magnetite Fe_3O_4 is a naturally occurring iron oxide mineral and a member of the spinel group. It is highly magnetic [873,874]. It is one of three common naturally occurring iron-rich oxide. The others are haematite and ilmenite. It is also found in many sedimentary rocks, including banded iron formations. It may sometimes be found in beach sand from rivers as a depository of eroded soil from various catchment areas.

5.24.2 Beneficiation of magnetite

Beneficiation of magnetite follow the conventional process route of crushing and grinding, cyclone classification and magnetic separation.

5.24.3 Magnetite value addition

Table 5.24 Uses of magnetite

Magnetite attributes and properties	highly magnetic (lodestone); black or grey in colour [873]
Magnetic property	lodestone used as magnetic compass; a tool in paleomagnetism to understand plate tectonics and historic data for magnetohydrodynamics
Source of iron and other applications	important iron ore [875]; in audio recording using magnetic acetate tape; catalyst for the industrial synthesis of ammonia [876]; removes arsenic ions form water; coating industrial watertube steam boilers after treatment with hydrazine
Iron-metabolising bacteria [476]	redox reactions in microscopic magnetite particles; reduction of Cr(VI) to less toxic Cr(III) in solutions

5.25 Mica

5.25.1 Overview of mica

Micas are a group of phyllosilicate minerals with muscovite and phlogopite being the commercially important ones [878] having the following general chemical formula [877]:

$$X_2Y_{4-6}Z_8O_{20}(OH,F)_4$$

Where X maybe K, Na, or Ca or less commonly Ba, Rb, or Cs
Y maybe Al, Mg, or Fe or less commonly Mn, Cr, Ti, Li, etc.
Z maybe Si or Al, but also include Fe^{3+} or Ti
If Y = 4, the mica is classed dioctahedral and if Y = 6, it is classed trioctahedral. A mica is *common* if the X ion is K or Na; but it is *brittle* if X ion is Ca.

Mica is found in igneous, metamorphic and sedimentary regimes, whereas large crystals are usually mined from granitic pegmatites. A variety of sources have been identified for flake mica. Schist (metamorphic rock) may be a by-product of processing feldspar and kaolin resources, placer deposits and pegmatites. Sheet mica may be recovered from mining scrap and flake mica from pegmatite deposits. Thin sheets of mica caused by foliation are chemically inert, dielectric, elastic, flexible, hydrophilic, insulating, lightweight, platy, reflective, refractive, resilient, and range in opacity from transparent to opaque.

5.25.2 Beneficiation of mica

Mica is pulverised dry to required size of finely divided particles prior to sheet forming. A sheet is formed by pouring a colloid mixture of the ground mica, water and a colloid agent onto a mesh screen. The formation of a sheet which includes a resin binder is completed by means of vacuum and a hydraulic press.

5.25.3 Mica value addition

Table 5.25 Uses of mica

Mica attributes and properties	high dielectric breakdown; thermally stable; resistant to corona discharge; insulating properties; supports electrostatic field while dissipating minimal energy in the form of heat; good electrical insulator at the same time a good thermal conductor; impervious to most gases
Muscovite	electrical industry in capacitors that are ideal for high frequency and radio frequency; used in sheet and ground forms [878]
Phlogopite mica	remains stable at high temperatures; used when a combination of high-heat stability and electrical properties is required; used in sheet and ground forms [878]
Dry-ground mica	in joint compound for filling and finishing seams and blemishes in gypsum wallboard (dry wall); filler and extender; provides a smooth consistency; improves the workability of the compound and provides resistance to cracking
Paint industry	pigment extender; facilitates suspension; reduces chalking; prevents shrinking and shearing of the paint film; increases resistance of the paint film to water penetration and weathering; brightens the tone of coloured pigments; provides paint adhesion
Well drilling	ground mica used as an additive to drilling fluids as the coarsely ground mica flakes helps prevent the loss of circulatory by sealing porous sections of the drill hole
Plastics industry	extender and filler; lightweight insulation to suppress sound and vibration; reinforcing material in plastic vehicle fascia and fenders; provides improved mechanical properties and increased dimensional stability, stiffness, and strength; imparting high-heat dimensional stability; reduces warpage; best surface properties of any filled plastic composite
Rubber industry	used as inert filler; mould release compound in the manufacture of moulded rubber products such as tyres and roofing; this is due to its platy texture which acts as an antiblocking, antisticking agent; as a rubber additive, mica reduces gas permeation and improves resiliency [878]
Surface coating	dry-ground mica is used as coating of rolled roofing and asphalt singles; prevents sticking of adjacent surfaces; decorative coating on wallpaper, concrete, stucco, and tile surfaces; flux coating welding rods, special greases, and core and mould release compounds, facing agents, mould washes in foundry industry
Insulating material	sound-absorbing insulation for coatings in polymer systems; decreases the permeability of moisture and hydrocarbons in industrial coating; insulator in concrete block and home attics; can be poured into walls usually in retrofitting uninsulated open top walls; electrical insulation; high-temperature and fire-resistant power cables in aluminium plants, blast furnaces, critical wiring circuits (e.g. defence systems, fire and security alarm systems, surveillance systems), heaters and boilers, lumber kilns, metal smelters; tanks and furnace wiring; bonding material; flexible, heater, moulding, and segment plates; mica paper and tapes [878]; insulation for electric motors and generator armatures, field coil, magnetic and commutator core; V-rings cut and stamped from copper segments; insulation of tubes and rings in armatures, motor starters, and transformers
Brake and clutch linings	dry-ground phlogopite mica is a substitute to asbestos in vehicle brake linings and clutch plates to reduce noise and vibrations

(Continued)

Table 5.25 (Continued)

Polymer industry	reinforces additives for polymers to increase strength and stiffness; improves stability to heat, chemicals, and UV radiation; heat shields and temperature insulation; in polar polymer formulations it increases strength of epoxies, nylons and polyesters
Paint industry	wet-ground mica is used in vehicle industry to retain the brilliancy of cleavage faces in pearlescent paints; substrates of mica with TiO_2 compose metallic-looking pigments; reflective colour in vehicle paint, shimmery plastic containers, high-quality inks used in advertising and security applications
Cosmetics industry	reflective and refractive properties of mica in blushes, eye liner, eye shadow, foundations, hair and body glitter, lipstick, lip gloss, mascara, moistening lotions, nail polish; powdered white mica included in some brands of tooth paste; acts as a mild abrasive to aid polishing on the tooth surface, adds cosmetically pleasing, glittery shimmer to the paste; provides a coloured shiny surface to latex balloons
Soil conditioner	potting soil mixes; gardening plots
Greases	mica, tar or graphite added to greases composed of fatty oils to increase durability and give better surface to axles
Atomic force microscopy	used as clean imaging [497] substrate; imaging of bismuth films, plasma glycoproteins, membrane bilayers, DNA molecules [879]
Various other uses	in electrical components; electronics, isinglass, atomic force microscopy; in diaphragms for oxygen-breathing equipment; marker dials for navigation compasses; optical filters; pyrometers; thermal regulators; stove and kerosene heater windows; radiation aperture covers for microwave ovens; mica thermic heater elements; birefringent used to make quarter and half wave plates; aerospace components in air-, ground- and sea-launched missile systems; laser devices, medical electronics and radar systems; mechanically stable in micrometer-thin sheets which are relatively transparent to radiation (e.g. alpha particles); in radiation detectors such as Geiger-Müller tubes; mica is used as a substrate in the production of ultraflat, thin-film surfaces (e.g. gold surfaces)

5.26 Natron

5.26.1 Overview of natron

Natron is an admixture of sodium carbonate decahydrate $Na_2CO_3 \cdot 10H_2O$, sodium bicarbonate $NaHCO_3$ together with small quantities of sodium chloride and sodium sulphate. It occurs in saline lake beds in association with thermonatrite, nahcolite, trona, halite, mirabilite, gaylussite, gypsum, and calcite. It is the source of soda ash (sodium carbonate anhydrate Na_2CO_3) as a result of calcination.

5.26.2 Beneficiation of natron

Natron is formed geologically as a transpiro-evaporite mineral by crystallising during the drying up of lakes that are rich in sodium carbonate which itself is formed by absorption of atmospheric carbon dioxide by a highly alkaline, sodium-rich lake brine:

$$NaOH_{(aq)} + CO_2 \rightarrow NaHCO_{3(aq)} \tag{81}$$

$$NaHCO_{3(aq)} + NaOH_{(aq)} \rightarrow Na_2CO_{3(aq)} + H_2O \tag{82}$$

5.26.3 Natron value addition

Table 5.26 Uses of natron

Natron attributes and properties	its colour ranges from white to colourless when pure, varying to grey or yellow with impurities; stable at room temperature
Ceramics	used along with sand and lime to make glass; flux to solder precious metals together
Detergent	used along with other chemicals

5.27 Nepheline syenite

5.27.1 Overview of nepheline syenite

Nepheline and alkali feldspar constitute nepheline syenite [880] which is a holocrystalline plutonic rock that is generally equigranular, equidirectional and gross with grain size of 2 mm to 5 mm. Nepheline syenite has a high ratio of $(Na_2O + K_2O)/SiO_2$ and $(Na_2O + K_2O)/Al_2O_3$ representing the existence of nepheline and alkaline mafic minerals, respectively. Nepheline is a feldspathoid which reacts with quartz to produce alkali feldspar and the nepheline syenites are distinguished from ordinary basic syenites by the presence of nepheline and the occurrence of many other minerals rich in alkalis and in rare earths and other incompatible elements. The alkaline feldspar that dominates them is represented by orthoclase and the exsolved lamellar albite and form perthite. Sodalite and nepheline are the principal feldspathoid minerals which are hosts to triclinic aeigmatite, sodium-rich pyroxene, biotite, titanite, iron oxides, apatite, fluorite, melanite garnet, cancritinite and zircon. Magnetite, ilmenite, apatite and titanite are accessory minerals.

The chemical peculiarities of nepheline syenites include richness in alkalis and in alumina with silica ranging between 50 and 56%, while lime, magnesia and iron are not present in great quantities.

5.27.2 Beneficiation of nepheline syenite

Nepheline syenite ore is first crushed and milled. It is then sent to a separation process which is determined according to the accompanying minerals. As nepheline syenite usually contains magnetite and haematite, magnetic separation to remove such particles is needed.

5.27.3 Nepheline syenite value addition

Table 5.27 Uses of nepheline syenite

Nepheline syenite attributes and properties	geochemically an alkaline rock which is mostly pale coloured, grey or pink and similar to granites
Glass and ceramics	as nepheline syenite does not co-exist with quartz and is rich in feldspar and nepheline, it is used in the manufacture of glass and ceramics [881,882]; especially container glass and fibre glass
Industrial uses	refractories; glass making; ceramics; pigments and fillers; construction façade; interior wall texture; counter tops
Source of rare minerals	unusual minerals; rare earth elements
Environment	clue to environment formation

5.28 Ochre

5.28.1 Overview of ochre

Ochre is a natural earth pigment containing hydrated iron oxide. Its colour ranges from yellow to deep orange or brown [883–885]. It is a family of earth pigments whose major ingredient is iron(III) oxide-hydroxide, limonite, responsible for the yellow colour. When natural sienna and umber pigments are heated, they are dehydrated and some of the limonite is transformed into haematite, giving them more reddish colours, called burnt sienna and burnt umber [1247]. Ochres are non-toxic and can be used to make an oil paint that dries quickly and covers surfaces thoroughly.

5.28.2 Beneficiation of ochre

Ochre pigments can be made using synthetic iron oxide. Sienna, which contains both limonite and a small amount of manganese oxide, can be heated to dehydrate it during which some of the limonite is transformed into haematite, giving them more reddish colour, called burnt sienna and burnt umber.

5.28.3 Ochre value addition

Table 5.28 Uses of ochre

Ochre attributes and properties	a natural earth pigment containing hydrated iron oxide which is responsible for the variants in colours; non-toxic
Pigment	mix of ochre and animal fat for body decoration and hair after braiding; oil paint that dries quickly and covers surfaces thoroughly

5.29 Pegmatites

5.29.1 Overview of pegmatites

Pegmatites are holocrystalline, intrusive igneous rocks made manifest by interlocking phaneritic crystals [886] and composed of quartz, feldspar and mica. They may contain amphibole, Ca-plagioclase feldspar, pyroxene, feldspathoids and other unusual minerals occurring in recrystallised zones and apophyses associated with large layered intrusions. They have similar basic composition as granite. Crystals found within these pegmatites include spodumene, microcline, beryl and tourmaline.

Spodumene $LiAl(SiO_3)_2$, petalite $LiAl(Si)_4O_{10}$ and lepidolite $K(Li,Al,Rb)_3(Al,Si)_4O_{10}(F,OH)_2$ are the principal lithium pegmatite minerals with the last one being a lithium mica containing minor quantities of caesium, rubidium and fluorine. The Bikita pegmatite hosts the largest single accumulation of petalite, bikitaite $|Li(H_2O)|$ $[AlSi_2O_6]$, yielding large petalite crystals. These complex pegmatites, one of the largest in the world, are rich in exotic constituents such as lithium, beryllium, tantalum and niobium. Bikitaite is a mineral within the fractures in the lithium-rich pegmatites and is associated with the minerals that include the following: eucryptite, quartz, petalite, feldspar, calcite, stibnite, allophone, albite and fairfieldite.

5.29.1.1 Strategic pegmatites

Some granitic pegmatites may contain diverse minerals like ZrO_2, Nb_2O_5 and BeO bearing rare metals considered "critical" or "strategic resources" such as Ta, Nb, Be, Sb, W, rare earth elements (REE) and Co. Strategic elements are those that are of greatest risk to supply disruptions or are important to a country's economy or defence.

The complex-type pegmatites of the LCT (Li-Cs-Ta) is a particularly important class of rare-element pegmatites which contain extremely high concentration of Rb, Cs, Be, Ta, Nb, and Sn, as well as elevated levels of fluxing components (Li, P, F, and B).

5.29.2 Beneficiation of pegmatites

The beneficiation of pegmatites is specific to the pegmatite mineral of interest: lepidolite, lithium, petalite, spodumene. However, the main objective of their beneficiation is to extract lithium.

Separation of lithium minerals can be efficiently achieved by taking advantage of their physical, electrical and magnetic properties. Physical separations are performed by wet and dry screening, tabling and magnetic, electromagnetic, electrostatic, magnetohydrostatic bullet and heavy media separation. Gravity separation is feasible only if the spodumene is coarsely grained.

Flotation is used to generate high grade spodumene concentrate suitable for lithium extraction. Depending on objective, the flotation concentrate can then be further processed by high-temperature techniques (pyrometallurgy) or chemical techniques (hydrometallurgy) to produce lithium carbonate or other desirable lithium compounds. Lithium can be extracted from spodumene concentrates after roasting. A concentrate with a least 6% Li_2O (approximately 75% spodumene) is suitable for roasting which is performed at about 1050° C. During the roasting, spodumene goes through a phase transform α-spodumene to ß-spodumene. The α-spodumene is virtually refractory to hot acids but as a result of the phase transformation, the spodumene crystal structure expands by about 30% and become amenable to hot sulphuric acid attack. After roasting, the material is cooled and then mixed with sulphuric acid and roasted again at about 200° C. An exothermic reaction starts at 170° C and lithium is extracted from the mixture. Due to this expansion, the specific gravity of the spodumene decreases from 3.1 g/cm³ (natural α-spodumene) to around 2.4 g/cm³ (ß-spodumene). After roasting, the material is cooled and then mixed with sulphuric acid (95–97%). The mixture is roasted again at about 200°C. An exothermic reaction starts at 170°C and lithium is extracted from ß-spodumene to form lithium sulphate, which is soluble in water.

Working with the lithium concentrate, one can use a standardised flowsheet to produce high grade lithium products such as lithium carbonate or lithium hydroxide.

5.29.3 Pegmatites value addition

Table 5.29 Uses of pegmatites

Pegmatites attributes and properties	large size crystal components; inhomogeneous; exhibit zones with different mineral assemblages
Imbended minerals	rare earth minerals; gemstones such as aquamarine, tourmaline, topaz, fluorite, apatite, corundum, tin, tungsten
Lithium	primary source of lithium either as spodumene, lithiophyllite or lepidolite
Source of other minerals	beryl, beryllium; bismuth, molybdenum, tantalum, niobium, REE

5.30 Petalite

5.30.1 Overview of petalite

Petalite $LiAlSi_4O_{10}$, castorite, is a lithium aluminium phyllosilicate mineral which has been found to crystallise in the monoclinic system. It belongs to the feldspathoid group and is found in the lithium-bearing pegmatites co-located with spodumene, lepidolite, tourmaline and as an important ore of lithium. If heated to about 500° C it under 3 Kbar of pressure in the presence of a dense hydrous alkali borosilicate fluid with a minor carbonate component [887] petalite converts to spodumene and quartz. Abuyelite Li_2CO_3 is another petalite in spodumene-hosted fluid inclusions.

5.30.2 Beneficiation of petalite

The same as for pegmatites in section 5.29.2.

5.30.3 Petalite value addition

Table 5.30 Uses of petalite

Petalite attributes and properties	colourless, grey, yellow, yellow grey, white tabular crystals
Ceramics	raw material for the glass-ceramic cooking ware; raw material for ceramic glazes
Gemstones	colourless variety used as gemstones
Mineral	source of lithium

5.31 Phosphate

5.31.1 Overview of phosphate

Phosphate rock consists mainly of apatite minerals. Sedimentary marine phosphorites are the principal deposits of phosphate rock. Sedimentary marine phosphorites are the major deposits of phosphorate rock which consists mainly of apatite minerals. The inorganic chemical and salt of phosphoric acid is a phosphate PO_4^{3-}. In organic chemistry it is a phosphate, or organophosphate and an ester of phosphoric acid. The element phosphorus naturally occurs in the form of phosphates found in many phosphate minerals some of which deposits may include significant quantities of radioactive uranium isotopes and naturally occurring heavy metals.

5.31.2 Beneficiation of phosphate

Conventional crushing, grinding and screening is done to phosphate rock in preparation of various uses. The specific refining process is chosen to suite the mineralogical, textural and chemical characteristics of the ore. Large amounts of waste phosphogypus are created in the course of processing phosphate rock concentrates. High levels of cadmium, lead, nickel, copper, chromium and uranium may be present in the tailings of the mining operations. If not carefully managed these wastes products can leach these heavy metals into groundwater leading to concentration of toxic heavy metals in food products [888].

5.31.3 Phosphate value addition

Table 5.31 Uses of phosphate

Phosphate attributes and properties	phosphate PO_4^{3-} is an inorganic chemical and salt of phosphoric acid, and in organic chemistry it is an ester of phosphoric acid
Biochemistry	vital part of plant and animal nourishment; production of calcium phosphate feed supplements for animals
Biogeochemistry (ecology)	sedimentary phosphate enriched in cadmium and uranium
Agriculture	component of nitrogen-phosphorus-potassium fertiliser for food crops; no alternatives; two phosphate compounds are present in REACH substances of very high concern (SVHC) list: trilead dioxide phosphate and tris(2-chloroethyl)phosphate
Phosphate rock	only source for phosphorus production
Other uses	dishwasher powders and detergents; animal feedstocks

5.32 Pollucite

5.32.1 Overview of pollucite

Pollucite $(Cs,Na)_2Al_2Si_4O_{12} \cdot 2H_2O$ is a zeolite mineral whose substituting elements are iron, calcium, rubidium and potassium. It is a major source of caesium and sometimes rubidium. It is found in lithium-rich granite pegmatites in association with albite, ambylgonite, apatite, cassiterite, columbite, eucryptite, lepidolite, microline and muscovite.

5.32.2 Beneficiation of pollucite

Pollucite is mined mostly on a small-scale metal operation. The ore is crushed, hand-sorted and then ground. Caesium is extracted from pollucite by acid digestion, alkaline decomposition, or direct reduction.

5.32.2.1 Acid digestion

HCl, H_2SO_4, HBr, or HF acid is used to dissolve the silicate rock. With HCl acid, a mixture of soluble chlorides is produced, and the insoluble chloride double salts of caesium are precipitated as caesium antimony chloride Cs_4SbCl_7, caesium iodide chloride Cs_2ICl, or caesium hexachlorocerate $Cs_2(CeCl_6)$. The pure CsCl is obtained after separation, decomposition of the pure precipitated double salt and evaporation of the water. With sulphuric acid the insoluble double salt caesium alum $CsAl(SO_4)_2 \cdot 12H_2O$ is produced directly. The aluminium sulphate present is converted to the insoluble aluminium oxide by roasting the alum with carbon and leaching the resulting product with water to yield a Cs_2SO_4 solution.

Alternatively, caesium is recovered as caesium chloride and other alkali metal and poly-valent metal chlorides by use of HCl. The iron and aluminium present can be precipitated as their hydroxides and separated from the solution of the alkali metal chlorides. On addition of potassium permanganate to the solution, caesium permanganate is selectively precipitated and separated from the residual solution containing the metal chlorides. Reacting the caesium permanganate with a reducing agent results in the production of caesium carbonate and caesium delta manganese dioxide.

5.32.2.2 Alkaline decomposition

Pollucite is roasted with calcium carbonate and calcium chloride to yield insoluble calcium silicates and soluble caesium chloride which is then leached with water or dilute ammonia NH_4OH to yield a dilute solution of caesium chloride $CsCl$. On evaporation the solution produces caesium chloride or transformed into caesium alum or caesium carbonate.

5.32.2.3 Direct reduction

In this method, caesium metal may be obtained from the purified compounds of the ore. Reduction of caesium chloride, and other caesium halides, is achieved at 700 to 800° C with calcium or barium, followed by distillation of the caesium metal. Similarly, magnesium can be used to reduce aluminate, carbonate, or hydroxide. The metal can also be recovered by electrolysis of the fused caesium cyanide $CsCN$. Thermal decomposition of caesium azide CsN_3, which is produced from aqueous caesium sulphate and barium azide, at 390° C yields exceptionally pure and gas-free caesium metal.

5.32.3 Pollucite value addition

Table 5.32 Uses of pollucite

Pollucite attributes and properties Mineral resource	zeolite mineral; brittle fracture but no cleavage; heavy source of caesium

5.33 Potash

5.33.1 Overview of potash

Potash is any potassium compounds, potassium-bearing materials, mined and manufactured salts that contain potassium in water-soluble form [889,890]. Potassium reacts violently with water [893] and therefore it does not occur in nature. The most common salt is potassium chloride KCl. Potash deposits of commercial quantities come originally from evaporite deposits which are usually buried deep in the earth. They are usually rich in potassium chloride and sodium chloride.

5.33.2 Beneficiation of potash

Conventional shaft mining is used to obtain potash ores which are ground into a powder [891]. Alternative methods include dissolution mining and evaporation from brines. With the evaporation method, hot water is poured into the potash which gets dissolved and then pumped to the surface for concentration by solar induced evaporation. Amine reagents added to either the mined or evaporated solutions coating KCl and not NaCl. The amine and KCl are floated to the surface while the NaCl and clay settle at the bottom. The amine + KCl can now be skimmed, dried and packaged for use as a K rich fertiliser. Potassium chloride become available quickly for plant nutrition as it readily dissolves in water.

5.33.3 Potash value addition

Table 5.33 Uses of potash

Potash attributes and properties	water soluble
Agriculture	fertiliser; K is the 3rd major plant and crop nutrient after nitrogen and phosphorus [892]; potash improves water retention, yield, nutrient value, taste, colour, texture, and disease resistance of food crops; application to fruit and vegetables, rice, wheat and other grains, sugar, corn, soybeans, palm oil and cotton
Industry	used in aluminium recycling by the chloralkaline industry to produce potassium hydroxide; in metal electroplating; oil-well drilling fluid; snow and ice melting; steel heat-treating; in medicine as a treatment for hypokalaemia; water softening. Potassium hydroxide is used for industrial water treatment; precursor to potassium carbonate and several forms of potassium phosphate, many other potassic chemicals; soap manufacturing; potassium carbonate used to produce feed supplements, cement, fire extinguishers, food products, photographic chemicals, and textiles; in brewing beer; pharmaceutical preparations; catalyst for rubber manufacturing [889]

5.34 Pozzolan

5.34.1 Overview of pozzolan

A pozzolan is a naturally occurring siliceous or siliceous and aluminous material of volcanic origin. It reacts chemically with calcium hydroxide at ordinary temperature to form compounds possessing cementitious properties [895].

5.34.2 Beneficiation of pozzolan

Thermal activation of kaolin-clays to obtain metakaolin produces artificial pozzolans which may also be waste or by-products from high-temperature process such as fly ashes from coal-fired electricity production. Other pozzolans are produced as by-products of silica fume from silicon smelting, highly reactive metakaolin and burned organic matter residues rich in silica such as rice husk ash.

5.34.3 Pozzolan value addition

Table 5.34 Uses of pozzolan

Pozzolan attributes and properties	a siliceous material of volcanic origin; reacts with calcium hydroxide in the presence of water to form compounds possessing cementitious properties; supplementary cementitious materials
Cement industry	used in cement and concrete; cheaper, pollution free; increased durability; less greenhouse gases emitted than from Portland cement; higher compressive strength performance; reaction with calcium hydroxide produces additional C-S-H and C-A-H reaction products which fill pores and lower permeability of the binder
Chemical	lowering solution alkalinity and increases alumina concentrations decreases or inhibits the dissolution of the aggregate aluminosilicates [897]

5.35 Pumice

5.35.1 Overview of pumice

Pumice is a volcanic rock that consists of highly vesicular rough textured volcanic glass [898,899]. In composition it is commonly [900] silicic or felsic and composed of highly microvascular glass pyroclastic where the following may occur: andesite, dacitic, pantellerite, phonolite, rhyolitic and trachyte. It is a common product of explosive eruptions [901] with 90% porosity.

5.35.2 Beneficiation of pumice

When super-heated, highly pressurised rock is violently ejected from a volcano, pumice is created with a foamy configuration as a result of simultaneous rapid cooling and rapid depressurisation. Bubbles in the matrix are frozen as a result of the simultaneous cooling and depressurisation.

5.35.3 Pumice value addition

Table 5.35 Uses of pumice

Pumice attributes and properties	highly vesicular rough textured volcanic glass
Insulating material	lightweight concrete or insulative low-density cinder blocks
Construction material	additive for cement; plaster-like concrete; aqueducts
An abrasive	in polishes, pencil erasers, cosmetic exfoliants, production of stone-washed jeans; pumice stones in beauty saloons during pedicure process to remove dry and excess skin from the bottom of foot as well as calluses; finely ground pumice is added to some toothpastes and heavy-duty hand cleaners; some brands of chinchilla dust bath
Horticulture	growing substrate for growing horticultural crops; pumice mixed with soil provides good aeration for plants; used for hydroponics as a soilless growing medium
Other uses	water filtration, chemical spill containment, cement manufacturing, pet industry

5.36 Salt

5.36.1 Overview of salt

Common salt is composed principally of sodium chloride NaCl. It is a crystalline mineral whose small amounts of trace elements are present in sea salt and freshly mined salt.

5.36.2 Beneficiation of salt

Evaporation is commonly used to process common salt from salt mine water, seawater [902] or mineral-rich spring water in shallow pools. To produce refined table salt, the salt rock is dissolved in water, purified by precipitating out other minerals and re-evaporation.

Large-scale extraction of salt is from vast sedimentary deposits which have been laid down over very long periods from evaporation of seas and lakes. The salt rock is directly mined or the salt is extracted in solution by pumping water into the deposits. Mechanical evaporation of brine in pans under vacuum is the common purification process for salt. Recrystallisation is usually done in further purification during which the brine is treated with chemicals that precipitate out most impurities such as magnesium and calcium salts.

5.36.3 Salt value addition

Table 5.36 Uses of salt

Salt attributes and properties	common crystalline mineral salt
Trace elements in salt	good for plant, human and animal health
Table salt	used for food seasonings as well as for food preservation; in the preservation of meat and fish and the canning of meat and vegetables; may be iodised to prevent iodine deficiency; edible salt usually contains anti-caking agents, such as sodium aluminosilicate or magnesium carbonate to make it free-flowing, and may be iodised to prevent iodine deficiency
Industrial	major source of caustic soda and chlorine; used in many industrial processes and in the manufacture of polyvinyl chloride (PVC), plastics, paper pulp and many other products
Manufacturing industry	production of aluminium involves salt as a flux; manufacture of soaps and glycerine; an emulsifier in the manufacture of synthetic rubber; firing of pottery when added to the furnace where it vaporises before condensing onto the surface of the ceramic material, forming a strong glaze
Stabilising material	added to drilling fluid to provide stable "wall" to prevent holes collapsing when drilling through loose materials such as gravel or sand
Other uses	water conditioning processes; de-icing highways and agricultural use; mordant in textile dying; to regenerate resins in water softening; for the tanning of hides

5.37 Selenium

5.37.1 Overview of selenium

Selenium (Se) is a non-metal which occurs in metal sulphide ores where it partially replaces sulphur [620,621]. It is usually an amorphous, brick-red powder. It exists in several allotropes [619] that depend on heating and cooling at different temperatures and rates. Many Se forms are insulators [458] except grey Se which is a semiconductor showing appreciable photoconductivity. It is also insoluble in CS_2 [619] and resists oxidation by air, and is also resistant to attacks by nonoxidizing acids. It forms polyselenides with strong reducing agents. Selenium does not exhibit unusual changes in viscosity when gradually heated as happens with sulphur [618]. Selenium most often occurs in soils in soluble forms such as selenate which are leached into rivers easily by run-off [620,621] and into oceans [622]. Selenium may be the result of burning and the mining and smelting of sulphide ores.

5.37.2 Beneficiation of selenium

Commercially, elemental selenium is produced from selenide *anode mud* as a by-product in the electrolytic refining of sulphide ores such as copper, nickel and lead. The industrial production of selenium is via the extraction of selenium dioxide from the residues during the purification of copper. The residue is oxidised by sodium carbonate to produce the selenium dioxide which is then mixed with water and the solution is acidified to form selenous acid in the oxidation step. Selenous acid is reduced to elemental selenium [624,625] by bubbling it with sulphur dioxide. The power necessary to operate the electrolysis cells is significantly decreased during the electrowinning of manganese by addition of selenium dioxide [626]. The current production of copper is a combination of solvent extraction and electrowinning (SXEW) altering the availability of selenium as only a comparably small part of the selenium in the ore is con-leached with copper [623].

Melting selenium rapidly results in the formation of a black, vitreous form, which is an industrial production of beads [618]. Black selenium is a brittle and lustrous solid which is structurally irregular and complex consisting of polymeric rings. It is soluble in CS_2.

When rapidly melted, it forms the black, vitreous form, which is usually sold industrially as beads [618]. The structure of black selenium is irregular and complex and consists of polymeric rings with up to 1000 atoms per ring. Black Se is a brittle, lustrous solid that is slightly soluble in CS_2. On heating, black selenium softens at 50° C and is transformed to grey selenium at 180° C. The conversion temperature is reduced by the presence of halogens and amines [619].

5.37.3 Selenium value addition

Table 5.37 Uses of selenium

Selenium attributes and properties	non-metal which does not occur in its elemental state in nature; has photovoltaic and photoconductive properties
Glassmaking	glassmaking and pigments; Se compounds confer a red colour to glass cancelling out the green and yellow tints that arise from iron impurities; various selenite selenate salts are added; CdSe and CdS are added if red colour is desirable [627]

(Continued)

Table 5.37 (Continued)

Electronics	used in photocells; uses in electronics have be supplanted by silicon semiconductor devices; cadmium selenite for production of fluorescent quantum dots; copper indium gallium selenide is material used in the production of solar cells [632.1136]; in photocopying [633–636]; photocells; light metres; was once used widely in selenium rectifiers; power DC surge protection; zinc selenite was first material for LEDs; sheets of amorphous selenium converts x-ray images to patterns of charge in xeroradiography, solid-state, flat-panel x-ray cameras [637]; in x-ray crystallography, incorporating Se atoms in place of S helps with single and multiple-wavelength anomalous dispersion phasing [638]
Alloys	used with bismuth in brasses [628] to replace toxic lead; improves machinability of steel [629,630] and copper [631]
Rubber	small amounts of organoselenium compounds used to modify the vulcanisation catalysts used in the production of rubber [623]
Photography	used in toning of photographic prints; its use intensifies and extends the tonnage range of black-and-white images; improves the permanence of prints [639–641]

5.38 Silica

5.38.1 Overview of silica

Silicon (Si) occurs only in silicates and oxides (silica) which account for 75% of the Earth's crust [903]. Abundant free silica is in the form of quartz, extremely high purity of which is required for the production of silicon metal.

5.38.2 Beneficiation of silica

Silicon production for the photovoltaics industry starts with carbothermic reduction of quartz into metallurgical grade (MG) silicon which is then purified into polysilicon of suitable purity for use in solar cells [1136]. This entails high cost of production as the process is very energy intensive [904–906]. The carbothermic reduction takes place in a furnace containing quartz and carbon materials such as coke and charcoal. Molten silicon metal of 98.5% purity settles at the bottom of the furnace; this constrains impurities that include iron, aluminium and calcium. Further purification is done by treating the molten silica with oxidative gas and slag forming additives in the Siemens process. Thereafter thin silicon wafers are produced.

5.38.3 Silica value addition

Table 5.38 Uses of silica

Silica attributes and properties	semiconductor; its electrical resistivity decreases with increasing temperature and with increasing concentrations of elements such as B, Al, Ga and P
Silicone products	aluminium alloys; silicone and silane chemicals; electronic and solar uses; semiconductors [498]; explosives; refractories and ceramics [903,907,908]

(Continued)

Table 5.38 (Continued)

Metallurgical grade silicon (MG-Si)	used in aluminium; silicone is dissolved in molten aluminium to improve the viscosity of the liquid Al and mechanical properties of Al alloys; chemical applications; raw material for solar and electronics grade silicon but must be refined further to ultrahigh-purity grades
Chemical industry	used to produce silicones, synthetic silica and silanes [904,909]; silicone products are used as surfactants, lubricants, sealants, and adhesives; insulating silicone rubber; increases mechanical strength and the elasticity of elastomers; silanes used in glass, ceramic, foundry and painting industries
Semiconductivity	solar grade silicon (polysilicon) is ultra-high purity (between 6N and 9N) to ensure semiconducting properties; in Siemens process; used in electronic devices such as transistors, printed circuit boards, integrated circuits and solar panels; semiconductor-grade silicon metal used in making computer chips [910–912]

5.39 Soda ash

5.39.1 Overview of soda ash

Soda ash is the water-soluble sodium carbonate Na_2CO_3 which can be found as a crystalline heptahydrate. It readily effloresces to form a white powder, the monohydrate. It forms a strongly alkaline water solution. It is white, odourless and hygroscopic.

5.39.2 Beneficiation of soda ash

Sodium carbonate can be produced from sodium chloride and limestone by the Leblanc process [913] involving the boiling of sea water (sodium chloride) in sulphuric acid to produce sodium sulphate and hydrogen chloride gas:

$$2NaCl + H_2SO_4 \rightarrow Na_2SO_4 + 2HCl \tag{83}$$

The sodium sulphate is subsequently blended with crushed limestone (calcium carbonate) and coal, and burning the mixture to give sodium carbonate and calcium sulphide with the evolution of carbon dioxide:

$$Na_2SO_4 + CaCO_3 + 2C \rightarrow Na_2CO_3 + 2CO_2 + CaS \tag{84}$$

The ashes are washed, dissolving the sodium carbonate into a solution from which it can be collected on evaporation of the water.

By the Solvay process, sodium chloride can be converted to sodium carbonate using ammonia. The process centres around a large hollow tower at the bottom of which calcium carbonate (limestone) is heated to release carbon dioxide:

$$CaCO_3 \rightarrow CaO + CO_2 \tag{85}$$

The carbon dioxide is allowed to bubble up through the top where a concentrated solution of sodium chloride and ammonia enters the tower. This results in the precipitation of sodium bicarbonate.:

$$NaCl + NH_3 + CO_2 + H_2O \rightarrow NaHCO_3 + NH_4Cl \tag{86}$$

Heating the sodium bicarbonate converts it to sodium carbonate accompanied by the release of water and carbon dioxide:

$$2NaHCO_3 \rightarrow Na_2CO_3 + H_2O + CO_2 \tag{87}$$

Simultaneously, the ammonium chloride by-product regenerates ammonia when treated with the residual lime (calcium hydroxide) left over from carbon dioxide generation:

$$CaO + H_2O \rightarrow Ca(OH)_2 \tag{88}$$

$$Ca(OH)_2 + 2\,NH_4Cl \rightarrow CaCl_2 + 2\,NH_3 + 2\,H_2O \tag{89}$$

The Solvay process is substantially more economical than the Leblanc process because the former recycles its ammonia while consuming only brine and limestone with calcium chloride as its only waste product.

By the Hou's process carbon dioxide is pumped through a saturated solution of sodium chloride and ammonia to produce sodium bicarbonate:

$$NH_3 + CO_2 + H_2O \rightarrow NH_4HCO_3 \tag{90}$$

$$NH_4HCO_3 + NaCl \rightarrow NH_4Cl + NaHCO_3 \tag{91}$$

Due to its low solubility the sodium bicarbonate is collected as a precipitate and then heated to yield pure sodium carbonate.

$$2NaHCO_3 \rightarrow Na_2CO_3 + H_2O + CO_2 \tag{92}$$

As more sodium chloride is added to the remaining solution of ammonium and sodium chlorides ammonium chloride is precipitated in a sodium chloride solution by temperature adjustments and exploiting the common-ion effect.

Hou's process is coupled to the Haber process and offers better atom economy by eliminating the production of calcium chloride since ammonia no longer needs to be regenerated. Ammonium chloride, the by-product, is saleable as a fertiliser.

5.39.3 Soda ash value addition

Table 5.39 Uses of soda ash

Soda attributes and properties	water-soluble forming strongly alkaline solution; hygroscopic
Industrial	used as a water softener; in laundering; in manufacture of glass; as a flux for silica by lowering the melting point of the mixture; can remove grease, oil, and wine stains; descaling agent in boilers such as coffee pots, espresso machines; used in dyeing industry; wetting agent in brink industry

(Continued)

Table 5.39 (Continued)

Buffer	buffer in photographic film developing agents, swimming pools; food additive for acidity regulating; anti-caking agent; raising agent; stabilizer
Electrolyte	very good conductor and not corrosive to anodes; good primary standard for acid-base titrations
Other uses	wetting agent in brick industry; foaming agent and abrasive in toothpaste where it temporarily increase mouth pH; used to neutralise the H_2SO_4 needed for acid de-linting of fuzzy cottonseed; used together with salt to clean silver

5.40 Spodumene

5.40.1 Overview of spodumene

Spodumene $LiAl(SiO_3)_2$ is a lithium-bearing pyroxene mineral composed of lithium aluminium inosilicate. It is found in various colours from colourless to yellowish, purplish, or like kunzite, yellowish-green or emerald-green hiddenite, prismic crystals, often of great size. The normal α-spodumene converts to β-spodumene at temperatures above 900° C [914]. Spodumene is associated with minerals that include quartz, albite, petalite, eucryptite, lepidolite and beryl [915] in lithium-rich pegmatites and aplites. Hiddenite, a variety of spodumene, is a pale emerald green gem, the presence of chromium being responsible for the colour. Kunzite is also another variety of spodumene which is a pink to lilac coloured gemstone. Minor to trace amounts of manganese is responsible for the colour which is frequently enhanced by irradiation.

5.40.2 Beneficiation of spodumene

Lithium is extracted from spodumene by fusing in acid as described in sections 3.16.2 and 5.29.2 prior. However, spodumene may be viewed as a less important source of lithium compared to alkaline brine lake sources which produce lithium chloride directly. In this case, lithium chloride is converted to lithium carbonate and lithium hydroxide by reaction with sodium carbonate and calcium hydroxide, respectively. The advantage of the pegmatite-based production is that it is quicker to set up than from brines. Additionally, the lithium from spodumene has higher purity than that from brines which may contain high amounts of iron, magnesium or other deleterious materials.

5.40.3 Value addition of spodumene

Table 5.40 Uses of spodumene

Spodumene attributes and properties	various colours from colourless to yellowish, purplish, or like kunzite, yellowish-green or emerald-green hiddenite, prismic crystals, often of great size
Gemstone	varieties kunzite and hiddenite noted for their strong pleochroism
Source of lithium	Li is used in ceramics, mobile phones and automotive batteries, medicine, pyroceram; fluxing agent

5.41 Talc

5.41.1 Overview of talc

Talc is hydrated magnesium silicate mineral $H_2Mg_3(SiO_3)_4$ or $Mg_3Si_410(OH)_2$ which is found as foliated to fibrous masses. It is insoluble in water and slightly soluble in dilute mineral acids. It is greasy to the touch. It is the main constituent of soapstone. It is the result of the metamorphism of magnesium minerals [916] such as serpentine, pyroxene, amphibole, olivine in the presence of carbon dioxide.

5.41.2 Beneficiation of talc

Talc can be formed from the reaction between serpentine and carbon dioxide with the release of water and formation of magnesite.

Serpentine + carbon dioxide → talc + magnesite + water

$$2Mg_3Si_2O_5(OH)_4 + 3CO_2 \rightarrow Mg_3Si_4O_{10} + 3MgCO_3 + 3H_2O \tag{93}$$

Alternatively, talc is formed when dolomite reacts with silica and water.

Dolomite + silica + water → talc + calcite + carbon dioxide

$$3CaMg(CO_3)_2 + 4SiO_2 + H_2O \rightarrow Mg_3Si_4O_{10}(OH)_2 + 3CaCO_3 + 3CO_2 \tag{94}$$

or as a result of a metamorphic reaction between magnesium chlorite and quartz.

chlorite + quartz → kyanite + talc + water

5.41.3 Talc value addition

Table 5.41 Uses of talc

Talc attributes and properties	greasy feel; water-insoluble hydrated magnesium silicate mineral
Industrial uses	filler in paper manufacturing; plastic, paint and coatings; rubber; food; electric cable; pharmaceuticals; cosmetics (talcum powder); ceramics; stoves; sinks; electrical switchboards; crayons; soap; used for surfaces or lab counter tops; in lubricating oils
Powder	baby powder for preventing rashes on areas covered by diaper; in basketball to keep player's hands dry; tailor's chalk; chalk for welding and metalworking
Health	food additive in pharmaceutical products as a glidant; used as pleurodesis agent to prevent recurrent pleural effusion or pneumothorax
Ceramics industry	used in bodies and glazes; in low-fire art-ware bodies it imparts whiteness and increases thermal expansion to resist crazing; improves strength and vitrification in stone wares; a source of MgO flux in high temperature glazes; matting agent in earthenware glazes; in production of magnesia mattes at high temperature
Other uses	in production of materials used in the building interiors such as content paints in wall coatings; organic agriculture; food industry; cosmetics; hygiene products such a baby detergent powder
Drugs	an adulterant to illegal heroin; with intravenous use, it may lead to talcosis, a granulomatous inflammation in the lungs

5.42 Vermiculite

5.42.1 Overview of vermiculite

Vermiculite is a hydrous phyllosilicate mineral formed by the weathering or hydrothermal alteration of biotite or phlogopite. The weathered micas in which the potassium ions between the molecular sheets are replaced by magnesium and iron ions take the form of vermiculite clays. The interface between felsic and mafic or ultramafic rocks such as pyroxenites and dunites is the space where vermiculite occupies as the alteration product as in carbonatites and metamorphosed magnesium rich limestone. It is found interlayered with chlorite, biotite and phlogopite. The mineral phases that are associated with vermiculite include apatite, corundum, serpentine and talc. It undergoes exfoliation which is the phenomenon of expansion when the mineral is heated sufficiently. This effect is routinely produced in commercial furnaces.

5.42.2 Beneficiation of vermiculite

The weathering or hydrothermal alteration of biotite or phlogopite results in the formation of vermiculite which is an alteration product at the contact between felsic and mafic or ultramafic rocks such as pyroxenites and dunites, when heated. Carbonatites and metamorphosed magnesium rich limestone may also be the source of vermiculite with associated mineral phases that include corundum, apatite, serpentine and talc. It is found interlayered with chlorite, biotite and phlogopite.

5.42.3 Vermiculite value addition

Table 5.42 Uses of vermiculite

Vermiculite attributes and properties	ion exchange properties
Agricultural	soilless growing media for horticulture; growing media for hydroponics; seed germination aid; storing bulbs and root crops; soil conditioner; carrier for dry handling and slow release of agricultural chemicals; substrate for cultivation of fungi
Vermiculite material	for open fireplaces; high-temperature or refractory insulation; acoustic panels; fireproofing of structural steel and pipes
Structural material	calcium silicate boards; brake linings; roof and floor screeds and insulating concretes; insulator; sealant for pores and cavities of masonry construction; moulds that can stand high temperatures; used in aluminium smelting industry; flat boards [917]; brake linings; thermal resistance; rheological aids
Refractory material	combined with alumina cements and other aggregates to produce refractory/insulation concretes and mortars
Coatings and binders	coatings and binders for construction materials, gaskets, specialty papers/textiles, oxidation resistant coatings on carbon-based composites; packing material
Chemical operations	in fluid purification processes for wastewater; chemical processing and pollution control of air in mines and gases in industrial processes

(Continued)

Table 5.42 (Continued)

General purpose material	substrate for various animals and incubation of eggs; lightweight aggregated for plaster, proprietary concrete compounds; firestop mortar and cementitious spray fireproofing; resistance to chipping, cracking, shrinkage; additive to fireproof wallboard; interior fill for firestop pillows
Steel industry	hot topping for transporting material to next production process; permits slow cooling of hot pieces in glassblowing, lampwork, steelwork, glass beadmaking
Swimming pools	smooth pool base
Domestic uses	hand warmers; in AGA cookers as insulation; explosives storage as a blast mitigant; absorbs hazardous liquids for solid disposal; in gas fireplaces to simulate embers

5.43 Wollastonite

5.43.1 Overview of wollastonite

Wollastone $CaSiO_3$ is a white calcium inosilicate mineral whose calcium may be substituted by small amounts of iron, magnesium and manganese. It is obtained by reacting calcium oxide and silica.

$$CaO + SiO_2 \rightarrow CaSiO_3 \tag{93}$$

It may also be formed by the reaction between calcite and silica with the release of carbon dioxide.

$$CaCO_3 + SiO_2 \rightarrow CaSiO_3 + CO_2 \tag{94}$$

Calcite, diopside, epidote, feldspar, garnets, plagioclase, pyroxene, tremolite and vesuvianite are some of the minerals associated with wollastonite. It is commonly found as a constituent of a thermally metamorphosed limestone.

5.43.2 Beneficiation of wollastonite

Wollastonite is beneficiated when dry starting with primary and secondary crushing of the ore. The crushed material is then subjected to X-ray fluorescent separation and classification of the material according to grain size. This is followed by magnetic separation on a conveyor belt and further magnetic separation on a roll-type separator to remove impurities from the concentrate of various grain sizes. Subsequently the ore is impact ground, dried and followed by electrostatic separation and air sizing.

5.43.3 Wollastonite value addition

Table 5.43 Uses of wollastonite

Wollastonine attributes and properties	has high brightness and whiteness; low moisture and oil absorption; low volatile content; acicular products; resistance to chemical attacks; inertness; stability at high temperatures, flexural and tensile strength
Ceramics	source of calcium and silicon [918]; application due to its fluxing properties, freedom from volatile constituents, whiteness, and acicular particle shape [919]

(Continued)

Table 5.43 (Continued)

Friction products	brakes and clutches; substitute for asbestos in floor tiles, friction products; insulating board and panels; roofing products
Metalmaking	serves as flux for welding; source of calcium oxide; a slag conditioner; protects surface of molten metal during continuous casting of steel
filler material	fabricator of mineral wool insulation; ornamental building material [920]
Plastics industry	improves tensile and flexural strength; reduces resin consumption; improves thermal and dimensional stability at elevated temperatures
Wollastonite products	acicular nature provide improvement in dimensional stability in plastics, flexural modulus, and heat deflection
Paint industry	as an additive; improves the durability of the paint film; acts as a pH buffer; improves its resistance to weathering; reduces gloss; reduces pigment consumption; acts as a flatting and suspending agent

5.44 Zeolites

5.44.1 Overview of zeolites

Zeolites, such as bikitaite $|Li(H_2O)|[AlSi_2O_6]$ which is mined in Zimbabwe and yields large petalite crystals, are a large group of natural synthetic hydrated nonstoichiometric aluminium silicates. Characteristically, they are 3-dimensional structures with large, cage-like cavities large enough to accommodate Na, Ca, or other cations, water and even small organic molecules. The cations housed in the cage-like cavities balance the negative charge of the aluminosilicate framework of a zeolite. About 45 natural zeolites are known to exist, the more important ones being the fine-grained zeolites such as clinoptilolite $(Na,K)AlSi_5O_{111}H_2O$ which is formed by the alteration of fine-grained volcanic deposits by underground water. Zeolites are found in low-temperature metamorphic rocks and are often contaminated by other minerals such as ferrous, sulphates, quartz, other zeolites, and amorphous glass.

5.44.2 Beneficiation of zeolites

Natural zeolites occur offering a limited range of atomic structures and properties. Synthetic zeolites are now routinely produced [921] with wider range of properties and larger cavities. An important method of zeolite production is by mixing sodium, aluminium and silica chemicals with steam to create a gel. When the gel has aged, it is heated to about $90°C$. An alternative method involves kaolin clay which has been heated in a furnace until it begins to melt, then is chilled and ground to powder. The powder is subsequently mixed with sodium salts and water, aged, and heated again.

The zeolites produced synthetically are dependent on the composition of the starting materials and the reaction conditions including acidity, temperature and water pressure.

5.44.3 Zeolites value addition

Table 5.44 Uses of zeolites

Zeolites attributes and properties	complex 3-dimensional structures with large cavities that can accommodated cations, atomic clusters, water molecules, and small organic molecules; can interact with water [922] to absorb or release ions (ion exchange); can selectively absorb ions that fit their cavities (molecular sieves); can hold large molecules and break them into smaller pieces (catalytic cracking); structurally stable; high heat of water adsorption
Water treatment	water softeners to remove Ca ions, which react with soap to form scum; clinoptilolite used to clean ammonium ions (NH_4^+) from sewage and agricultural wastewater
Industrial processes	ion-exchanges agents; catalysts; molecular filters
Air filters	natural zeolites absorb SO_2 from waste gases; purify gases from power plants
Industrial applications	synthetic zeolites absorb or hold molecules as molecular sieves to remove water and nitrogen impurities from natural gas; interact with organic molecules in refining and purifying natural gas and petroleum chemicals; catalysts; used to help break down large organic molecules in petroleum into smaller molecules that make up gasoline (catalytic cracking); used in hydrogenating vegetable oils and in many other industrial processes involving organic compounds
Molecular sieving and adsorptive applications	adsorption of chemical species by zeolites [923,924]; used to produce sharp separation of molecules by size and shape; strong electrostatic field within a zeolite results in very strong interaction with polar molecules such as water; non-polar molecules also strongly adsorbed due to the polarising power of these electric fields; zeolites re-usable many times, cycling between adsorption and desorption
Ion exchange	exchange their cations for those of surrounding fluids; Na zeolite A is a remover of water hardness ions; i.e. detergent builder
Shape selective catalysts	extremely selective reactions can occur over zeolites when certain products and reactants or transition state are kept from forming with the pores because of size and shape
Laundry detergent	largest outlet for zeolites; does not contribute to the eutrophication of lakes, streams and bays
Solar-powered refrigerators	high heat of water adsorption; high adsorption capacity; undergo reversible adsorption/desorption; store energy during off-peak periods and release it during peak periods; in refrigeration and air-cooling systems to reduce water in the air

Chapter 6

Precious stones

6.1 Diamonds

6.1.1 Overview of diamonds

Diamonds are the only gemstone made of only one element: carbon [482]. Invariably, however, small amounts of carbon neighbours in the periodic table, such as boron and nitrogen are also found in diamonds accounting for the different colours of diamonds. Lattice defects are responsible for brown colouration while the green colour is the result of radiation exposure. Purple, pink, orange or red may also arise from lattice defects.

Carbon-containing minerals at high temperature and pressure at depths of 140 to 190 kilometres in the Earth's mantle over periods of 1 to 3.3 billion years are the sources for the formation of natural diamonds. Deep volcanic eruptions result in a magma bringing diamonds close to the Earth's surface. This magma cools into igneous rocks known as kimberlites and lamproites where the diamonds crystals grow larger with longer residence in the cratonic lithosphere [929].

6.1.1.1 Synthetic diamonds

Synthetic diamonds [935] can be produced in a high-pressure high-temperature (HPHT) method [930,931] which approximately simulates the conditions in the Earth's mantle. They can also be produced by chemical vapour deposition [932,933]. Silicon carbide and cubic zirconia are diamond simulants that resemble diamonds in both appearance and the other many properties [934].

Diamond enhancements may be made by certain specific treatments designed to better the gemmological characteristics of the diamond stone. These include laser drilling to remove inclusions, applications of sealants to fill cracks, treatments to improve a white diamond's colour grade or give a fancy colour to a white diamond [936].

6.1.1.2 Industrial diamonds

A significant part (80%) of global diamonds are brown-coloured and are termed industrial diamonds as they are predominantly used for industrial purposes due to their hardness and thermal conductivity. Market conditions partly determine the ill-defined distinction between gem-quality diamonds and industrial diamonds. Bort diamonds are the lowest-quality industrial diamonds which are mostly opaque.

6.1.2 Beneficiation of diamonds and other gemstones

The process of beneficiating gemstones from their ores varies from place to place and is dependent on the nature and quality of the gemstone. In general, those from alluvial deposits are assorted by washing with sieves. Slurry containing the rough gemstone along with mud and other materials is washed with water on a sieve. A vibratory motion in applied to the sieve so as to separate the slurry quicker. The gem rough is then hand-picked from the sieves. Depending on the size of the rough, different sieve sizes are employed.

Diamond beneficiation process from the mother rock utilises the properties of diamonds to separate them. *Crushing* is done to reduce the size of the rocks in the ore. *Heavy media separators* make use of the property of specific gravity wherein a heavy stone will sink in a liquid of lower specific gravity. Either one or a series of heavy media separators are used to separate the diamonds. A centrifugal force is applied to the liquid inside, for faster separation. A *grease table or belt* uses the property that diamonds stick to grease. In this process, ore is poured over a table and washed with water. A vibratory motion is given to the table where the diamonds with a few other minerals will stick while the rest is washed off. Subsequently, heat is used to separate the diamonds from the grease [937]. *X-ray separators* use the principle that diamonds fluoresce to X-rays. In this method, the ore is dropped in a fine stream through a channel and an X-ray beam is directed at it. As a diamond fluoresces, detector activates an air jet and the diamond is directed into another channel. *Magnetic separators* in the form of belts are used where there are magnetic minerals present in the ore. *XRT (X-ray transmission)* is a unique sorting algorithm that allows for extremely high accuracy in the detection and ejection of diamonds and thereby able to separate minerals from gangue waste. XRT detects different X-ray absorption levels of different material types.

6.1.2.1 Diamond cutting and polishing

Mined rough diamonds are converted to diamond gems through a multi-step process called "*cutting*" [938–941] to produce a faceted jewel where the specific angles between facets would optimise the diamond lustre. The following considerations need to be taken into account in cutting: the crystal orientation and crystallographic structure by X-ray diffraction to choose the optimal cutting directions must be analysed; inclusions and crystal flaws that are visible on-diamond to be removed or retained must be identified; the method or splitting or cutting the diamond must be determined.

After cutting there are numerous stages of polishing the shapes of the diamond to remove material by gradual erosion. After polishing the diamond is re-examined for possible flaws, either remaining or induced by the process. The identified flaws may be concealed through various diamond enhancement techniques. The diamond cutting process include the following sequential steps: *planning, cleaving or sawing, bruting, polishing* and *final inspection.*

6.1.3 Diamonds value addition

Table 6.1 Uses of diamonds

Diamond attributes and properties	hardness 10.0 on Mohs scale; gemstone made of only one element: carbon; hardness and thermal conductivity [925–927]; diamond is able to disperse light of different colours [928]; maintains its polish; toughness; resists breakages from forceful impact [482]; withstand crushing pressures; diamond is *hydrophobic* but *lipophilic*
Gemstone	graded using four basic parameters – the four Cs of connoisseurship: *colour*, *cut*, *clarity* and *carat* (weight); resistant to scratching and therefore well-suited to daily wear; wedding rings; used to polish other diamonds; regular diamonds may be treated under combination of HPHT to produce diamonds that are harder than diamonds used in hardness gauges
Diamond extraction	diamonds can easily be wet and stuck by oil or grease
Industrial diamonds	used to polish, cut, or wear any material including diamonds; used as diamond-tipped drill bits and saws; diamond powder as an abrasive; used in laboratories as containment for high pressure experiments; high-performance bearings; heat sink for integrated circuits in electronics due to their thermal conductivity; diamond knives and anvil cell
Synthetic diamonds	industrial applications; 90% of diamond grinding is of synthetic origin

6.2 Emerald

6.2.1 Overview of emerald

Emerald is a cyclosilicate green [943] variety of the mineral beryl $Be_3Al_2(SiO_3)_6$ which has a green colour because of the inclusions of trace elements of chromium and vanadium [942]. There is a tendency of emeralds having numerous inclusions and surface-breaking fissures [886] which are imperfections unique to each emerald and can be used to identify a particular stone.

6.2.2 Beneficiation of emeralds

Beneficiation from ores is as described in section 6.1.2. In order to fill in surface-reaching cracks and improve their clarity and stability, most emeralds are oiled as part of the post-lapidary process. Cedar oil or other liquids including synthetic oils and polymers of similar refractive index, such as *opticon*, is used. Fractures on emerald stones are often waxed with glass or resin to stabilise the stone and make the fractures and surface-reaching inclusions less visible. Reduction of visibility of inclusions is often done by heating and drilling.

6.2.3 Emeralds value addition

Table 6.2 Uses of emeralds

Emeralds attributes and properties	hardness of 7.5–8 on the Mohs scale; fine emerald possesses verdant green hue and high degree of transparency [943]
Gemstone	blue aquamarine; yellow heliodor; pink morganite; red beryl or bixbite; colourless goshenite

6.3 Ruby

6.3.1 Overview of ruby

Aluminium oxide occurs in a variety of forms of the mineral corundum, one of which is the gemstone ruby [944], which has a pink to blood-red colour. Ruby is α-alumina (the most stable form of Al_2O_3) in which a small fraction of the Al^{3+} ions are replaced by Cr^{3+} ions. The presence of the element chromium is the cause of the red colour. Sapphires are the other varieties of gem-quality corundum.

Imperfections are always found in natural rubies due to colour impurities and inclusions such as rutile needles known as "silk". These needle inclusions found in natural rubies distinguish them from synthetics, simulants, or substitutes. Fusing alumina with little chromium as a pigment [893] is one way of making synthetic rubies. They can also be produced through the Czochralski's pulling process, flux process and the hydrothermal process. Red spinels, red garnets and coloured glass are some of the imitation rubies that are marketed by being falsely claimed to be rubies.

6.3.2 Beneficiation of ruby

Beneficiation from ores is as described in section 6.1.2.

The rough ruby stone is usually heated before cutting. Almost all rubies today are treated in some form, with heat treatment being the most common practice [896]. This improves the quality of the gemstone by colour alteration. The transparency is improved by dissolving rutile inclusions, healing for fractures (cracks) or even completely filling them. Improving the quality of gemstones by treating them, commonly by application of heat [896], include colour alteration, improving transparency by dissolving rutile inclusions, healing of fractures (cracks) or even completely filling them. Another treatment is by filling the fractures inside the ruby with lead glass (or similar material) [945]. Copper or other metal oxides as well as elements such sodium, calcium, potassium etc. can enhance glass powder [946] by adding colour.

6.3.3 Ruby value addition

Table 6.3 Uses of ruby

Ruby attributes and properties	hardness 9.0 on Mohs scale
Gemstones	ruby prices determined by colour; red, blood-red or "pigeon blood" commands premium over other rubies; after colour follows clarity; cut and carat (weight) are other important factors; ruby is always lighter red or pink than garnet; ruby without any needle-like rutile inclusions may indicate that the stone has been treated
Other uses	applications where high hardness is required; at wear exposed locations in modern mechanical clockworks; as scanning probe tips in coordinate measuring machines; ruby lasers and masers are made from rods of synthetic ruby [947].

6.4 Sapphire

6.4.1 Overview of sapphire

Sapphire is blue gemstone variety of the mineral corundum α-Al_2O_3. Trace amounts of iron, titanium, chromium, copper, or magnesium are responsible for the gemstone's colours of blue, yellow, purple, orange, or green, respectively. The sapphire with pink or red tint is due to chromium impurities and it is called ruby which is the first variety of corundum discussed in 6.4. Padparadscha is the third gem-variety of corundum with a pinkish orange colour. Sapphire and rubies often occur together.

Sapphires occurs in alluvial deposits or primary underground workings. They may also occur in rock formations. Sapphires are divided into three broad categories that indicate types of constituent microscopic inclusions arising from different geographical locations and manifested in different appearances or chemical-impurity concentrations. These are classic metamorphic, non-classic metamorphic or magmatic, and classic magmatic [948].

Sapphires may be manufactured for industrial or decorative purposes from agglomerated aluminium oxide, sintered and fused in an inert atmosphere, resulting in transparent but slightly porous polycrystalline product.

6.4.2 Beneficiation of sapphires

Beneficiation from ores is as described in section 6.1.2. Sapphires' clarity and colour may be enhanced by treating them by several methods which include heating in furnaces to temperatures between 500 and 1,800° C for several hours. This can also be achieved by heating in a nitrogen-free atmosphere oven for several days. Colour enhancement by addition of impurities to them is called diffusion treatment. As an example, beryllium, chromium, iron, nickel or titanium may be absorbed and diffused into the crystal structure of the sapphire under very high heat.

6.4.3 Sapphire value addition

Table 6.4 Uses of sapphire

Sapphire attributes and properties	harness 9.0 on Mohs; high thermal conductivity; very high melting temperature (2030° C); colours blue, yellow, purple, orange, or green due to traces of iron, titanium, chromium, copper or magnesium, respectively; when pink or red tint its ruby; able to handle very high power densities in the infrared or UV spectrum without degrading due to heating; Electrical insulating properties
Gemstone	Jewellery
Non-ornamental applications	in some infrared optical components; high-durability windows; wristwatch crystals and movement bearings; very thin electronic wafers which are used as the insulating substrate of very special-purpose solid-state electronics (especially integrated circuits and GaN-based LEDs)
Industrial use of synthetic sapphires [949–951]	material used in laser and nanotechnologies; shatter resistant windows in armoured vehicles and various military body armour suits

(Continued)

Table 6.4 (Continued)

Blue glass or sapphire glass	is a synthetic sapphire which is both transparent and extraordinarily scratch-resistant [952]; wide optical transmission band from UV to near-infrared (0.15–5.5 μm); very strong optical material
Applications dependent on hardness and toughness	used in high pressure chambers for spectroscopy, crystals in various watches, and windows in grocery store barcode scanners [950]; end windows on some high-powered laser tubes
High-power radio-frequency (RF) applications	CMOS chips on sapphire used in cellular telephones, public-safety band radios, and satellite communications systems
Silicon on sapphire "SOS"	allows for the monolithic integration of both digital and analogue circuitry all on one IC chip; construction of extremely low power circuits; crystal sapphire boules are core-drilled into cylindrical rods, and wafers are then sliced from these cores
Semiconductor industry	uses wafers of single-crystal sapphire as substrate for growth of devices based on gallium nitride (GaN); reduces costs to a third of germanium ones; used in blue LEDs [953]

Chapter 7

Semi-precious stones

7.1 Agate

7.1.1 Overview of agate

Agate is a cryptocrystalline variety of silica which occurs in various kinds of rock but is associated with volcanic rocks and common in certain metamorphic rocks [954]. Most agates may be found as nodules in volcanic rocks or ancient lavas, in former cavities produced by volatiles in the original molten mass. These may have been filled by siliceous matter deposited in regular layers upon the walls. Veins or cracks in volcanic or altered rock underlain by granitic intrusive masses may be filled with agate giving rise to *banded agate, riband agate* and *stiped agate*.

7.1.2 Beneficiation of agate

Beneficiation from ores is as described in section 6.1.2.

7.1.3 Agate value addition

Table 7.1 Uses of agate

Agate attributes and properties	fine grain and bright colour; extremely resistant to weathering; hardness; ability to retain highly polished surface finish; resistant to chemical attack
Industrial use	used to make knife-edge bearings for laboratory balances and precision pendulum; used to make mortars and pestles to crush and mix chemicals; used for leather burnishing tools
Decorative applications	ornaments such as pins, brooches or other type of jewellery; paper knives, inkstands, marbles and seals; decorative displays, cabochons, beads, carvings, and Intarsia art; face-polished and tumble-polished specimens of varying sizes and origin

7.2 Amethyst

7.2.1 Overview of amethyst

Amethyst [955], one of several forms of quartz (SiO_2), is a semiprecious stone whose violet colour is due to irradiation, iron impurities, and the presence of trace elements, which result in complex crystal lattice substitutions [956–958,960]. It exhibits primary hues from light pinkish to deep purple [959].

Production of synthetic amethyst is possible by gamma ray, X-ray or electron beam irradiation of clear quartz which has been first doped with ferric impurities. If the synthetic amethyst is exposed to heat, the irradiation effects can be partially cancelled and the amethyst generally becomes yellow or even green [955,961].

7.2.2 Beneficiation of amethyst

Beneficiation from ores is as described in section 6.1.2.

7.2.3 Amethyst value addition

Table 7.2 Uses of amethyst

Amethyst attributes and properties	same hardness as quartz
Gemstone	Jewellery

7.3 Aquamarine

7.3.1 Overview of aquamarine

Aquamarine is a blue or cyan variety [965] of beryl $Be_3Al_2Si_6O_{18}$ which occurs at most deposits that yield beryl. The yellow beryl is *aquamarine chrysolite* [962] while the deep blue version of aquamarine is called *maxixe* whose colour fades to white when exposed to sunlight or is subjected to heat treatment. The colour may return with irradiation. The ferrous ions, Fe^{2+} is responsible for the blue colour of aquamarine while the ferric Fe^{3+} is responsible for the golden-yellow colour. When both are present the colour is a darker blue as in maxixe. Charge transfer between Fe^{3+} and Fe^{2+} due to heat or light may cause of decolouration of maxixe [963–966]. Irradiation with high-energy particles (gamma rays, neutrons or even X-rays) [967] can transform the dark-blue maxixe into green, pink or yellow beryl.

Aquamarines are explored in areas that contain Precambrian metamorphic rock which may contain many minerals such as biotite, felspar, garnet, mica, quartz and many gemstones within it. The metamorphic rocks comprise some pegmatitic veins that host aquamarine crystals. These veins maybe sheared and varies from 10 m to 40 m in thickness. The aquamarine crystals are of cabochon to facet grade varieties becoming visible bearing few inclusions. Their mineralisation is found within the pegmatic veins.

7.3.2 Beneficiation of aquamarine

As the centre of the pegmatite is massive it is fractured and banded; the veins are fractured through opencast mining method using pneumatic tools such as compressor, jack hammer, picks, shovels, chisels, ladders etc. Bulldozers may also be used to aid the mining of the aquamarine. On freeing the gemstone from the rock, it is collected and cleaned off with water.

After that the rough aquamarines are sorted for size, shape and colour followed by gem processing during which the rough gemstones are transformed into beautiful and attractive gems.

7.3.3 Aquamarine value addition

Table 7.3 Uses of aquamarine

Aquamarine attributes and properties	possesses all the different shades of sea water – light blue colour to dark with a touch of green in all possible variations
Gemstone	jewellery item

7.4 Beryl

7.4.1 Overview of Beryl

Beryl $Be_3Al_2Si_6O_{18}$ is a beryllium aluminium cyclosilicate mineral which is colourless when pure, but frequently tinted by impurities to green, blue, yellow, red and white. Beryl of various colours commonly occurs in granite, rhyolite and granitic pegmatites, but also in mica schists and is associated with tin and tungsten ore bodies. It may also occur where carbonaceous shale, limestone and marble have been acted upon by regional metamorphism. The carbonaceous material provides the chromium or vanadium responsible for the colour to the emerald. It is thought that aquamarine is a blue or cyan variety of beryl while emerald is green beryl. Beryl contains a significant amount of beryllium [968].

7.4.1.1 Golden beryl

Golden beryl or *heliodor* ranges in colours from pale yellow to a brilliant gold with very few flaws. The Fe^{3+} ions are responsible for the golden-yellow colour [969,970].

7.4.1.2 Goshenite

Goshenite is the colourles variety of beryl. It can be coloured yellow, green, pink, blue and in intermediate colours by irradiating it with high-energy particles with resultant colours dependent on the content of Ca, Sc, Ti, V, Fe and Co impurities [970]. Since colours in beryl are caused by impurities it may be assumed that *goshenite* is the purest variety of beryl. However, there are several elements that can act as colour inhibitors in beryl rendering the assumption not always true.

7.4.1.3 Morganite

"Pink beryl", "rose beryl" and "pink emerald" are synonymous with *morganite* which is a rare light pink to rose-coloured gem-quality variety of beryl as are orange/yellow varieties. It can be treated by heat to remove patches of yellow or by irradiation to improve its colour. Mn^{2+} ions are thought to be responsible for the pink colour [969].

7.4.1.4 Red beryl

A red variety of beryl may be known as "red emerald" or "scarlet emerald". Mn^{3+} ions [969] are thought to be responsible for the dark red colour.

Red beryl is found in topaz-bearing rhyolites while beryl gems are ordinarily found in pegmatics and certain metamorphic stones. Bixbyite, haematite, orthoclase, pseudobrookite, quartz, spessartine and topaz are some of the minerals associated with red beryl [970,971].

There is also synthetic red beryl that can be produced.

7.4.2 Beneficiation of beryl

Beneficiation from ores is as described in section 6.1.2.

7.4.3 Beryl value addition

Table 7.4 Uses of beryl

Beryl attributes and properties	occurs in a wide variety of colours: emerald (green), aquamarine (greenish blue to blue), morganite (pink to orange), red beryl (red), heliodor (yellow to greenish yellow), maxixe (deep blue), goshenite (colourless), green beryl (light green)
Gemstone	occurs in a wide variety of colours
Chatoyant specimens	can be cut into cabonchons to produce interesting *cat's eyes*;
Goshenite use	used for manufacturing eyeglasses and lenses
Mineral	source of beryllium

7.5 Chrysoberyl

7.5.1 Overview of chrysoberyl

Chrysoberyl $BeAl_2O_4$ is an aluminate of beryllium in the form of both mineral and gemstone in three varieties: ordinary yellow-to-green chrysoberyl, cymophane and alexandrite. It is formed as a result of pegmatitic processes. Due to the fact that it is a hard and dense mineral which is resistant to chemical alteration, it may be weathered out of rocks and deposited in river sands and gravels in alluvial deposits with other gem minerals such as corundum, diamond, garnet, spinel topaz and tourmaline. Crystals of beryl or chrysoberyl can form if the pegmatite fluid is rich in beryllium. Chrysoberyl has a high ratio of aluminium to beryllium.

7.5.1.1 Alexandrite

Some chromium would have to be present in chrysoberyl for alexandrite to form. This situation arises only when Be-rich pegmatite fluids react with Cr-rich country rock. Synthetic laboratory-grown stones with the same chemical and physical properties as natural alexandrite can be produced [973]. However, some gemstones falsely described as lab-grown synthetic alexandrite may actually be corundum laced with trace elements such as vanadium or colour-change spinel indicating that they are not actually chrysoberyl. As such they are better described as simulated alexandrite rather than synthetic ones [974].

7.5.1.2 Cymophane

Cymophane exhibits chatoyancy or opalescence similar to the eye of a cat, hence it is known as "*cat's eye*". There are varieties where microscopic tubelike cavities or needle-like inclusions [975] of rustle occur in an orientation parallel to the c-axis, producing a chatoyant effect visible as a single ray of light passing across the crystal. The Fe^{3+} impurities are responsible for the colour in yellow chrysoberyl.

7.5.2 Beneficiation of chrysoberyl

Beneficiation from ores is as described in section 6.1.2.

7.5.3 Chrysoberyl value addition

Table 7.5 Uses of chrysoberyl

Chrysoberyl attributes and properties	hardness 8.5 on Mohs scale i.e. between corundum (9) and topaz (8) [972]; yellowish-green and transparent to translucent
Gemstone	jewellery; when the mineral exhibits good pale green to yellow colour and is transparent, it is used as a gemstone

7.6 Garnet

7.6.1 Overview of garnets

Garnets are a group of nesosilicate minerals with the following different species: almandine, andradite, grossular (variety of which are hessonite or cinnamon-stone and tsavorite), pyrope, spessartine and uvarovite, making up two solid solution series: pyrope-almandine-spessartine and uvarovite-grossular-andradite. They occur in many colours including black, blue, brown, green, orange, pink, purple, red, yellow and colourless.

Garnets can be represented by the following general formula:

$$X_3Y_2(SiO_4)_3$$

where X = divalent cations (Ca, Mg, Fe, Mn)$^{2+}$
       Y = trivalent cations (Al, Fe, Cr)$^{3+}$

Almandine, $Fe_3Al_2(SiO_4)_3$, carbuncle is a deep red transparent stone which can be found in metamorphic rocks like mica schists. It is associated with minerals such as staurolite, kyanite, andalusite.

Pyrope $Mg_3Al_2(SiO_4)_3$ has a colour that varies from deep red to black. The magnesium can be replaced by calcium and ferrous ions. As a mineral it is essentially an isomorphous mixture of pyrope and almandine in proportion 2:1. It is an indicator mineral for high-pressure rocks and is often contained in mantle-derived rocks, peridotites and eclogites.

Spessartine or spessartite is manganese aluminium garnet $Mn_3Al_2(SiO_4)_3$. It is often found in granite pegmatite and allied rock types and in certain low grade metamorphic phyllites.

Andradite is a calcium-iron garnet $Ca_3Fe_2(SiO_4)_3$ of variable composition in the following colours: black, brown, green, red or yellow. It has the following varieties: topazolite (yellow to green), demantoid (green) and melanite (black). It occurs both in deep-seated igneous rocks like syenite as well as serpentine, schists and crystalline limestone.

Grossular is a calcium-aluminum garnet $Ca_3Al_2(SiO_4)_3$. The calcium may in part be replaced by ferrous iron and the aluminium by ferric iron. It occurs in contact metamorphosed limestones with diopside, vesuvianite, wernerite and wollastonite. Tsavorite is its name used in Kenya and Tanzania.

Uvarovite is a calcium-chromium garnet $C_3Cr_2(SiO_4)_3$ which is a rare garnet with a bright green colour, usually occurring as small crystals associated with chromite in kimberlites, peridotite, and serpentinite and in crystalline marbles and schists.

Knorringite is a magnesium-chromium garnet $Mg_3Cr_2(SiO_4)_3$ whose pure endmember never occurs in nature. Pyrope containing a significant component of knorringite is only formed under high pressure and often occurs in kimberlites.

An expansion of the crystallographic structure of garnets from the prototype can be considered to include chemicals with the general formula $A_3B_2(CO_4)_3$ where C represents a large number of elements including Al, Ge, Ga, Fe and V besides silicon.

Yttrium aluminium garnet (YAG), $Y_3Al_2(AlO_4)_3$ is an illustrative synthetic gemstone with such a varied crystallographic structure. Yttrium aluminium garnet (YAG), $Y_3Al_2(AlO_4)_3$, is used for synthetic gemstones.

A strong neodymium magnet can separate garnets from all other natural transparent gemstones commonly used in the jewellery industry [976].

7.6.2 Beneficiation of garnet

Beneficiation from ores is as described in section 6.1.2.

7.6.3 Garnets value addition

Table 7.6 Uses of garnets

Garnet attributes and properties	range of hardness on the Mohs scale of about 6.5 to 7.7
Pyrope	an indicator mineral for high-pressure rocks in search of diamonds
Gemstones	pure crystals with varieties occurring in shades of green, red, yellow, and orange
Abrasives	harder species like almandine; garnet sand is good abrasive with alluvial grains, which are rounder, being more suitable for sand blasting treatments; cutting steel and other materials in water jets with high pressure; garnet from hard rock more suitable because it is angular and more efficient in cutting; garnet paper used for finishing bare wood [977]; glass polishing and lapping; larger grain sizes used for faster work and the smaller ones for finer finishes
Gadolinium gallium garnet $Gd_3Ga_2(GaO_4)_3$	used as a substrate for liquid-phase epitaxy of magnetic garnet films for bubble memory and magneto-optical application
Geothermobarometry	used in the interpretation of the genesis and temperature-time histories of many igneous and metamorphic rocks; useful in defining metamorphic facies of rocks
Water filtration media	garnet sand
Placer deposit	river sand garnet occurs as a placer deposit
Synthetic gemstones	yttrium aluminium garnet (YAG), $Y_3Al_2(AlO_4)_3$, is used for synthetic gemstones; with neodymium (Nd3+) as a dopant, YAl-garnet may be used as the lasing medium in lasers; due to its high refractive index YAG used as a diamond simulant

7.7 Howlite

7.7.1 Overview of howlite

Howlite which is a calcium borosilicate hydroxide $Ca_2B_5SiO_9(OH)_5$, a borate mineral that occurs in evaporite deposits [978]. Its most common form is irregular white nodules with fine grey or black veins in an erratic, often web-like pattern, opaque with a sub-vitreous lustre. Its crystals are colourless, white or brown, prismic and flattened [978] and are often translucent or transparent.

The most common form of howlite is irregular nodules, sometimes resembling cauli-flower. Crystals of howlite are rare, having been found in only a couple localities worldwide [978]. The crystals are colourless, white or brown.

7.7.2 Beneficiation of howlite

Beneficiation from ores is as described in section 6.1.2.

Boric acid is prepared from howlite minerals by first treating the mineral with sulphuric acid in order to dissolve boron compounds. The solution when separated from the solids in suspension is reacted with hydrogen sulphide in order to precipitate arsenic and iron impurities contained in it. When the impurities are removed the solution is again reacted with ammonia to precipitate out aluminium impurities after which the resultant solution is reacted with hot sulphuric acid in order to generate boric acid. The reaction mixture is cooled in order to precipitate boric acid and separate it from the remaining solution. It is susceptible to be recycled to the sulphuric acid treatment state thereby concentrating the mineral. Ammonia is similarly regenerated to be used again in the process.

7.7.3 Howlite value addition

Table 7.7 Uses of howlite

Howlite attributes and properties	monoclinic structure with 3.5 Mohs hardness; porous texture
decorative applications	small carvings or jewellery components; easily dyed to imitate other minerals, especially turquoise; dyed howlite (or magnesite) is marketed as *turquenite*
Mineral	source of boric acid

7.8 Jade

7.8.1 Overview of jade

Jade refers to two different metamorphic rocks constituted of different silicate minerals: *nephrite* and *jadeite*.

Microcrystalline interlocking fibrous matrix of calcium, magnesium-iron rich amphibole mineral series tremolite (calcium-magnesium)-ferroacinolite (calcium-magnesium-iron) make up *nephrite*. Actinolite (one form of the silky fibrous asbestos mineral) is the middle member of this series with an intermediate composition. The iron content determines the intensity of the green colour.

Jadeite is a microcrystalline interlocking crystal matrix pyroxene mineral rich in sodium and aluminium and exhibits various colours including blue, lavender-mauve, pink, and emerald green.

7.8.2 Beneficiation of jade

Mining of jade is done from large boulders that contain significant deposits of the mineral using a diamond core drill in order to extract samples. On removing the boulders and accessing the jade, it is broken into more manageable 10-tonne pieces using water-cooled diamond saws. Jade may be enhanced by three main methods referred to as the ABC Treatment System [979].

No treatment takes place in any way in *Type A* except surface waxing. In *Type B* treatment, a promising but stained piece of jadeite is exposed to chemical bleaches and/or acids and impregnating it with a clear polymer resin. This improves the transparency and colour of material. *Type C* involves artificially staining and dying a piece of the mineral resulting in a dull colour and losing translucency. *Type B + C* is a combination of *B* and *C* meaning that the piece has been impregnated and artificially stained. *Type D* refers to a composite stone such as a doublet comprising a jade top with a plastic substrate.

7.8.3 Jade value addition

Table 7.8 Uses of jade

Jade attributes and properties	jadeite hardness on Mohs scale is between 6.0 and 7.0; nephrite measures between 6.0 and 6.5
ornamental and decorative	hardstone carvings; historical ornaments with beads, button, and tubular shapes [980]; adze heads, knives, and other weapons

7.9 Mtorolite (chrome chalcedony)

7.9.1 Overview of mtorolite

The green variety of the mineral chalcedony [989] is called chrome chalcedony [981], called so presumably because it is coloured by small quantities of chromium [982]. It was discovered in the 1950s in Zimbabwe [990] where it is known as Mtorolite [983], mtorodite [984], or matorolite [985], principally near to the mining town of Mtoroshanga, located on the Great Dyke geological feature [983] well known for hosting vast deposits of chromium.

Mtorolite and chrysoprase are similar in appearance but differ in that the former is coloured red by chromium (as chromium(III) oxide) with tiny black specks of chromite [987] while chrysoprase is coloured green by aluminium [981,986]. Other varieties of chalcedony are agate, carnelian, chrysoprase, heliotrope, and onyx, being cryptocrystalline forms of silica, consisting of fine intergrowths of the minerals quartz and moganite [988].

7.9.2 Beneficiation of mtorolite

Beneficiation from ores is as described in section 6.1.2.

7.9.3 Mtorolite value addition

Table 7.9 Uses of Mtorolite

Mtorolite attributes and properties	coloured due to small amounts of chromium
Main application	jewellery

7.10 Opal

7.10.1 Overview of opal

Opal is a hydrated amorphous mineraloid of silica deposited at a relatively low temperature found in the fissures of almost any kind of rock in association with basalt, limonite, marl, rhyolite and sandstone. It takes on many colours depending on the conditions in which it is formed [994]. Other than the gemstone varieties there are other kinds of opal. Synthetic opals can be produced experimentally and commercially [993], but they are distinguishable by their regularity and porosity.

7.10.2 Beneficiation of opal

Beneficiation from ores is as described in section 6.1.2.

7.10.3 Opal value addition

Table 7.10 Uses of opal

Opal attributes and properties	diffract light due to its internal structure [991–993]
Gemstone use	cut and polished to form cabochon; opal doublet consists of thin layer of opal backed by a layer of dark-coloured material (ironstone, potch, onyx or obsidian); an opal triplet is similar to the doublet
Jewellery application	limited due to opal's sensitivity to heat and predisposition to scratching

7.11 Pearl

7.11.1 Overview of pearl

Pearls are produced by a living shelled mollusc within its soft tissue (especially the mantle). Pearls are hard objects composed of calcium carbonate in minute crystalline form. While natural pearls occur spontaneously in the wild, these are extremely rare, the majority being *cultured* or *farmed* from pearl oysters and freshwater mussels. The unique lustre of pearl is a result of its reflection, refraction and diffraction of light from its translucent layers the number of whose layers in the pearl determine the fineness of the lustre. The iridescence that pearls display is caused by the overlapping of successive layers, which breaks up light falling on the surface. Pearls can be dyed black, blue, brown, green, pink, purple or yellow.

7.11.2 Beneficiation of pearls

Pearls are produced by natural processes when an irritating microscopic object becomes trapped within the mantle folds of shelled molluscs. However, the majority of these "pearls" are not valued as gemstones. The most commercially significant are nacreous pearls. These are produced by mollusc and bivalves or clams. Nacreous pearls are made from layers of nacre similar to the process used in the secretion of the mother pearl which lines the shell.

The mollusc's mantle, which is the protective membrane, deposits layers of calcium carbonate ($CaCO_3$) in the form of the mineral aragonite or a mixture of aragonite and calcite [995,996] held together by an organic horn-like conchiolin. The mother-of-pearl is made up of the combination of aragonite and conchiolin which is nacre. The stimuli are organic material, parasites, or even damage that displaces mantle tissue to another part of the mollusk's body. Entry is gained when the shell valves are open for feeding or respiration.

Cultured pearls are a response to a tissue implant to a shell and may be produced by a number of methods [997] which include using freshwater or seawater shells, transplanting the graft into the mantle or into the gonad, and adding spherical bead as a nucleus.

7.11.3 Pearl value addition

Table 7.11 Uses of pearl

Pearl attributes and properties	gem-quality pearls are nacreous and iridescent
Applications	jewellery, adorning clothing, in cosmetics, medicine and paint formulations

7.12 Quartz

7.12.1 Overview of quartz

Quartz SiO_2 is an abundant mineral in the Earth's continental crust [985] with a crystal structure that is a continuous framework of SiO_4 silicon-oxygen tetrahedra. Several varieties of quartz [998,999] are semi-precious gemstones. There are many different varieties of quartz [998–999], several of which are semi-precious gemstones. Common coloured varieties include amethyst, citrine, milky quartz, rose quartz, smoky quartz and others. Pure quartz is colourless and transparent or translucent.

Quartz is present in granite and other felsic igneous rocks and also common in sedimentary rocks such as sandstone and shale as well as in various amounts as an accessory mineral in most carbonate rocks. It is a constituent of schist, geneiss, quartzite and other metamorphic rocks. Because it is not easily weathered, it is common as a residual mineral in stream sediments and residual soils.

Quartz crystallises from molten magma and also chemically precipitates from hot hydrothermal veins as gangue sometimes with minerals such as copper, gold and silver. Large crystals of quartz may occur in magmatic pegmatites.

High-temperature polymorphs of SiO_2 that are found in high-silica volcanic rocks include tridymite and cristobalite. Some meteorite impact sites and metamorphic rocks formed at pressures greater than those typical of the Earth's crust have been found to host coesite and stishovite which are denser polymorphs of SiO_2. Lightning strikes in quartz sand may cause the formation of lechatelierite which is an amorphous silica glass SiO_2.

Amethyst is formed when there is iron in the area where it is formed. It is sometimes found to grow together in the same crystal with citrine and is referred to as ametrine.

Critine, which is a variety of quartz, has colour ranges from pale yellow to brown due to ferric impurities. As natural critines are rare, the ones commercially available are heat-treated amethysts or smoky quartz.

The colour varieties of quartz generally due to trace amounts of metals such as iron, manganese, titanium, rutile, phosphate, aluminium and others. The white colour in milk quartz or milky quartz thought to be caused by minute fluid inclusions of gas, liquid, or both, trapped during crystal formation.

7.12.2 Beneficiation of quartz

Beneficiation from ores is as described in section 6.1.2.

Some clear quartz crystals are treated using heat or gamma-irradiation to induce colour. Prasiolite and the majority of citrine are produced by heat treatment of amethyst. Deepening of its colour is achieved by heat treament of carnelian. Quartz is sometimes coated with metal vapour to give it an attractive sheen.

7.12.3 Quartz value addition

Table 7.12 Uses of quartz

Quartz attributes and properties	pure quartz is colourless and transparent or translucent; piesoelectric properties; (they develop an electric potential upon the application of mechanical stress)
Gemstone	varieties of quartz used in jewellery and hardstone carvings
High-quality quartz crystals	crucibles and other equipment for growing silicon wafers in the semiconductor industry
Photograph pickups	use of piezoelectic properties; crystal oscillator; quartz clock; quartz crystal microbalance; in thin-film thickness monitors

7.13 Tanzanite

7.13.1 Overview of tanzanite

Tanzanite $Ca_2Al_3(SiO_4)_3(OH)$ belongs to the epidote group and is a blue/violet variety of zoisite, a calcium aluminium hydroxyl sorosilicate mineral. It has been mined since 1967 in the Mererani Hills [1000] near the city of Arusha and Mount Kilimanjaro, Tanzania. Tanzanite appears to have different colours when viewed under alternate lighting conditions. Several colours have been observed in various specimens: blue, brown, cyan, green, shades of purple, red, violet and yellow. It is usually reddish brown in its rough state. Heat treatment enhances its dichroic colours ranging from bluish-violet to violetish-blue [1001,1002].

7.13.2 Beneficiation of tanzanite

Beneficiation from ores is as described in section 6.1.2.

Tanzanite is heat-treated in a furnace at temperatures 370–390° C for 30 minutes. After that the stone shows no cracks, bubbles [1004] with the source of the heating gemologically undetectable [1005] and of no apparent consequence.

7.13.3 Tanzanite value addition

Table 7.13 Uses of tanzanite

Tanzanite attributes and properties	remarkably strong trichroism, appearing alternately sapphire blue, violet and brown depending on crystal orientation [1003]
Gemstone	extremely rare [1000]
Collector market	green tanzanite

7.14 Topaz

7.14.1 Overview of topaz

Topaz $Al_2SiO_4(F,OH)_2$ is a silicate mineral containing aluminium and fluorine. It is colourless and transparent but normally tinted by metal impurities to the following colours: blue brown, pale grey, reddish-orange, wine-red or yellow. Other colours it can assume are blue, pale green, white, gold, pink (rare) reddish-yellow or opaque to transparent/translucent. Treatment of brown or pale topazes makes them bright yellow, gold, pink or violet coloured. Heat treatment and irradiation of colourless, grey or pale yellow and blue material produce darker blue. Artificially coating the mystic topaz which is colourless gives it a rainbow effect. However, exposure of some imperial topaz stones to sunlight for extended time may cause them to fade.

The mineral is usually associated with silicic igneous rocks of the granite and rhyolite type. It crystallises in pegmatites or in vapour cavities in rhyolite lava flows. It can occur with fluorite and cassiterite in various areas.

7.14.2 Beneficiation of topaz

Beneficiation from ores is as described in section 6.1.2.

7.14.3 Topaz value addition

Table 7.14 Uses of topaz

Topaz attributes and properties	colourless and transparent but usually tinted by impurities
Gemstone	various colours due to impurities and heat treatment

7.15 Tourmaline

7.15.1 Overview of tourmaline

Tourmaline is a crystalline boron silicate mineral that incorporates elements such as aluminium, iron, lithium, magnesium, potassium or sodium. It is a semi-precious stone which exhibits a wide variety of colours including dichroism, in that they change colour when viewed from different directions. Some tourmaline gems are altered by heat treatment or irradiation to improve their colour. Additionally, some heavily included tourmalines, such as rubellite, are sometimes clarity-enhanced [1007].

The composition of the tourmaline mineral group varies considerably because of isomorphous replacements (solid solution). Its general formula can be written as

$$XY_3Z_6(T_6O_{18})(BO_3)V_3W$$

Where [1006]:
X = Ca, Na, K, □ = vacancy
Y = Li, Mg, Fe^{2+}, Mn^{2+}, Zn, Al, Cr^{3+}, V^{3+}, Fe^{3+}, Ti^{4+}, vacancy
Z = Mg, Al, Fe^{3+}, Cr^{3+}, V^{3+}
T = Si, Al, B
B = B, vacancy
V = OH, O
W = OH, F, O

Tourmaline occurs in granite and granite pegmatites and in metamorphic rocks such as schist and marble. It is a durable mineral to the extent that it can occur in minute amounts as grains in sandstone and conglomerate, and is part of the ZTR for highly weathered sediments [1008].

7.15.2 Beneficiation of tourmaline

Beneficiation from ores is as described in section 6.1.2.

7.15.3 Tourmaline value addition

Table 7.15 Uses of tourmaline

Tourmaline attributes and properties	wide variety of colours; durable mineral found in minor amounts as grains in sandstone and conglomerate; part of the ZTR index for highly weathered sediments
Gemstone	Jewellery

Chapter 8

Dimension stones

8.0 Dimension stone

8.0.1 Overview of dimension stone

A natural stone or rock that has been selected and trimmed, cut, drilled, ground to specific sizes and shapes is defined as a dimension stone if it meets the other requirements of colour, texture and pattern, and surface finish. The duration of the ability of the dimension stone to endure and maintain its essential and distinctive characteristics of strength, resistance to decay, and appearance is also an important selection criterion for a dimension stone [1009].

A variety of igneous, metamorphic, and sedimentary rocks such as granite, limestone, marble, travertine, quartz-based stone and slate are the raw material from which structural and decorative dimension stones are made. Special minor types [1010] that may be considered for the same function are alabaster (massive gypsum), soapstone (massive talc), serpentine and various products fashioned from natural stone. Geological phenomena such as mineral grains, inclusions, veins, cavity fillings, blebs, and streak create an extraordinary wide range of colours available in thousands of patterns.

8.0.2 Beneficiation of dimension stones

The original rock is first shattered by heavy and indiscriminate blasting. Precise and delicate techniques, such as diamond wire saws, diamond belt saws, burners (jet-piercers), or light and selective blasting with Primcord are used to separate the dimension stone. A variety of finishes that can be applied to dimension stone are applied to achieve diverse architectural and aesthetic effects. They include a finished slab of stone (usually granite) for the production of countertops and bathroom vanities; reciprocating gangsaws use steel shot as abrasive for sawing countertop slabs from rough blocks of stone. Resin is often used to fill microfissures and surface imperfections due to the loss of poorly bonded elements such as biotite to end up with finished (i.e. polished and honed) slabs.

8.0.3 Dimension stone value addition

Table 8.1 Uses of dimension stone

Dimension stone attributes and properties	produced from igneous, metamorphic, and sedimentary rocks such as granite, limestone, marble, travertine, quartz-based stone; durability
Construction industry	buildings and veneers in place of solid stone blocks; limestone, quartz-based stone (sandstone), marble or granite; tiles with common colours of white and light earth colours; roofing slate

(Continued)

Table 8.1 (Continued)

Other applications	stone monuments (grey, black and mahogany granite); tombstones; grave markers; mausoleums
Marble	military cemeteries (usually white)
granite and quartz	for durability against rain which is naturally acidic because of the CO_2 present in the atmosphere; acidity can be due to oxides of sulphur and nitrogen due to anthropogenic emissions
Limestone and sandstone	more susceptible to rapid erosion due to dissolution of acid-vulnerable carbonates by acidic rainfall
Curbing	traffic-related use in curbing (vehicular) and flagstone (pedestrian); curbing is made of thin stone slabs used along streets or highways to maintain integrity of sidewalks and borders; granite for curbing and quartz-based stone for flagstone [1011]
Recycling dimension stone	aggregate in the form of concrete when structures are demolished; fireplace mantels; benches, veneer, landscaping (such as for retaining walls)

8.1 Granite

8.1.1 Overview of granite

Granite is a holocrystalline felsic intrusive igneous rock of granular and phaneritic nature. It is composed mainly of alkali feldspar, quartz, mica and amphibole minerals [1012]. Granites are sometimes found in circular depressions surrounded by a range of hills, formed by the metamorphic aureole of hornfels. It usually occurs in the continental plates of the earth's crust with its outcrops tending to form tors and rounded massifs.

8.1.2 Beneficiation of granite

Skills are required to carve granite by hand because it is a hard stone. Computer-controlled rotary bits and sandblasting over a rubber stencil is one of the modern methods of carving granite. Any kind of artwork or epitaph can be created leaving the letters, numbers and emblems exposed on the stone.

8.1.3 Granite value addition

Table 8.2 Uses of granite

Granite attributes and properties	massive; hard and tough; poor primary permeability
Construction industry	gravestones and memorials; flooring tiles in public and commercial buildings and monuments; polished granite for kitchen countertops due to its high durability and aesthetic qualities
Granite tables	base for optical instruments due to granite's rigidity, high dimensional stability and excellent vibration characteristics
Polished granite surfaces	used by engineers to establish a plane of reference granites are relatively impervious and inflexible
Substituted granite	sandblasted concrete with heavy aggregate content has similar appearance as rough granite
Granite processing	granite blocks are cut into slabs which can be cut and shaped by a cutting centre

8.2 Marble

8.2.1 Overview of marble

Marble is composed of recrystallised carbonate minerals, mostly calcite or dolomite as a non-foliated metamorphic rock of interlocking mosaic of crystals [1013]. Various mineral impurities such as chert, clay, iron oxides, sand or silt are responsible for the varieties in the colours of marble which is white when pure. Serpentine with silica impurities give the marble a green colouration.

Some bacteria species and fungi can cause structural degradation of marble [1014–1016].

8.2.2 Beneficiation of marble

Early mining of marble involved an exploitation of the natural fissures of the rock, where fig wood wedges were inserted and inflated with water so that the natural expansion caused the detachment of the block.

Current method involves the use of a helical wire and the penetrating pulley based on a 4 mm to 6 mm diameter steel wire combined with the abrasive action of silica sand and an abundant amount of water as a lubricant. The helical wire is configured to be a continuous loop of tensioned steel that moves at a speed of 5 to 6 metres per second, cutting the marble at a rate of 20 cm per hour.

8.2.3 Marble value addition

Table 8.3 Uses of marble

Marble attributes and properties	composed mainly of calcite, dolomite or serpentine; varied and colourful patterns
Building material	sculpture
Decorative material	computer displays; marble dust combined with cement and synthetic resin make *reconstituted* or *cultured marble*; appearance of marble can be simulated with faux marbling

8.3 Serpentine

8.3.1 Overview of serpentine

Serpentine [1017] is a group of common rock-forming hydrous magnesium iron phyllosilicate $(Mg,Fe)_3Si_2O_5(OH)_4$ minerals which are greenish, brownish, or spotted and commonly occurring in serpentinite rocks. Minor amounts of other elements including chromium, cobalt, manganese or nickel may be present in the minerals.

8.3.2 Beneficiation of serpentine

Same as for dimension stone in section 8.0.2. Beneficiation may be for the production of asbestos and magnesium.

8.3.3 Serpentine value addition

Table 8.4 Uses of serpentine

Serpentine attributes and properties	serpentine soil is toxic to plants because of high levels of Ni, Cr, and Co; opaque to translucent; specific gravity 2.2–2.9; hardness 2.5–4 Mohs; infusible and susceptible to acids; lamellated antigorite has a hardness of 3.5–4 and its lustre is greasy; bowenite is especially hard serpentine (5.5)
Mineral	source of magnesium and asbestos
Decorative stone	gemstones and ornamental carvings; jewellery and hardstone carving [1018]
Industrial application	railway ballasts; building material; asbestiform types used as thermal and electrical insulation (chrysotile asbestos); fire retardant devices and heat protection
Verd antique	ornamental green marble replacement; countertops, sculptures, plaques, and tiling; polished as cabochons or beads
Antigorite	ornamental carvings; soft and easy to work with; create exceptional art; used as book ends
Health hazard	when used as a road surface, forming a long-term health hazard by breathing; asbestos content can be released to the air; water supplies normal weathering processes may contain asbestos from serpentine at low levels

8.4 Slate

8.4.1 Overview of slate

Slate is a fine-grained, foliated, homogeneous metamorphic rock that is derived from an original shale-type sedimentary rock composed of clay or volcanic ash [1019]. It is frequently grey in colour, especially when seen, en masse, covering roofs but it also occurs in a variety of colours even from a single locality.

Slate is a composite of several minerals that include quartz and muscovite or illite, often along with biotite, chlorite, haematite, and pyrite. Less frequently it is also composed of apatite, graphite, kaolinite, magnetite, tourmaline or zircon as well as feldspar.

8.4.2 Beneficiation of slate

Slate can be produced by quarrying from a *slate quarry* or by reaching through tunnel in a *slate mine*. The slate can be made into roofing slates which are a type of roof shingle or a type of roof tile. They can be split into thin sheets which remain relatively flat and easy to stack.

8.4.3 Slate value addition

Table 8.5 Uses of slate

Slate attributes and properties	flat sheets of stone; beauty and durability; low water absorption; resistant to frost damage and breakage due to freezing; fire resistant and energy efficient
Industrial use	roofing; floor tiles [1019]; used for interior and exterior flooring, stairs, walkways and wall cladding; chemical sealants used on tiles to improve durability and appearance; increase stain resistance; reduce efflorescence; increase surface smoothness; set into walls to provide rudimentary damp-proof membrane; small offcuts used as shims to level floor joists; building walls and hedges, sometimes combined with other kinds of stone; table coasters

(Continued)

Table 8.5 (Continued)

Electric and heat insulator	used to construct electric switchboards and relay controls for large electric motors
Other uses	whetstone to hone knives; laboratory bench tops and for billiard table tops due to its thermal stability and chemical inertness; blackboards and individual writing slates using slate or chalk pencils; high quality slate for tombstones and commemorative tablets

8.5 Soapstone

8.5.1 Overview of soapstone

Soapstone *or steatite* is dominantly composed of the mineral talc and is thus rich in magnesium but with varying amounts of chlorite and amphiboles and traces of FeCr oxides. Dynamothermal metamorphism and metasomatism are thought to be responsible for its production. These occur in areas where tectonic plates are subducted with influx of fluids.

Low-cost biaxial porcelains of nominal composition $(MgO)_3(SiO_2)_4$ make up steatite ceramics which are approximately 67% silica and 33% magnesia. They may contain minor quantities of other oxides such as CaO or Al_2O_3.

8.5.2 Beneficiation of soapstone

Soapstone is easily carved and can be pried by large crude stone tools, chisel and scrape from the face of an outcrop.

8.5.3 Soapstone value addition

Table 8.6 Uses of soapstone

Soapstone attributes and properties	relatively soft because of its high talc content; feels soapy to the touch; dielectric and thermal insulating properties; durability; can be pressed into complex shapes before firing; undergoes transformations when heated to temperatures of 1000–1200° C into enstatite and cristobalite; an increase in Mohs to 5.6–6.5 [1020]
Insulating material	insulator or housing for electrical components
'Whisky stones'	soapstone can be put in a freezer and later used in place of ice cubes to chill alcoholic beverages without diluting
Architectural applications	countertops; interior surfacing; used as marker by welders and fabricators; remains visible when heat is applied; used by seamstresses, carpenters, and other craftsmen as a marking tool; moulds for casting objects from soft metals such as pewter or silver; not degraded by heat; the slick surface allows the finished object to be easily removed
Other uses	tile; substrates; washers; bushings; beads; pigments; inlaid designs; sculpture; coasters; kitchen countertops; sinks; bowls; cooking slabs [1021]; gravemarkers; cooking-pots [1022]; construction of fireplace surroundings; cladding on metal woodstoves; woodburning masonry heaters because it can absorb, store and evenly radiate heat; countertops and bathroom tiling

Chapter 9

Energy sources

9.1 Coal

9.1.1 Overview of coal

Coal is a sedimentary rock which is found in rock strata in layers and veins called beds or coal seams. It is black or brownish-black in colour. It is combustible and is formed from vegetation which has been consolidated between other rock strata and altered by the combined effects of pressure and heat over millions of years.

Anthracite coal, which is very hard, is a metamorphic rock which has been exposed to elevated temperature and pressure. Carbon is the primary constituent of coal incorporating variable quantities of other elements mainly hydrogen, nitrogen, oxygen and sulphur [1023]. Coal's metamorphic grade increases successively into peat, lignite, sub-bituminous coal, and lastly anthracite when geological processes are applied to dead biotic material over a long period of time, under suitable conditions [1024].

9.1.2 Beneficiation of coal

When coal is brought up to the surface it undergoes separation from gangue material. This is achieved by immersing the coal in a solution of water and tiny magnetite particle as the *dense media separator*. The coal floats while the other materials sink to the bottom.

In the next step the coal goes through crushing and screening to produce appropriate size fractions for the intended uses and/or markets. The coal is also processed to remove impurities and reduce ash and sulphur. It may also be charred to remove hydrogen and oxygen.

9.1.3 Coal value addition

Table 9.1 Uses of coal

Coal attributes and properties	largest global anthropogenic source of carbon dioxide; extraction of coal, its use in energy production and its by-products are all associated with severe environmental and health effects
Source of energy	coal is the largest source of energy for the generation of electricity; source of heat; refining of metals
Bituminous coal	fuel in steam-electric power generation [1025]; heat and power application in manufacturing; used to make coke; domestic water heating
Anthracite	commercial and residential space heating [1026,1027]
Integrated gasification combined cycle (IGCC)	coal is first gasified to create syngas which is burned in a gas turbine to produce electricity; hot exhaust gases from the turbine are used to raise steam in a heat recovery steam generator which powers a supplementary steam turbine; use pulverised coal

9.1.3.1 Coking coal

9.1.3.1.1 OVERVIEW OF COKING COAL

The use of coal is strongly conditioned by the composition, rank and amount of impurities (e.g. sulphur and phosphorus), the content of volatile matter and ash. Coking coal or metal-lurgical coal which is used industry for the production of steel is derived from coal after special treatment.

9.1.3.1.2 BENEFICIATION OF COKING COAL

The run-of-mine coal is treated before marketing in order to remove impurities such as waste rock. The quality of the coal and the intended use determine the required processing. One method, known as washery, is for the coal to be crushed, separated by size and subsequently treated in an oscillating column of water, in which the unwanted fragments sink faster than coal.

The final stage of preparation is the conversion of coal to coke by driving off impurities to leave almost pure carbon. In the process the coal softens, liquefies and then re-solidifies into hard but porous lumps. This coking process involves heating coking coal to about 1000° C in the absence of oxygen to drive off volatile compounds. This is pyrolysis.

9.1.3.1.3 COKING COAL VALUE ADDITION

Table 9.2 Uses of coking coal

Coking coal attributes and properties	almost pure coal from which all impurities and volatiles have been removed
Industrial use	production of steel; used to make furnace coke or metallurgical coke [1028]; (need to meet maceral composition and reactivity constituents, CSR index – coke strength after reaction)
Other uses	in alumina refineries; paper manufacturing; chemical and pharmaceutical industries
By-products of coke ovens	ammonia salts; nitric acid and agriculture fertilisers
Coal ash	potential source of rare earth elements, gallium, and several other strategic materials

9.1.3.2 Coal to liquid fuel (CTL)

9.1.3.2.1 OVERVIEW OF COAL TO LIQUID (CTL)

Coal to liquid technology can be utilised to produce fuels (diesel, petrol, jet fuel and paraf-fin), fertilisers, explosives, power, polymers, synthetic waxes, lubricants, chemical feed-stocks and alternative liquid fuels such as methanol and dimethyl aether (DME), etc. The starting material/feedstock in the CTL process is coal which is gasified into syngas. The produced syngas is first purified by the removal of sulphur before it goes for the Fischer-Tropsch process for making fuel [452,469].

9.1.3.2.2 BENEFICIATION OF CTL

Coal liquefaction refers to the conversion process of coal into liquid hydrocarbons, namely: liquid fuels and petrochemicals. The process is commonly referred to as "Coal to Liquid" (CTL).

$$n\,C + (n+1)\,H_2 \rightarrow CnH_2\,n_{+2} \tag{95}$$

The specific technologies may be direct (DCL) or indirect liquefaction (ICL) processes. The ICL involves gasification of coal to a mixture of carbon monoxide and hydrogen (syngas) and then using a process such as Fischer-Tropsch [452,469] process to convert the syngas mixture into liquid hydrocarbons. The DCL converts coal into liquids directly, without the intermediate step of gasification, by breaking down its organic structure with the application of solvents or catalysts in a high pressure and temperature environment [1029–1031]. Hydrogenation is involved in both ICL and DCL.

9.1.3.2.3 CTL VALUE ADDITION

Production of fuels, power, fertiliser, explosives and other polymeric materials.

9.2 Coal Bed Methane (CBM)

9.2.1 Overview of coal bed methane

Coal bed methane (CBM) is a form of natural gas extracted from coal beds where methane CH_4 is adsorbed into the solid matric of the coal. The methane, in a near-liquid state, lines the inside of pores within the coal. Free gas may be present in open fractures *(cleats)* in the coal which may also be saturated with water. Very little heavier hydrocarbons such as propane or butane are present in coalbed methane. It may however contain trace quantities of ethane, nitrogen, carbon dioxide and few other gases.

9.2.2 Beneficiation of CBM

For the production of the methane gas, water in the coal fracture spaces is pumped out first, leading to a reduction of pressure and enhancing desorption of gas from the matrix. The fracture permeability acts as the major conduit for the gas flow. Higher permeability results in higher gas production. As coal displays a stress-sensitive permeability changes applied to fractured reservoirs play an important role during stimulation and production operations.

9.2.3 CBM value addition

Table 9.3 Uses of CBM

CBM attributes and properties	its occurrence in underground coal mining presents a serious safety risk
CBM uses	an important source of energy for power generation, heating, boiler fuel, town gas; may be sold to natural gas pipeline systems
End-use specifications	depends on the gas quality, especially the concentration of the methane, the presence of other contaminants
Other uses	coal drying; heat source for mine ventilation air; supplemental fuel for mine boilers; vehicle fuel as compressed or liquefied natural gas (LNG); manufacturing feedstock; fuel source for fuel cells [468,481,486]

9.3 Natural gas

9.3.1 Overview of natural gas

Natural gas is formed as a fossil fuel when layers of decomposing organic matter are exposed to intense heat and pressure over thousands of years [1034]. It is a non-renewable [1032] hydrocarbon gas mixture comprising principally of methane and varying amounts of other higher alkanes, ethane, propane, butane and pentane. It is found in deep underground rock formations and may be associated with other hydrocarbon reservoirs in coal beds and as methane clathrates.

9.3.2 Beneficiation of natural gas

Hydraulic fracturing or "fracking" is a method of releasing natural gas from subsurface porous rock formations [1035]. By-products sulphur, ethane, natural gas liquid (NGL) propane, butanes and pentanes are yielded from the processing of raw natural gas using a typical natural gas processing plant that is made up of various unit processes. Some gases and other hydrocarbons with molecular weights above methane (CH_4) are removed mid-stream to produce a natural gas fuel which is used to operate the natural gas engines for further pressurised transmission.

9.3.3 Natural gas value addition

Table 9.4 Uses of natural gas

Natural gas attributes and properties	it is also the main source of helium (2–7%) and sometimes a usually lesser percentage of carbon dioxide, nitrogen, and/or hydrogen sulphide [1033]
Chemical industry	consumer fuel; chemical plant feedstock in the manufacture of plastics; production of ammonia via Haber process for use in fertiliser production; production of hydrogen; hydrogenerating agent; commodity for oil refineries; in manufacture of fabrics, glass, steel, paint
Energy power	to power natural gas-powered engines; heating; cooking [1036]; electricity generation; fuel for vehicles

9.3.3.1 Gas to liquids (GTL)

In the GTL process no gasification step is necessary since natural gas is used as the primary feedstock. The natural gas is converted into syngas which is a mixture of carbon monoxide and hydrogen by a steam reforming process:

$$CH_4 + H_2O \rightleftharpoons CO + 3H_2 \tag{96}$$

In the gas to liquids refinery process natural gas or other gaseous hydrocarbons are converted into longer-chain hydrocarbons such as diesel fuel. As in the CTL described earlier methane-rich gases are converted into liquid synthetic fuels either by direct conversion – using non-catalytic processes that convert methane to methanol in one step – or through syngas as an intermediate as in the Fischer-Tropsch process [452,469].

Natural gas may be converted to methanol which provides an alternative route to fuel via the *methanol to gasoline* process (MTG). The methanol polymerises over a zeolite catalyst to form alkanes. A third gas-to-liquids process builds on this MTG technology by converting natural gas-derived syngas directly into liquid fuel [1037].

Table 9.5 Uses of GTL

GTL processes	refineries can convert some of their gaseous waste products (flare gas) [1033] into valuable fuel oils; sold as is or blended with diesel; use for the economic extraction of gas deposits in locations not economical to build a pipeline
Microchannel reactors	for the conversion of unconventional, remote and problem gas into liquid fuels [1034]
FPSO	for offshore conversion of gas to liquids such as methanol, diesel, petrol, synthetic crude, and naphtha

9.4 Peat

9.4.1 Overview of peat

Peat is made out of partially decayed organic matter in peatlands or mires [1038]. It forms when plant material, usually in wet areas, is inhibited from decaying fully by acidic and anaerobic conditions. Wetland conditions where flooding obstructs flows of oxygen from the atmosphere, slowing down the rates of decomposition [1039], are very conducive to the formation of peat whose most important source are peatlands. Other less common wetland types deposit peat, including fens, pocosins, and peat swamp forests. It is composed mainly of wetland vegetation: principally bog plants including mosses, sedges and shrubs in features that include ponds, ridges and raised bogs [1043]. Peat deposits provide records of past vegetation and climates [1040] especially as they are the first step in the geological formation of other fossil fuels such as coal which is a non-renewable source of energy [1042].

9.4.2 Beneficiation of peat

The site for new peat bog must contain horticulture grade sphagnum with a depth of at least 2 metres. When this is verified, ditches are built to drain surface water and access roads are built to allow heavy equipment to operate. The top inch or two of peat are loosened by milling machines. This way, the underlying bog is kept intact and not eroded away. A few days are provided to allow the top layer to dry for easy of harvesting which can now be done using vacuum harvesters. These collect fibres from about a quarter of an inch top layer without destroying the integrity of the peat fibres.

The harvested peat is emptied into large piles from where large truck-drawn hoppers carry the vacuum harvested peat to the processing facilities. Screening is done to remove impurities such as wood, roots or large clumps of peat. The peat is then packages "as is" or is blended with other materials. Wetting agents are added, along with limestone, to raise the peat pH to 6.0 and aggregates such as perlite or vermiculite to increase porosity.

9.4.3 Peat value addition

Table 9.6 Uses of peat

Peat attributes and properties	efficient carbon sink [1038]; highly compressible under even small loads; often stains the water yellow or brown due to the leaching of tannins
Harvested peat	source of fuel [1041]
Industrial use	precursor of coal; in dehydrated form, it is a highly effective absorbent for fuel and oil spill on land and water; insulating purposes
Agricultural application	soil conditioner in order to retain or slowly release water without killing roots when it is wet; mixed into soil to improve its structure and to increase acidity; dried peat used to absorb excrement from cattle that are wintered indoors; stores nutrients; important raw material in horticulture; it is recommended to treat peat thermally, e.g. through soil steaming, in order to kill inherent pests and reactivate nutrients
Household usage	for cooking and domestic heating
Fire hazard	can burn indefinitely; peat fires can burn underground; can smoulders undetected for very long periods in a creeping fashion through underground peat layers; global threat with significant economic, social and ecological impacts
Construction industry	pose difficulties to builders of structures, roads, and railways; peat deposits also pose major difficulties to builders of structures, roads, and railways as it is highly compressible
Freshwater aquaria	in soft water or blackwater river systems; suitable for demersal (bottom-dwelling) species such as *corydoras* catfish; softens water by acting as an ion exchanger; contains substances that are beneficial to plants and for the reproductive health of fishes; prevents algae growth and kill microorganisms; stains water yellow or brown due to the leaching of tannins
Water filtration	treatment of septic tank effluent and urban run-off; filter for septic tanks; used as a water purifier

9.4.4 Lignite

When peat is naturally compressed over a long period, it forms lignite [1044], which is a soft brownish-black combustible rock often referred to as brown coal. It has a relatively low carbon content [1044] and an ash content ranging from 6 to 19% compared to bituminous coal which has 6–12% [1045]. Its content of volatile matter [1046] is high making it easier to convert into gas and liquid petroleum products than higher-ranking coals.

Geologically, lignite begins as an accumulation of partially decayed plant material, or peat, whose temperature and pressure increases with local geothermal gradient and tectonic setting. The result is compaction of the material and loss of the water and volatile matter, principally methane and carbon dioxide. Through this process, coalification, the carbon content of the material is concentrated and resultantly the heat content as well. With the passage of time and deeper burial there is further expulsion of moisture and volatile matter. Eventually the material is transformed into higher rank of coal such as bituminous and anthracite coals [1047].

Table 9.7 Uses of lignite

Lignite attributes and properties	lowest rank of coal due to its relatively low heat content; carbon content around 60–70% [1044]; high moisture content and susceptibility to spontaneous combustion causing transportation and storage problems
Fuel usage	fuel for steam-electric power generation
Mineral	source of germanium
Chemical industry	source of light aromatic hydrocarbons for the chemical synthesis industry
Amine-treated lignite (ATL)	reacts with quaternary amine to form ATL which is used in drilling mud to reduce fluid loss during drilling

9.5 Petroleum (oil)

9.5.1 Overview of petroleum

Petroleum is a fossil fuel that is co-located with natural gas and formed when large quantities of dead organisms, such as zooplankton and algae, are beneath sedimentary rock under intense heat and pressure. It is made up of hydrocarbons of various molecular weights and other organic compounds. The subsurface conditions determine the proportions of gas, liquid and solid in an underground oil reservoir [1048]. Petroleum is composed of a very large number of different hydrocarbons and the unique mix of molecules defines its physical and chemical properties such as colour and viscosity [1049,1050]. Alkanes (paraffins), cycloalkanes (naphthenes), aromatic hydrocarbons, or more complicated chemicals like asphaltenes are the most commonly found molecules in petroleum.

Crude oil in the reservoir is usually found in association with natural gas which forms a "gas cap" over the petroleum and saline water which generally sinks beneath it. It may also occur in a semi-solid form mixed with sand and water.

9.5.2 Beneficiation of petroleum

Crude oil, with some natural gas dissolved in it, is produced from oil wells. The gas may contain heavier hydrocarbons such as pentane, hexane and heptane in a gaseous state because of the underground temperature and pressure that are higher than at the surface. Petroleum is recovered by oil drilling which is carried out after studies of structural geology, sedimentary basis analysis, and reservoir characterisation in terms of porosity and permeability of geologic reservoir structures have been completed [1051,1052]. It is then refined and separated, mostly by fractional distillation at an oil refinery to produce a large number of consumer products such as petrol, jet fuel, kerosene, asphalt and other chemical hydrocarbon reagents used to make plastics and pharmaceuticals.

9.5.3 Petroleum value addition

Table 9.8 Uses of petroleum

Petroleum attributes and properties	fossil fuel; consists of hydrocarbons of various molecular weights and other organic compounds
Chemical products	to make plastics and pharmaceutical products; solvents; fertilisers; pesticides
Fuel	petrol; diesel; jet fuel; kerosene; liquefied petroleum gas
Energy production	energy generation; heating

Part III

Policy issues about mining in Africa

Chapter 10

Mineral-driven industrialisation of Africa

10.1 Role of mineral-based materials in society

From time immemorial humankind has always relied on materials for every aspect of life and for the improvement of the quality of life of modern industry and society [381,1053–1061]. As direct inputs for many sectors of the economy [1062–1063] these materials play a vital role in all the stages of entire supply chains. The origins of these materials are mineral ores from which metals, industrial materials, stones and energy sources are produced to anchor a very wide variety of industrial activities.

Thus, materials derived from minerals are intrinsically linked to every aspect of our life. To illustrate, the health sector uses surgical equipment containing high-performance magnets from rare earth elements, electricity is generated from fossil fuels and distributed by pylons and cables constructed of aluminium and copper, the PGMs are used for catalytic convertors to reduce harmful emissions from vehicles; a broad range of metals have applications in the electronic industry essential for the communication and military uses; water purification processes rely on reagent chemicals from broad range of minerals ores. The rapid technological progress is reliant on access to a growing number of materials required for the hi-tech goods whose complexity and sophistication necessitate a concomitant increase in the number and breadth of these raw materials. To maintain and improve the quality of life, continued access to mineral-based materials is essential.

10.2 Important and critical materials for industrialisation

The access to mineral-based resources has been given great attention globally, and the concept of very important or 'critical' resources or materials has emerged in several studies in the EU, USA, Japan and elsewhere [1061,1064,1065]. The industrialised countries conduct regular studies to identify materials which they consider important for their economies. Fifty-four (54) non-energy, non-agricultural, abiotic and biotic materials were reportedly considered in 2014 as so [1066]. The list comprises the following:

aluminum, antimony, barytes, bauxite, bentonite, beryllium, borates, coking coal, chromium, clays (and kaolin), cobalt, copper, diatomite, feldspar, fluorspar, gallium, germanium, gold, gypsum, hafnium, indium, iron ore, limestone, lithium, magnesite, magnesium, manganese, molybdenum, natural graphite, natural rubber, nickel, niobium, perlite, phosphate rock, platinum group metals, potash, pulpwood, rare earth elements, rhenium, sawn softwood, scandium, selenium, silica sand, silicon metal, silver, talc, tantalum, tellurium, tin, titanium, tungsten, vanadium, zinc.

A combination of several complex factors linked to the importance of these raw materials and the changing supply conditions raises concerns for the industrialised countries. Growing competition for these materials from emerging economies and proliferation of both economic and "resource nationalism" is coupled to supply concentration. Growing economies in developing countries and the evolving materials markets are reflected in these factors triggering restrictions in supply from the world's most important suppliers, thus increasing risks across supply chains. This is accompanied by increases in the prices and price volatility of mineral-based raw materials, giving rise to continuing concerns in many countries and reducing the competitiveness of manufacturing. The impact on the whole value chain becomes inevitable.

Some minerals considered important to industrial countries are produced in only a few countries, perhaps marked by political and/or economic instability. If this factor is coupled to a low recycling rate and low substitutability the supply risk become very high. The combination of high risk associated with the supply of particular materials with high economic importance of those materials defines them as *critical*. Hence, some minerals become more critical than others [1065–1068]. Assessing the criticality of individual minerals is important for primary producers such as in Africa as well as the consuming countries. It serves as a supporting element when negotiating trade agreements, challenging trade distortions or measures or promoting research and innovation in beneficiation and value addition. The list of critical minerals helps to define the forward-looking policies in different areas including research and innovation, industrial policy, trade, development and in the communication and defence and security sectors.

> From the foregoing, the basic criteria for a 'critical' mineral are firstly, a relatively high supply risk due to worldwide production being concentrated in only a few countries with potential geopolitical constraints (compounded by low substitution and low recycling rates), and secondly, its economic importance which is a proportion of each mineral associated with specific industrial sectors [1068].

The criticality or strategic importance of these minerals may be continuously evaluated as these are largely determined by the demand depending on the new development, e.g. lithium has recently become critical since the advent of electrical cars and the exponential expansion of the electronics industry.

The industrialised countries have identified the following 27 materials as critical in 2017 [1066]:

> Antimony, baryte, beryllium, bismuth, borates, cobalt, coking coal, fluorspar, gallium, germanium, hafnium, helium, indium, magnesium, natural graphite, natural rubber, niobium, PGMs, phosphate rock, phosphorous, scandium, (light and heavy) rare earth elements, silicon metal, tantalum, tungsten, vanadium.

10.2.1 Economic importance of critical materials

The economic importance of a material may be calculated from a detailed assessment of the use of the material in specific industrial sectors as well as a raw material-specific substitution index [1064,1069–1071]. The following metals are commonly perceived as *'critical'* or '*strategic*': rare earth elements, tantalum, niobium, lithium, beryllium, gallium, germanium, indium, zirconia, and graphite.

10.2.2 Scarcity of supply of critical materials

The availability of raw materials is critical [1065,1067,1068,1072–1075] as they are essential to the global economy and for maintaining and improving our quality of life. There has been a rapid growth in the use and demand for metals or industrial minerals used in high-technology products. Hence, availability of these materials at competitive prices is essential for advances in high-technology products, clean-energy technology and commercialising new inventions. Conventional demand and supply analyses have emphasised technical, economic, environmental [1137] and social parameters, and technological breakthroughs. '*Criticality analysis*' on the other hand does the same, but also focuses on identifying and evaluating the risks and impacts of supply disruptions on the economy, national security, implementing green programmes, or other initiatives. For some materials (e.g. heavy rare earth elements, niobium, antimony, the assessment is more complex taking into consideration factors such as authoritarian regimes, monopoly-or oligopoly-type market conditions, political instability and potential or existing regional conflicts which can threaten reliable supply. Analyses may account for these risks. Many governments recognise the growing importance of raw materials to economic competitiveness [1064,1069–1071] and are taking an active role in mitigating supply risks.

Since some of the mineral reserves are concentrated in a very small number of countries, notably in Africa, providing a stable and affordable supply is an important issue. Africa should, therefore, note that consuming countries would wish to secure access to raw materials by ensuring undistorted world market conditions [1076–1080] through diplomacy with resource-rich countries and resource-dependent countries for cooperation, through international cooperation via G8, OECD, etc. to raise awareness about the issues and create dialogue, and by making access to primary and secondary raw materials a priority for trade and regulatory policy to ensure that measures that distort open market trade such as restrictions of exports and dual pricing are eliminated.

Supply risk is now also assessed by combining the level of governance in the country, and the risk of trade-related restrictions. The likely impact of supply risks may be greater than price volatility which can create an incentive to increase supply and avoid physical scarcity.

10.2.3 Material substitution for critical materials

Research leading to material and technology substitutes [1066,1081] may improve flexibility and help meet the material needs of the clean energy economy such as diverse battery chemistry and PV materials. Legislation requiring substitution may also lead to perverse outcomes. Illustratively, banning lead in solders increase demand for tin, silver and bismuth which are partly produced as by-products from lead production, thereby squeezing supply at the same time as increasing demand. It is difficult to assess potential for substitution ahead of actual supply constraints which generate innovation to provide the substitute. This requires expert judgement which may be arbitrary and qualitative suggesting that incorporating substitution effectively into the supply risk assessment remains a significant challenge.

10.2.4 Recycling of critical materials

The recycling and reuse of materials can significantly lower world demand for newly extracted material [1082–1089]. While consumer goods are an important source for recycling, the elements in the goods are distributed at low concentrations over a wide range of products which have to be collected for recycle. Sophisticated knowledge of the components present in the end-of-life (EoL) products stream is required for effective recycling especially from complex mixtures where some elements are difficult to separate.

10.2.5 Nanotechnology and mineral nanomaterials

Africa has vast mineral resources that are critical ingredients to the industrialisation of the continent which must critically consider nanotechnology [1253] and its significant application in mining exploration and material processing (Table 10.1). Nanotechnology refers to the manipulation of matter on an atomic, molecular and supramolecular scale [1090,1091,1253].

Minerals are very complex as their chemical properties vary as a function of particle size. The property variations maybe due to differences in surface and near-surface atomic structure in addition to crystal shape and surface topography making a difference in important geochemical and biogeochemical reactions and kinetics [1092].

Table 10.1 Nanotechnology applications in mining

Examples of mineral nanoparticles	sun creams; self-cleaning glass; scratch-resistant coatings; carbon nanotube reinforced composites
Occurrences	in ceramics; beer and wine production; cosmetics; paints; paper; plastics; rubbers; they reinforce, filter, toughen, stiffen materials in daily use
Specific applications in mining	reclamation [1093] of trace amounts of valuable or harmful materials from soil, water and industrial process streams, including tailing ponds [1094]; non-toxic method of extracting gold from consumer waste instead use of cyanides [1095–1097]
Interface and colloid science breakthroughs	graphene-based carbon nanotubes; other fullerenes; nanomaterials with fast ion transport; nanoparticles and nanorods [1098]
Mineral separation	hydrophobic nanoparticles [477,1090,1091,1099–1101] replacing conventional low molecular weight, water-soluble flotation collectors
Nanomaterial and nanocoating commodities	carbon nanotubes for conductive fillers or medical sensors; titanium dioxide for UV absorption in sun cream [1109]; nanoporous polymers for supercapacitors, or tough nanostructured ceramics; nanocoating such as diamond-like carbon for combating friction and wear; invisible conductive layers for switchable glass; improved solar cells; focussed ion beam etching and epitaxial growth in silicon wafer etching leading to higher processing power and the high capacity memories
Consumer goods	nanomaterials already incorporated [1102–1105]; car bumpers made lighter; stain repellant clothing; more radiation resistant sunscreen; stronger synthetic bones; cell phone screens of lighter weight; glass packaging for drinks leads to longer shelf-life; balls for various sports made more durable [1106]
Household application	self-cleaning or "easy-to-clean" surfaces on ceramics or glasses; nano ceramic particles have improved the smoothness and heat resistance of common equipment such as the flat iron

(Continued)

Table 10.1 (Continued)

Chemical catalysis	large surface area to volume ratio; fuel cells [461,468,481]; catalytic converters and photocatalytic devices; for production of chemicals e.g. nanoparticles with a distinct chemical surrounding (ligands), or specific optical properties
Heat Affected Zone (HAZ)	weld adjacent to welds can be brittle and fail without warning when subjected to sudden dynamic loading; addition of nanoparticles such as magnesium and calcium make the HAZ grains finer in plate steel; increases weld strength
Solar cells [1253]	nanopillars, which are cheaper replacing traditional silicon solar cells; semiconductor nanoparticles [490] for use in the next generation of products such as display technology, lighting, solar cells and biological imaging [497]; nanoelectronics with MOSFET's being made of small nanowires ~ mm in length
Nanoscale discoveries and inventions	nanoscale structures and processes in nano innovation [1253]; 2D materials that are one atom thick, such as graphene (carbon), silicone (silicon) and staphene (tin); smartphones; large screen TV sets; solar cells; batteries are nano-enabled products; nanocircuits and nanomaterials are creating a new generation of wearable computers and a wide variety of sensors
Technology driver	higher performance materials; intelligent systems; new production methods with significant impact for all aspects of society [1107,1108]

10.2.6 The Fourth Industrial Revolution (FIR or Industry 4.0): implications for the mining industry

As the world is moving forward at an alarming pace a term, *Fourth Industrial Revolution*, (FIR or Industry 4.0) has been coined to describe the current trend of automation and data exchange, particularly in the manufacturing industries [1110]. It encompasses *Internet of Things* which focusses on the integration of all data into platforms that allow real-time decision making. The Fourth Industrial Revolution is upon us, and the African mining industry will be left behind if it does not join this innovation curve.

There are technological advances that are predicted to herald the Fourth Industrial Revolution [111]. It will be noted that the Third Industrial Revolution used electronics and information technology to automate production. Building on this Third Industrial Revolution the Fourth Industrial Revolution is the digital revolution that has been occurring since the middle of the twentieth century. The fusion of technologies that is blurring the lines between the physical, digital and biological spheres characterises the Fourth Industrial Revolution. The prospect of this happening is heightened by current emerging technological breakthroughs in fields such as artificial intelligence, robotics, the Internet of Things, autonomous vehicles, 3-D printing, nanotechnology, biotechnology, materials science, energy storage and quantum computing. Nanotechnology and materials science are principally based on mineral-derived materials.

The inability to accurately monitor adequately in real time poses a significant challenge across the entire mining sector, even more so when using mechanised mining methods. Production-related issues have a direct impact on the efficiency of the mines and mining processes. The use of real-time information for monitoring and control allows proactive intervention that can correct deviations and unsafe conditions as they arise. Thus, the 'Internet of Things' becomes an imperative for the future of mining [1112–1114].

10.3 Mineral by-products

Many of the anticipated *critical* elements are associated in nature with 'attractor' or 'carrier' base metals (Table 10.2) and therefore can be co-products of primary metal smelters. Producers of base metals are key sources of critical elements so that their supply is dependent on taking a systems-integrated metal production approach. Their supply may be significantly increased by improving the extent to which these critical elements are separated from their carrier base metals. It is for this reason that many elements can be successfully extracted in combination with others, or as subsidiary processes. Many of the rarer elements are associated with 'attractor' or 'carrier' base metals most commonly aluminium, copper,

Table 10.2 Some elements and their carrier/attractor base metal or mineral ores

Element	'Attractor' or 'Carrier' base metals or mineral ores
Beryllium	Tin and tungsten ore bodies
Bismuth	Mineral ores of lead, copper, tin, molybdenum, tungsten, silver, zinc and antimony
Cadmium	Sulphide ores of zinc, lead, copper and phosphate rock
Caesium	Pollucite (main source of Caesium), found in zoned pegmatites associated with lithium minerals lepidolite and petalite
Cobalt	Nickel, silver, lead, copper and iron ores
Fluorspar	Quartz, baryte (barite), calcite minerals
Gallium	Bauxite, Sphalerite, Phosphate rock, fly ash of coal
Germanium	Zinc sulphide, coal fly ash
Gold	Nickel, copper, cobalt, iron ores
Hafnium	Zirconium ores
Indium	Zinc sulphide ores
Lithium	Pegmatite minerals lepitolites, petalite
Manganese	Iron ores, spodumene
Molybdenum	Copper and tungsten ores
Niobium	Tantalite, dolomite, tin ores
PGMs	Nickel, copper, cobalt deposits
Rare Earth Elements	Mineral deposits of bastnaesite, monazite; laterite and low-level thorium (TH) or uranium (U) deposits; copper, iron and gold ores
Rhenium	Copper ores
Rubidium	Zone pegmatite ore bodies containing mineable quantities of caesium as pollucite or the lithium minerals lepidolite
Selenium	Selenium is found in metal sulphide ores, such as copper sulphides, where it partially replaces the sulphur; it is produced as a by-product in the refining of these ores, most often during production
Silver	Silver is found in native form, as an alloy with gold (electrum), and in ores containing sulphur, arsenic, antimony or chlorine; the principal sources of silver are the ores of copper, copper-nickel, lead, gold, and lead-zinc
Tantalite	Iron, tin, titanium ores
Tellurium	Copper minerals
Thorium	Monazite
Titanium	Ilmenite
Tungsten	In 45 mineral ores including wolframite (iron- manganese tungstate, $((Fe,Mn)WO_4)$, scheelite (calcium tungstate, $(CaWO_4)$, ferberite $(FeWO_4)$, and hübnerite $(MnWO_4)$
Uranium	In hundreds of minerals including phosphate rock, lignite, monazite
Wollastonite	Garnets, epidote ores

iron, lead, nickel, tin and zinc and therefore can be co-products of primary metal smelters. The choice to extract them is dependent on whether it is economically viable to mine the main product and whether the co-product value is sufficient to influence the processing design for the base metal.

Furthermore, as some high-technology metals are derived as by-products of base metal extraction, supply of these metals is tied directly to production levels of related base metals. The level of their production cannot be easily varied independently of the main base metal co-products without a major upward variation in their prices.

Strategies must be found to increase production of rarer elements from their 'carrier' base metals and quantify potential supplies from this approach perhaps focussing on iron (rare earth elements), aluminium (gallium), copper (cobalt, rhenium, molybdenum, tellurium and selenium), zinc (germanium and indium), nickel (cobalt and PGM) and tin (niobium and tantalum).

10.3.1 Undeclared by-products

The export of minerals from Africa has been carried out for many years without taking into account the many co-deposited minerals or by-products to the major minerals of interest. This inadvertent and *undeclared* export of potentially critical minerals as co-products of major minerals has been depriving Africa of valuable wealth.

Table 10.3 below lists the possible mineral by-products that are likely to be co-deposited and mined together with the main minerals of interest. The table may be useful to either the mineral importers or exporters in determining which particular by-products need their presence confirmed, or otherwise, before the export of the main minerals of interest.

Table 10.3 Minerals and their possible by-products

Main Mineral	Possible by-products
Aluminium (bauxite)	Gallium
Antimony	Bismuth, silver
Arsenic	Silver
Bastnaesite	Rare earth elements
Baryte (barite)	Fluorspar
Calcite minerals	Fluorspar
Cobalt	Gold, PGMs
Copper	Bismuth, cadmium, cobalt, gold, molybdenum, PGMs, rare earth elements, rhenium, selenium, silver, nickel, tellerium
Dolomite	Niobium
Epidote	Wollastonite
Garnet	Wollastonite
Gold	Rare earth elements, silver
Iron	Cobalt, gold, manganese, rare earth elements, tantalite, tungsten
Lead	Bismuth, cadmium, cobalt, silver
Lepidolite	Caesium, rubidium
Lithium	Caesium, rubidium

(Continued)

Table 10.3 (Continued)

Main Mineral	Possible by-products
Molybdenum	Bismuth
Monazite	Rare earth elements, thorium, uranium
Nickel	Cobalt, gold, PGMs, silver
Petalite	Lithium, caesium
Phosphate	Cadmium, gallium, uranium
Pollucite	Caesium, rubidium
Quartz	Fluorspar
Silver	Bismuth, cobalt
Spodumene	Manganese
Tantalite	Niobium
Thorium	Rare earth elements
Tin	Beryllium, bismuth, niobium tantalite
Titanium	Tantalite
Tungsten	Bismuth, beryllium, calcium, iron, manganese, molybdenum
Uranium	Rare earth elements
Zinc	Bismuth, cadmium, gallium, germanium, indium, silver
Zirconium	Hafnium

10.4 Riverbed minerals

Many rivers in Africa have been massively silted but are also rich in minerals that have been drained into them from their catchment areas whose geological features and minerals potential can be established. Numerous dikes of pegmatites and aplites that cut the granite and veins of white quartz are common in many areas. There are dolerite outcroppings to the catchments that have intruded into the granite and have since stood out as large mafic massifs due to their resistance to erosion.

There are two main causes of the siltation. Virtually all mining methods to recover major mineral(s) of interest loosen the soil which result in siltation into drainages, creeks and rivers and ultimately seas and oceans. These contain unrecovered valuable mineral products in what is commonly referred to as mineral sands in rivers. Thus, the extensive mining (small and large) activities in the river's catchment loosen the soils. When the rains come, these soils are swept into the rivers or their many tributaries. Thus, the results of the general soil erosion in the catchment areas end up as silted material in the rivers.

The other cause of the siltation is the erosion of their banks. This is very evident especially where there are settlements and gardening activities along the rivers. Mineral sands contain suites of minerals with high specific gravity known as 'heavy minerals'. These minerals occur in very low concentrations in a variety of igneous and metamorphic rocks, but being chemically and physically resistant to weathering, and having comparatively high specific gravity, they tend to accumulate in placer deposits in river channels or along coastal shorelines.

The major minerals commonly found in the rivers include the following: gold, diamonds, rare earth elements, ilmenite (source of titanium), tantalite- columbite, tungsten, cassiterite (source of tin).

10.5 Obsolescent or near-obsolescent mineral materials

Some elements have been rendered obsolete for a variety of reasons. Indeed, some elements have been restricted for environmental [1137] or health reasons (e.g. mercury, cadmium, asbestos, lead). Obsolescence may be a result of technological development, as in platinum.

10.5.1 Asbestos

The desirable properties of asbestos made it a very widely used material, and its use continued to grow throughout most of the twentieth century until the knowledge of carcinogenic effects of asbestos dust caused its effective demise as a mainstream construction and fireproofing material in most countries. It is now known that prolonged inhalation of asbestos fibres can cause serious and fatal illnesses including lung cancer, mesothelioma and asbestosis (a type of pneumoconiosis) [1115,1116]. By the beginning of the twentieth century concerns were beginning to be raised, which escalated in severity during the 1920s and 1930s. By the 1980s and 1990s asbestos trade and use started to become banned outright, phased out, or heavily restricted in an increasing number of countries.

10.5.2 Cadmium

The use of cadmium is generally decreasing due to its toxicity [1117,1118]. That is why the nickel-cadmium batteries are replaced with nickel metal hydride and lithium-ion batteries. The bioinorganic aspects of cadmium toxicity have been reviewed. The most dangerous form of occupational exposure to cadmium is inhalation of fine dust and fumes, or ingestion of highly soluble cadmium compounds. Inhalation of cadmium-containing fumes can result initially in metal fume fever but may progress to chemical pneumonitis, pulmonary oedema and death.

Cadmium is also an environmental hazard. Human exposures to environmental cadmium are primarily the result of fossil fuel combustion, phosphate fertilisers, natural sources, iron and steel production, cement production and related activities, non-ferrous metals production and municipal solid waste incineration. The International Agency for Research on Cancer has classified cadmium and cadmium compounds as carcinogenic to humans. Cadmium exposure is a risk factor associated with early atherosclerosis and hypertension, which can both lead to cardiovascular disease.

Concern must, therefore, be raised about exposure to the major sources of cadmium which are sulphide ores of zinc, lead, copper and phosphate rock!

10.5.3 Lead

The concern about lead's role in cognitive deficits in children has brought about widespread reduction in its use (lead exposure has been linked to learning disabilities). Most cases of adult elevated blood lead levels are workplace related. High blood levels of lead are associated with delayed puberty in girls. Lead has been shown many times to permanently reduce the cognitive capacity of children at extremely low levels of exposure. Lead exposure is a global issue as lead mining and lead smelting are common in many countries. Most countries had stopped using lead-containing gasoline by 2007. Lead exposure mostly occurs through ingestion. Lead paint is the major source of lead exposure for children. As lead paint deteriorates, it peels off, is pulverised into dust and then enters the body through hand-to-mouth contact or through contaminated food, water or alcohol. The use of lead for water pipes is problematic in areas with soft or (and) acidic water. Hard water forms insoluble layers in the pipes while soft and acidic water dissolves the lead pipes.

If ingested, lead is poisonous to animals and humans, damaging the nervous system and causing brain disorders [1119]. Excessive lead also causes blood disorders in mammals. Lead is a neurotoxin that accumulates both in soft tissues and the bones. Lead is a highly poisonous metal (whether inhaled or swallowed), affecting almost every organ and system in the body. The main target for lead toxicity is the nervous system, both in adults and children. Long-term exposure of adults can result in decreased performance in some tests that measure functions of the nervous system. Long-term exposure to lead or its salts (especially soluble salts or the strong oxidant (PbO_2) can cause nephropathy and colic-like abdominal pains. It may also cause weakness in fingers, wrists, or ankles. Lead exposure also causes small increases in blood pressure, particularly in middle-aged and older people and can cause anaemia. Exposure to high lead levels can cause severe damage to the brain and kidneys in adults or children and ultimately cause death. In pregnant women, high levels of exposure to lead may cause miscarriage. Chronic, high-level exposure has been shown to reduce fertility in males. Lead also damages nervous connections (especially in young children) and causes blood and brain disorders. Lead poisoning typically results from ingestion of food or water contaminated with lead but may also occur after accidental ingestion of contaminated soil, dust, or lead-based paint. It is rapidly absorbed into the bloodstream and is believed to have adverse effects on the central nervous system, the cardiovascular system, kidneys and the immune system.

10.5.4 Mercury

Mercury and most of its compounds are extremely toxic and must be handled with care; in cases of spills involving mercury (such as from certain thermometers or fluorescent light bulbs), specific cleaning procedures are used to avoid exposure and contain the spill. Protocols call for physically merging smaller droplets on hard surfaces, combining them into a single larger pool for easier removal with an eyedropper, or for gently pushing the spill into a disposable container.

Mercury can be absorbed through the skin and mucous membranes and mercury vapours can be inhaled, so containers of mercury are securely sealed to avoid spills and evaporation. Heating of mercury, or of compounds of mercury that may decompose when heated, should be carried out with adequate ventilation in order to minimise exposure to mercury vapour. The most toxic forms of mercury are its organic compounds, such as dimethylmercury and methylmercury. Mercury can cause both chronic and acute poisoning [1120–1122].

Acute exposure to mercury vapour has been shown to result in profound central nervous system effects, including psychotic reactions characterised by delirium, hallucinations and suicidal tendency. Occupational exposure has resulted in broad-ranging functional disturbance, including erethism, irritability, excitability, excessive shyness and insomnia. With continuing exposure, a fine tremor develops and may escalate to violent muscular spasms. Tremor initially involves the hands and later spreads to the eyelids, lips and tongue. Long-term, low-level exposure has been associated with more subtle symptoms of erethism, including fatigue, irritability, loss of memory, vivid dreams and depression.

10.5.5 Platinum

Approximately 50% of the platinum, 80% of the palladium and 80% of the rhodium produced are used in the automotive industry for the manufacture of catalytic converters, which are fitted to the vehicle's exhaust system. The primary role of catalytic converters [115] is to covert the malignant exhaust gases from diesel and petrol engines into environmentally [1137] benign gases before they are discarded into the atmosphere. There are concerted global efforts to reduce and/or replace the use of petrol- and diesel-powered vehicles with electric vehicles. A fully electric vehicle does not need any PGMs since it does not burn any fuel and discharge exhaust gases, which harm the environment. The advent of electric vehicles will therefore have a negative impact on the global demand of PGMs. The consequence is that the aforementioned PGMs would be rendered obsolete in a few years. If South Africa and Zimbabwe do not optimally exploit their resources now, they could sit with a resource for which there is no demand and/or value in future. It is thus critical for these governments to ensure that there is no PGMs ground lock-up by current mining companies.

10.5.6 Diesel and petrol

As a consequence of the advent of electric vehicles with the concomitant replacement of diesel and petrol there is the potential of making petroleum obsolete in as far as vehicle usage is concerned as there are concerted global efforts to reduce and/or replace the use of petrol- and diesel-powered vehicles with electric ones. The advent of electric vehicles will therefore also have a negative impact on the global demand of petroleum.

10.6 Mineral-driven Africa's development

The emergence of industrial-scale mining forced major demographic shifts in Africa's population. During the early stages of mining in Africa, labour had been primarily provided by young men from rural areas who would travel to the mines to provide temporary labour. This system, though, was too unreliable to provide a permanent labour force and was not acceptable to the mining corporations. The need to create a fixed, permanent labour force at the mines became the primary objective of the mining corporations. This resulted in the mining corporations introducing a variety of schemes to keep workers on site for lengthy periods of time [20]. A more extreme measure was taken to introduce *corporate compounds*. These enclosed compounds were built in the style of open-air prisons but further developed into large communities, towns and even cities like Johannesburg. Mass migration to towns, urban growth and the increasing urban demand for rural produce prompted the development of transport and communications infrastructure [18,19]. Railways were greatly expanded to link towns to each other and to the countryside, and ports were expanded to cope with increasing immigration and commercial activity, greatly stimulating mineral-based development in Africa. This was as far as the European corporate companies would go. What was needed is industrialisation.

Several fora have been held at which the development of the African continent from its minerals has been discussed [8–14]. The current thinking of the African leaders on the role of the mining sector is well encapsulated in the African Mining Vision [5,1123–1127].

The African Mining Vision has stated that it is more desirable to export the mineral resources in a more value-added form. This is the route to Africa's industrialisation. However, there is an optimal amount of domestic processing that should take place, and that amount is conditional on the availability of domestic technologies and on the terms of trade. Anything that makes trade in raw materials more attractive, such as an improvement in the terms of trade or a drop in shipping costs, will result in reduced resource-based manufacturing. It is desirable for African countries to formulate policies that protect their mineral-based economies against the deindustrialisation associated with otherwise beneficial falling transportation prices or improved terms of trade. The strategies for such policies include taxes on exports of raw ores and concentrates and support for local value-added manufacturing. The relative shipping cost disadvantages faced by many African countries are also a natural deterrent to trade in unprocessed minerals. These shipping costs are most likely a result of imperfectly competitive shipping markets.

10.6.1 Mineral value addition: an imperative for Africa's modern industrialisation

Value addition of minerals involves disciplines of materials science and materials technology [1128]. These are processes which create materials at the molecular and/or atomic scale for the purpose of advancing technology, developing more efficient products and creating new manufacturing technologies (Table 10.4). They lie at the very heart of ALL engineering . . . everything has to be made of something and requires materials technology.

The many industrialisation problems that Africa currently faces are due to the limits of the materials that are available and how they can be used. In other words, materials from their minerals affect the continent's industrialisation processes significantly [24,25]. Considering the foregoing, it is evidently clear that value additions to minerals is not a trivial undertaking. It requires a variety of skilled workforce in materials science and technology ranking from technicians with basic training to scientists and technologists with specialised training in R&D at the most advanced levels capable of dealing with the most complex scientific and technological issues.

It is worth noting that for over a century and a half spanning the period of colonial occupation of the continent, the colonial governments and their corporate agencies did nothing to advance value addition to the minerals for the benefit of African countries [4,7,17–21]. Instead value addition was done in the colonising countries resulting in impressive industrialisation courtesy of minerals from Africa. Furthermore, there is no evidence to indicate that the investing companies that are currently operating in the mining industry in Africa have any serious intention of value adding to the minerals for the benefit of host African countries. It is clearly evident that countries and companies outside Africa have not and will not invest meaningfully in mineral value addition for the benefit of Africa. Africa must realise that the continent is on its own when it comes to value addition to its mineral resources. Africa must design for itself the processes of beneficiation and value addition that create employment and yield mineral-based materials and products, thereby advancing the process of industrialisation. These processes necessarily create upstream and downstream linkages [3,5,1132–1134] coupled to sideway linkages of an infrastructural nature such as power, water, transport networks and the required social services. The scope and intensity of these activities and the utilisation of mineral value-added products define the level and extent of industrialisation brought about by the mining sector in each country. Mineral beneficiation and value addition contribute to industrialisation by creating jobs and wealth.

Table 10.4 Aspects of value addition to minerals in Africa

Mineral value addition characteristics	involves the discovery and design of new materials; solids arising in metallurgy and mineralogy [22,1129]; processing of materials into shapes and forms needed for specific application; it is at the intersection of various fields such as, and a syncretic discipline hybridising, metallurgy, materials science, ceramics, solid-state physics, and chemistry, mineral resources, energy and environment [23,1137]; creating and customising materials for specific uses; involves the manufacturing and processing of different materials with a focus on metals and chemical reactions
Areas for which mineral materials are created and customized	computer technology; transport; telecommunications; medicine; production of oil and natural gas
Product groups in mineral-based materials	basic metals and fabricated metal products; manufactured machinery and equipment; manufactured transport equipment; glass and construction materials; office machinery and computers; electrical machinery; medical and precision optical instruments; motor vehicles; transport equipment and electricity and gas
Necessary R&D areas for mineral value addition	electrochemistry; inorganic materials and ceramics; physical metallurgy; mineral resources; energy and environment [1137]; F&D provides more insight into the occurrence and use of individual mineral-based materials for Africa's industrialisation [1126,1130,1131];
Some of the processing methods of mineral value addition	casting; rolling; welding; ion implantation; crystal growth, thin-film deposition; sintering, glassblowing
Analytical methods in mineral value addition	electron microscopy; X-ray diffraction; calorimetry; nuclear microscopy (HEFIB); Rutherford backscattering; neutron diffraction; small-angle X-ray scattering (SAXS)
Some of the knowledge required in mineral value addition	ingot casting; foundry methods; blast furnace extraction; electrolytic extraction
Required understanding in mineral value addition	structural and functional properties of materials related to the use of materials that have great significant for energy technology; industrial processes; new technologies such as fuel cells [461,468,481,496], solar cells; lowering emissions of greenhouse gases; refining and recycling metals; environmentally friendly batteries; new materials using nanotechnology
Materials science applications in mineral value addition	combination of different materials into composite materials with new material properties; structured materials composed of two or three macroscopic phases; structure elements such as steel-reinforced concrete; thermal insulting tiles; studies of metal alloys and semiconductors; superalloys for jet engines; materials for artificial knee

(Continued)

Table 10.4 (Continued)

Semiconductor [498,1259]	a material that has resistivity between a metal and insulator; electronic properties easily altered by intentionally introducing impurities or doping; semiconductor materials are diodes, transistors, light-emitting diodes (LEDs), analogue and digital electric circuits; manufactures both as single discrete devices and integrated circuits (ICs) of devices manufactured and interconnected on a single semiconductor substrate [26];

A Holistic National Frame Work for Value-Addition

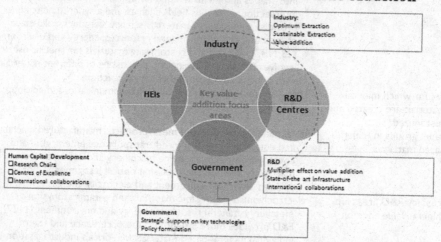

Figure 10.1 A national holistic framework for mineral value addition

Effective and efficient mineral value addition involves governments, industry, higher education institutions and R&D centres [32–34,1130–1139] as depicted in Figure 10.1.

It is important for Africa to realise that FULL beneficiation/extraction of every mineral must be done before its exporting, if the continent's mineral-driven industrialisation is to be achieved. This is particularly so for minerals that require various sequential stages of processing. For example, chromite requires concentration to make chrome concentrates which are exportable. But concentration is only the first stage of beneficiation. Historically chrome concentrates have been exported from African countries. Subsequent to that concentration the chrome concentrate can be processed to ferrochrome by smelting. Export of ferrochrome products is not to the full benefit of African countries. The ultimate product of chrome is stainless steel. This is achieved by further processing of the ferrochrome. For example, nickel or manganese may be added to produce superalloys with specific properties required for particular uses such as jet engines or gas turbines.

Similarly, with platinum group metals (PGMs) products which are exportable at various stages of beneficiation. PGMs are exported from many African countries to outside markets as concentrates. If these concentrates are smelted in the next step of extraction, they yield another exportable product called matte which contains PGMs and other by-products such as copper, nickel and cobalt. A next step in the extraction process is base metal refinery (BMR) which removes the by-products. The final step in the extraction of PGMs is the precious metal refinery (PMR) at which high purity of the PGMs is achieved. The succeeding steps increase the value of the metal. Africa, therefore, should insist on full beneficiation/extraction before allowing export of the particular minerals. To take a step further, for some minerals, granting of mining rights could be conditional on the commitment to establish full beneficiation/extraction facilities. Alternatively, tax disincentives could be introduced to discourage the export of raw concentrates.

Chapter 11

Science and technology and human resources

11.1 Science and technology and economic development

Developments in science and technology are driving rapid changes in the world resulting in benefits that are not equitably shared across the globe. It is noticeable that developing nations who are lagging behind in science and technology continue to fall farther and farther behind the industrialised nations that have resources to apply scientific advances and new technologies ever more intensively and creatively. As the industrialised nations continue to master the tools of science and technology, to vastly outspend the developing nations in R&D, and even to capture some of the developing nations' most precious human resources for their own use [1140], the disparity is predicted to grow ever wider. The reason for this expected phenomenon is that advances in science and changes in technology are main drivers of modern economic developments. Countries that are scientifically and technologically advanced are able to compete successfully in global markets and are able to generate income, higher wages and wealth. Their scientific and technological knowledge enables them to solve practical problems related to, for example, health, shelter, access to food and water, transportation and communication for the betterment of their countries and citizens [1141].

Without necessarily proving a direct causation between investment in science and technology and economic development, there is ample evidence of a clear relationship between a country's economic well-being and its ranking as a scientific and technological nation. It has been demonstrated that the countries of Africa rank poorly in science and technology compared to developed economies and other developing economies. African countries also rate relatively poorly.

Success of the newly developed countries, particularly in East Asia, is attributed to the significant investment in science, engineering and technology made in those countries. This has proved that capacity building through educational excellence is a critical prerequisite for sustainable economic and technological development in any nation and region. It is evidently clear that the weakness in the technological capacity of African countries is one of the factors affecting the continent's ability to harness its abundant natural resources, such as minerals, for socio-economic development. Lately, science, technology and innovation (STI) have been recognised by the African Union (AU) and the Regional Economic Communities (RECs) [3,5] to play an important role in Africa's efforts to eradicate poverty, achieve food security and fight diseases such as malaria, tuberculosis and HIV/AIDS. It is to state the obvious that value-added human capital [1142] is required to operate and maintain industries, build infrastructure, increase agricultural productivity and provide other valuable services. The contribution of science, engineering and technology (SET) in general, and research and training institutions in particular, are of critical importance in the efforts to attain the United Nations Sustainable Development Goals (SDGs) [1143,1144] and transform Africa's economies [1123–1127,1130–1134]. For Africa to transition from a resource-based economy to a knowledge-based economy demands significant investments in science and technology at levels unprecedented in the continent's history.

11.2 Higher education

The role of higher education, or tertiary education, and S&T in economic development are well documented [1142,1144]. Through training of competent and responsible professionals, higher education institutions assist in the improvement of the public sector and its institutions, provide crucial support for national innovation and constitute the backbone of a country's information infrastructure. Thus, higher education performs a multiplicity of societal functions, including developing human capital, building the knowledge base and disseminating, using and maintaining knowledge. It provides the advanced skills needed for a competitive work force and, particularly in the case of postgraduate education, the research capability necessary for innovation [1145] and utilisation of scientific and technological know-how. Higher education (together with key research institutes) is the engine room of the system of innovation [28]. Talented, educated people are central to the knowledge society and the higher education system is the wellspring of advanced skills and learning.

Notwithstanding freedom from conventional colonialism, poverty, hunger, diseases and civil wars are still prevalent in Africa. The continent has not been able to adequately harness its endowments of enormous resources, especially in minerals, for its sustainable development. This failure is attributable to lack of the required critical mass of human capital with appropriate knowledge and skills, technology, entrepreneurship and incentives to innovate [29]. There is a skills gap in the continent whereby the demand by industry and business for scientific and technological skills exceeds the supply of those skills currently provided by the existing higher education system [31,1142].

11.2.1 Pan African Minerals University of Science and Technology (PAMUST)

During the 4th Ordinary Session of the African Union Assembly, held in Abuja in January 2005, the African heads of state and government agreed to establish African Institutes of Science and Technology (AISTs) that can contribute to Africa's capacity for successful economic development [1146]. Realising that the African mineral industry [1–4] has evolved largely as a producer for foreign export markets for the industrialisation of other countries [3,5–7], one of these AISTs was to be for adding value to the continent's abundant mineral resources in order to effectively improve the lives of citizens in African countries. The continent's aspiration in this matter resonates very well with, and found expression in, the words of the former President of Zimbabwe, Robert Gabriel Mugabe, who on the occasion of his inauguration ceremony on 22 August 2013, made explicit the intentions of his government regarding the mineral sector:

> The mining sector will be the centrepiece of our economic recovery and growth. It should generate growth spurts across sectors, reignite that economic miracle which must now begin. The sector has shown enormous potential, but we are far from seeing its optimum. We have barely scratched our worth, even in the sense of merely bringing

above ground what we already know to be embedded in our rich soils. We need to intensify the exploitation of existing deposits. More mineral deposits remain unknown, unexplored. We need to explore new deposits, develop new greenfield projects in the mining sector. Above all, we need to move purposefully towards beneficiation of our raw materials [1147].

Hence the establishment of the Pan African Minerals University of Science and Technology (PAMUST) in Harare by an Act of Zimbabwe Parliament in 2016 [1148].

PAMUST has been established under the auspices of The Nelson Mandela Institution to become a pan-African centre of excellence that serves a comprehensive array of needs in the mining industry for the whole continent. It has been established as a postgraduate university offering master's degree courses and providing research at PhD and post-doctorate levels, initially in the following six areas: (i) Geosciences, (ii) Mining Engineering, (iii) Mineral Beneficiation and Extractive Metallurgy (iv) Materials Science and Engineering for mineral value addition, (v) Mineral Economics and (vi) ICT for the mining industry

11.3 Research and development institutions in mining

Presence of well-functioning R&D facilities to drive the process of technological innovation is a prerequisite to efficient mineral beneficiation and value addition. Even though most of the industrial countries do not have significant mineral resources they have well-established, well-functioning training and R&D institutions for mineral beneficiation/extraction and value addition [32]. Africa has very few such R&D institutions in mineral processing and metallurgy and their linkages to the fabrication sector (value addition) do not appear to be fully developed. Africa needs more of such world-class R&D institutions with linkages to the manufacturing sectors.

There are opportunities, if R&D can find the answers, of mining the extensive resources that still exist unmined, either in deep operations or in lower grades. These challenges can only be met if African countries establish collaborative and enabling environments for mining R&D, innovation and the development of world-class manufacturing outfits in the mining and beneficiation [33] industries. There are various focus areas requiring research and development such as in Table 11.1.

Of great importance for the development in the African mining sector is the financial commitment to the R&D programmes by government treasuries, Chambers of Mines and mining companies. There is need of co-investment by Chamber of Mines [1135] member companies. Mining companies should not 'go it alone' and set up their own R&D capacity and projects. Whilst this may be successful, these initiatives are likely to become the victims of fluctuating price cycles and budgetary constraints. Additionally, there must be commitments by mining departments at universities [22–31,1130] to participate in these programmes in a fully collaborative manner. This could be part of a strategic journey that aims to establish world-class organisations as a public-private partnerships (PPPs) [1149–1152] that will support the industry. For the preceding to happen meaningfully African countries must find innovative schemes of funding R&D for mineral beneficiation and value addition [32–34,1135–1139].

There is also need for providing platforms for dissemination of research and development information [1163], in helping to develop new R&D skills and competencies and in supporting R&D initiatives. There is a need as well to develop a new breed and critical mass of

Table 11.1 Areas for R&D in the mining sector

Geosciences	exploration geology, economic geology, GIS, surveying, geochemistry, mining geology, mineralogy, ore studies, geological environment, map interpretation, soil science, structural geology, geophysics, remote sensing, sedimentology, mineral deposits, area selection, ore genesis, pegging, basin modelling, tenements, risks, mapping, platform cover, mineral evaluation, mineralisation events, ground based geophysics, electromagnetic geophysics
Mining engineering	ore estimation, rock mechanics, eengineering geology, mine management, mine design, mine surveying, drilling engineering, geomechanics, hydrology, blasting, shaft sinking, ventilation systems, mine safety, soil mechanics, engineering geology, mine project evaluation, slope stability, mining machinery, mining transport systems, production operations management, mine planning, mine design, mine modelling and simulation, ore estimate, surface mining, mine surveying, closure and rehabilitation
Mineral beneficiation and extractive metallurgy	*Mineral Processing*: comminution, particle sizing, concentration, froth flotation, heavy media separation, sorting, dewatering, tailings disposal; *Hydrometallurgy*: leaching, separation, ion exchange, liquid-liquid solvent extraction, liquid-membrane processes, Solution concentration, precipitation processes, purification, metal recovery; *Electrometallurgy*: electrolytic processes for metal recovery, electrorefinery, electrowinning; *Pyrometallurgy*: calcining or roasting, smelting, refining
Materials science and engineering (value addition)	materials technology, materials characterisation, nanotechnology, fuel cells [461,468,496] technology, foundry, forging, drawing, gemmology, electroplating, tribology, corrosion and degradation of materials, steel making, PGM catalytic converters, craft jewellery, metal fabrication, structural materials, ceramics, physical metallurgy, mechanical properties, electronic materials, optoelectronic, materials, advanced materials, materials for energy and environment, solar cells, superconductors, photovoltaics, alloys, polymer science; cement, construction materials, industrial materials
Mineral economics	mineral resources policy, interpretation of statutes, environmental law and policy for mining [1137], mining agreements, mining and sustainable development, pricing issues, hedge funds, venture capital finance in mining, international mining laws and trading, mineral policy studies, ownership models, taxes and royalties, investment models, entrepreneurship, management, international business transactions, commercial dispute resolution in mining
ICT in mining	dispatch systems; modelling and simulation; GIS; underground operations; data, voice and video traffic; mechatronics systems; radio-frequency identification (RFID); GPS technologies; tracking the movement of minerals and equipment; planning and monitoring software, mapping, spatial concepts; software and mining technical systems; supporting optimisation; improve productivity; online weighment; sampling systems; sophisticated surveying equipment such as terrestrial laser scanners, slope stability radars, blasting accessories and modern vibration metres; mineral exploration; mine planning; voice communications; safe levels of hazardous gases; water levels in sump and pumping systems; running of idle conveyors; monitoring and control of various aspects of safety and production from the pithead; communication solutions for collision avoidance; localisation of mining machinery; wireless transmission

researchers, to take the programmes forward [1141,1142]. These researchers can be drawn from research organisations, universities, industry and interns.

Successful research and development produces new technologies that reduce production costs; enhances the quality of existing mineral commodities; reduces adverse environmental, health, and safety impacts; and creates or makes available entirely new mineral commodities.

11.4 Professional organisations in mining

In many mining countries and regions there are professional mining organisations whose membership include geologists, mining engineers, metallurgists and material scientists. These organisations are institutions whose activities encompass the whole materials spectrum, from exploration and extraction, through mineral characterisation, processing, forming, finishing and application, to product recycling and land reuse. They exist to promote and develop all aspects of materials science and engineering, geology, mining and associated technologies, mineral and petroleum engineering and extractive metallurgy, as leading authorities in the worldwide materials and mining communities. African mining and materials professionals are encouraged to join these organisations which are usually in joint collaboration and linkages with industry, government and academia [22–33, 1130]. These organisations are usually linked to a university and assist in monitoring and ensuring that courses affecting the mining industry conform to the standards for professional registration with the organisations, establish codes of practice and monitor legislative matters affecting their members' professional interests. The professional development programmes run by the organisations help to contribute to members' career development towards senior grades of membership.

The organisations provide a range of activities and initiatives for the benefit of the mining community as well as embracing professional codes of ethics. Their educational activities aim at promoting professional disciplines to younger generations by allowing access to a range of educational resources and materials. They may have close links with schools and colleges and are responsible for accrediting university and college courses and industrial training schemes. Their local and international links assist members to be abreast of technological development in the mining and related sectors.

11.5 African scholars and African academies of sciences

African scholars and academics, especially social and political scientists, made immense contribution in the decolonisation of Africa. Since then there have been numerous African scholars in diverse disciples including the mining sector. The expertise of these academics working in Africa's universities and research institutions can, and should, be harnessed for the benefit of the continent especially in the exploitation of African natural resources such as minerals. In this context the African science academies consists of Fellows who by virtue of their respective achievements in the field of science and technology are regarded as being of exceptional merit and distinction and who can be expected to act at all levels – national, continental and global – in the interest of Africa's development by maintaining quality and encouraging disciplinary inclusiveness that is relevant to the exploitation of the African mineral resources. They are expected to be at the forefront of co-creating knowledge with diverse communities of practice and expertise, engaging multiple perspectives, and ultimately helping to design Africa's industrialisation strategies. They are encouraged to develop strategic relationships with key influencers and establish credibility as honest and inclusive brokers.

As in other regions, national and continental academies of sciences [1164,1165] and their networks [1163,1166] have been established in most African countries. Academies of sciences primarily address policy-relevant scientific items on a very wide range of issues within the broad categories of materials, environment, energy and biosciences [1167]. It has also become increasingly obvious that key science-based issues with major policy ramifications may also include important aspects that are best addressed from the perspective of the social sciences, particularly economics.

These academies have a tripartite mandate [1069,1070,1142,1164,1165] of pursuing excellence by recognising scholars and achievers; providing advisory and think-tank functions for shaping the continent's strategies and policies; and implementing key science, technology and innovation programmes [1167]. The academies utilise their membership pool [1166], which consists of a community of scientists to engage with governments and policy makers on the continent. The membership comprises individuals who have reached the highest level of excellence in their field of expertise and have made contributions to the advancement of the field on the continent. The African scholars and their science academies mobilise and strengthen the African scientific community [1166] and publicise and disseminate scientific information, research and development results and policy on capacity building in science and technology [1166].

The African governments and their agencies are strongly advised to recognise the think-tank functions of their national and continental academies and their increasing role in setting the research agenda for the future development of the continent. The African mining sector stands to benefit from engagements with these academies of sciences in the continent.

11.6 Diaspora Africans

The term African diaspora is used to refer to people of African origin living outside the continent, irrespective of their citizenship and nationality. Appeal is made to the African diaspora who are willing to contribute to the development of the continent [1168,1169].

In the mining, as in other sectors, the African diaspora is increasingly skilled, educated and wealthy as the globalisation of trade, capital and labour has taken hold. Africa should benefit tremendously through these people in luring large multinational companies as well as entrepreneurial ventures. The diaspora population is positioned to bring technical and domain expertise to domestic research and development. Diaspora African academics abroad could volunteer time and resources to help African campuses improve the quality of education. African governments could do well to harness the energy and assets of Africans by communities around the world contributing money, time and technical expertise to their programmes. Overseas African business networks can be 'one of the main forces driving the dynamic growth required in the continent' [1170]. Africa needs to move towards a more knowledge-intensive economy with ever denser cross-border interactions. The changing nature of overseas African communities and their relationship to Africa is a subject that deserves further exploration [1170]. The role of science in the diaspora needs to be further highlighted. The diaspora's scientific achievements can lead to a direct and mutually reinforcing synergy between African science abroad and at home. The fear of a 'brain drain' is perhaps misguided. A 'brain' sitting in an uncongenial environment can drain away faster than if it goes to a stimulating interference-free environment abroad.

Every year, innumerable young, educated Africans leave home to seek their fortunes abroad. Most of these emigrants never return home, though their abilities and knowledge are a priceless commodity that could contribute to development of their home countries. In addition to this group, millions of people of African descent, located in communities around the world, are actively seeking ways to re-engage with their ancestral homeland.

The Africa Region of the World Bank launched the African Diaspora Programme [1140] in September 2007. The key elements of the strategy and programme were in the areas of policy development and implementation; finance, particularly the leveraging of remittances for development; and human capital development for 'brain gain'. The strategy also called for: working with the African Union (AU) in support of its global diaspora policies, programmes and projects; working with partner countries to obtain improvements in the enabling policy and institutional environments for engaging their diaspora in economic and social development activities; working with the partner donors to understand better the flow of remittances, including issues related to the costs of, and approaches to the leveraging and securitisation of remittances for development; and working with diaspora professional networks to increase knowledge sharing and transfer (brain gain) between the diaspora and their home countries [1170].

Chapter 12

Mineral economics

12.1 Africa's minerals in the global economy

The European conquest of Africa had its historical culmination with the Berlin Conference of 1884 that formalised the scramble for Africa and the establishment of extractive institutions that regarded Africa as a source of raw materials for European industries [17]. This imperial strategy involved the killing of millions of Africans and allowed the European countries to grab Africa's abundant mineral riches cheaply. A century later the European countries together with the USA were continuing to exploit Africa's mineral resources under more sophisticated and disguised strategies such as Structural Adjustment Programmes (SAPs) that compelled many African countries to sell their mineral resources to Western multinational corporations, who now own significant percentages of the large-scale mining industry in Africa. These corporations repatriate high percentages of their profits into foreign accounts and pay no, or little, income tax or duties. This means that the Western companies virtually monopolise Africa's mineral wealth, which now contributes little to African economies. The European/ USA strategy aims at extracting the maximum amount of mineral wealth from Africa for the West at the lowest cost, through the perpetration of a myriad of holocausts created by civil wars and SAPs imposed on many African countries. Most African exports to the West are raw mineral ores and concentrates whose prices have been kept low by wars since African armies need to sell these minerals for whatever money they can get in order to buy weapons [17], most of which are also from the West. Like the wars, the SAPs also contribute to the keeping of raw mineral prices low by enforcing the expansion of such exports to the West. Four hundred years of slave trade and a century of Western colonialism in African helped the building of USA and European economies. Currently, newer mines are being opened in Africa to keep pace with the burgeoning world market mineral demands.

The European colonial scramble for Africa was primarily for mineral exploitation that has been taking place during the colonial and post-colonial periods. The products from these minerals have been used for the infrastructural development of European cities [17–19] and industrialisation of their countries. Now the North American and Asian countries have joined the list of those countries whose new colonialism is driven by a determination to plunder the natural resources of Africa, especially its strategic and mineral resources. It is estimated that over 100 companies listed on the London Stock Exchange currently have mining operations in 37 African countries. These include the most prestigious multinational mining conglomerates in the world [4], making Africa a hive of activities related to mining. Through their operations they collectively control over US$1 trillion worth of Africa's most valuable resources [4,7]. By design, mining revenues do not stay in the countries where they are mined. In addition, the raw mineral ores and concentrates are not processed in the

African countries where they are mined to promote maximum value addition. The African governments must do more to protect the rights of the people affected by mining rather than protecting the profit margins of the corporations exploiting them. Across the continent multinational companies have left in their wake trails of breakdown and impoverishment of communities, environmental [1137] degradation and industrial injury, sickness and death. In contrast, as Africa is well endowed with mineral resources that are precursors to the materials required by the rest of the world, especially the industrialised countries; the use of these materials in people's everyday lives illustrates how world citizens are directly and indirectly linked to the fortunes of the African mining sector (**Tables 1.1** and **1.2**).

Given the staggering mineral export figures, one is left with the uncomfortable question as to why in such a wealthy continent have the issues of unemployment, poverty, disease, homelessness and crime assumed such equally staggering proportions? This is due to a variety of reasons ranging from historical, social, financial and even geo-politics, ethics, transparency (more lack of it), etc. These reasons invariably point to the weak integration of Africa's mining sector into national economic and social activities. These weaknesses need to be corrected by clear and appropriate mineral resources policies that are undergirded by robust legislative frameworks. This chapter discusses some of the issues of mineral economics as they relate to Africa.

12.2 Rates of depletion of finite mineral resources

The rates of depletion of their non-replenishable mineral resources [35] should be of great concern to African countries. The increase in populations and rate of economic growth of industrialised countries necessitates their need to ensure adequate supplies of raw materials that satisfies their high standard of living. It is recognised that Africa is the region that still has high-grade, easy-to-mine mineral deposits that account for ~ 44% of its total exports [1171–1186]. Notwithstanding the cost advantages in mineral resources that many African countries have, they are often not concerned with profits. Some, in fact, even take losses for two main reasons. The first is the need to keep people employed [1187–1189]. The second is that African countries need to generate foreign currencies to pay for the development of the nations [5,1123–1125,1127,1171]. Having taken these on board, it still is a fact that minerals are exhaustible commodities and only a finite supply exists. African countries must therefore concern themselves with the issue of how fast they wish to deplete their mineral resources. Production in the present reduces production that is possible in the future.

12.3 Critical infrastructure issues in the mining industry

12.3.1 Energy

Access to reliable and cost-effective forms of energy is a strategic priority for any mining operation globally. Traditionally the mining industry has relied on conventional fossil-based fuel sources – diesel, oil, coal and natural gas – to meet its energy demands [1190]. The mining industry is now facing the challenges of increasing fuel prices while mineral commodity prices tighten, resulting in ever-narrowing profit margins and increased opposition from communities to new conventional energy sources [1191]. Growing emerging markets are piling up pressure on the mining sector to expand into new and often remote locations as a response to increasing demand. For these reasons, mining companies are having to deal with unreliable grid power supply and uncertain power prices. This often means that grid-connected electricity needs to be supplemented with on-site generation, typically large-scale diesel generation, resulting in a dependency on diesel fuel. The sector is often challenged with meeting growing demand for mineral resources often located in countries and sites where the supply of energy is not always available, reliable or cost-effective. Off-grid power solutions are more likely to be required for the more remote mines.

The mining sector often experiences volatility in commodity prices [1179] and rising fossil fuel prices. This places margins under immense pressure, compelling mining companies to manage costs sustainably. As the mining sector is driven by a number of strong converging trends, including energy security concerns, it also faces growing demands from governments, customers, communities and other key stakeholders to operate in a sustainable manner [1143].

12.3.1.1 Renewable energies

The mining sector as whole is evaluating greater use of renewable energy plants – a trend set to intensify rapidly – as part of a broader strategy to lock in long-term fixed electricity prices and availability while minimising exposure to regulatory changes, market pricing and external fuels. Site-appropriate renewable energy solutions may provide cost-competitive energy while delivering greater energy supply reliability and consistency as a strategic imperative. For many mining companies this is essential for their bottom lines and increasingly, their licenses to operate.

Like others, the mining sector is often confronted with load-shedding, having to face the unenviable choice of either reducing operations or turning to costly diesel-powered generators. With energy already accounting for ~ 20% of grid-connected and 30% of off-grid operating costs, the pressure of this additional cost can substantially erode profit margins and place pressure on the price of commodities.

From the foregoing, it is evident that mining has a complicated relationship with renewable energy. The two may be viewed as traditional rivals, with renewables such as solar and wind power vying for a slice of the African energy grid, which is dominated by coal-powered and hydroelectric plants. As energy-intensive mining operations find themselves in an African energy grid that is struggling to meet demand, and as mining sites often find themselves isolated from the electric grid, there is increasing incentive for mining firms to venture into private production of electricity, in which renewable power generation can play a vital role.

Capital costs for renewables have dropped significantly, so renewable solutions are in many instances cheaper than diesel. The pressure of inadequate supply has arguably shifted the dynamic of the mining industry's relationship with renewable energy, as firms increasingly turn to renewables as one component of a basket of energy options used to maintain stable power at mining sites.

12.3.1.2 Other energy alternatives

While many advocates in the renewable energy sector see the reliance on diesel generators as making a clear case for a shift to renewables, this picture is incomplete. There are other players to be considered in the energy mix such as the future development of large hydropower projects in the DR Congo and Angola. Although currently speculative, the vast potential of 'bridging' alternatives to coal such as nuclear and shale gas can be seriously considered

[1192]. With impacts of global warming and the fast depletion of non-renewable energy sources, there is growing interest in biofuels, such as biodiesel fuel from jatropha seed oil, as alternatives to fossil fuels [165].

12.3.2 Transportation

The mining industry faces logistics challenges of moving mined products from remote mining sites or requiring transport for heavy mining equipment to and from remote sites, which add expense and complications that some other industries don't have to deal with. Transportation within the mining industry also commonly involves managing shipping by rail, which can be costly and difficult without a relationship with rail companies. Rail cars with the product need to be properly tracked to ensure that they are not sitting for long periods of time. It is also necessary to find the best cargo insurance to minimise an expensive risk of theft.

As African has 15 landlocked states, and as many of the railway lines they use have been damaged or destroyed in civil wars, the logistics of global trade has been rendered extremely difficult. Available funding for re-investment in the mining sector is reduced as average freight costs are raised with distance from ports. Landlocked countries face further disadvantages of having to rely on the political stability and good institutional structures in transit countries. A multitude of skilled planning disciplines for the development of improved freight movement and logistics operations for the mining sector in Africa is in great demand.

The auxiliary part of the transport system is for moving ore from the face to a chute or mine cars for transferring ore from a series of stope raises, through a sublevel drift, to an ore pass. The nature of transport system used to transport personnel and materials into the mine (shaft hoisting, drift haulage or rubber-tyred vehicle access) is determined by the mode of access (e.g. vertical shaft, inclined drift, in-seam access via outcrop, etc.).

12.3.3 Water

There are several issues [464] which are pertinent with respect to water management in the mining industry including the following: provision of secure and reliable supplies, dewatering of the mine, hydrology and water balance of the tailings storage facilities, stormwater management, protection of underlying surface waters and ground waters, use of water, environment impact and the need to drain water to allow mining operations. Water is used for the extraction of minerals that may be in the form of solids, such as coal, iron, sand, and gravel; liquids, such as crude petroleum; and gases, such as natural gas. Hence the mining sector is a large industrial user of water. In most mining operations, water is sought from groundwater, streams, rivers and lakes, or through commercial water service suppliers. But often mine sites are located in areas where water may already be scarce, and understandably, local communities and authorities often oppose mines using water from these.

The use of water in mining has the potential to affect the quality of surrounding surface and ground water. In response to environmental concerns and government regulations, the mining industry worldwide increasingly monitors water discharge from mine sites, and has implemented a number of strategies to prevent environmental damages [1137]. In well-regulated mines, hydrologists and geologists take careful measurements of water and soil to exclude any type of water contamination that could be caused by the mine's operations.

If protective measures are not taken the result can be unnaturally high concentrations of some chemicals, such as arsenic, sulphuric acid and mercury over a significant area of surface or subsurface. Run-off of river soil or rock debris – although non-toxic – also devastates the surrounding vegetation. There is potential for massive contamination of the area surrounding mines due to the various chemicals used in the mining processes as well as the potentially damaging compounds and metals removed from the ground with the ore. Large amounts of water produced from mine drainage, mine cooling, aqueous extraction and other mining processes increases the potential for these chemicals to contaminate ground and surface water. Dissolution and transport of metals and heavy metals by run-off and ground water is another example of environmental problems with mining [1137]. Water in the mine containing dissolved heavy metals such as lead and cadmium can leak into local groundwater, contaminating it. Long-term storage of tailings and dust can lead to additional problems, as they can be easily blown off site by wind.

Acid rock drainage occurs naturally within some environments as part of the rock weathering process but is exacerbated by large-scale earth disturbances characteristic of mining and other large construction activities, usually within rocks containing an abundance of sulphide minerals. In many localities, the liquid that drains from coal stocks, coal handling facilities, coal asherites and coal waster tips can be highly acidic, and in such cases, it is treated as acid mine drainage (AMD).

There are five principal technologies that are used to monitor and control water flow at mines sites: diversion systems, containment ponds, groundwater pumping systems, subsurface drainage systems and subsurface barriers. Contaminated water is generally pumped to a treatment facility that neutralises the contaminants, in the case of AMD.

Dewatering, pumping out groundwater out of a mine, may be required in order for the mine to operate. As subsurface mining progresses below the water table, water must be constantly pumped out of the mine in order to prevent flooding. On abandoning a mine, the pumping ceases, and water floods the mine. The initial step in most acid rock drainage situations starts with the introduction of water.

12.3.4 Information and Communication Technologies (ICT)

Information & Communication Technology (ICT) has become one of the most important tools for enabling increased transparency, process efficiency, improved productivity, optimisation of operational costs and decision making along with increasing employee, customer and investor satisfaction, etc. in every aspect of business process [1162]. Expectedly, ICT also now has a significant impact on mine operations [1154]. ICT systems are involved in process and machinery monitoring, fault detection and isolation of processes and machinery, and assessment of risk and hazards in the mining industry. Modernisation through automation and mechanisation of mining processes have significant implications on the number of people employed in the industry as well as the required skills levels. This necessarily requires significant attention to change management issues. The requirements for the upstream and downstream processes associated with mechanisation have to be understood.

Modernisation via automation and mechanisation of mining processes have significant implications on the number of people employed in the industry as well as the required skills level. This necessarily requires significant attention to change management issues. The understanding of the requirements for the upstream and downstream processes associated with mechanisation is made easier with ICT [11156]. The African mining industry may recognise, albeit in a general sense, the efficiency improvements and cost benefits associated with the implementation of ICT in its operations (Table 12.1) [1153].

Table 12.1 Uses of ICT in the mining sector

Time/space operations technologies	radio-frequency identification (RFID); GPS technologies; tracking the movement of minerals and equipment; planning and monitoring software, mapping, spatial concepts
Data collection	real time data collection and analyses; systems integration; integrated application software for uniform data and file structures, enhancement in management information system (MIS) etc.
Automation	software and mining technical systems; supporting optimisation; improve productivity
Dispatch systems	of trucks and shovel in open-pit mine operations
Modelling and simulation	condition monitoring; reduced running costs; improvements in operations; greater ROI
Measurements	online weighing; sampling systems; sophisticated surveying equipment such as terrestrial laser scanners, slope stability radars, blasting accessories and modern vibration metres.
GIS	mineral exploration; mine planning; mine management; social and environmental management; storage of all mine maps in digitised format making updating of the map easier, easy location/access of various installations and its shifting; social impact assessment at mines like resettlement and rehabilitation, provision of basic infrastructure in resettlement villages, general guidance on environmental monitoring etc. [1160]
In underground operations	voice communications; safe levels of hazardous gases; water levels in sump and pumping systems; running of idle conveyors; levels of stock in bunker; health monitoring of UG equipment; monitoring and control of various aspects of safety and production from the pithead
Employee biometric system	multipurpose electronic digital card (smart card) to each employ with biometric identification; information on employee personnel details, contribution towards social security, salary earnings and deductions, leave details, health information; attendance recordings, personal ID, availing medical facility; settlement of terminal dues [1155–1159,1162]
Governance issues	sales and marketing through e-auction; corporate e-banking for all payments; e-procurement; e-tendering document management system (for proper storage and quick retrieval of information/data); influence of prices; control of costs in order to remain competitive
Data, voice and video traffic	enables people to work together and share experiences; improves operational services and financial efficiencies in all areas; mines stores, weighbridges, and hospitals
Mechatronics systems	provide communication solutions for collision avoidance; localisation of mining machinery; wireless transmission
Global environment	centralisation of purchasing; R&D activities; product innovation; market development; deployment of systems integration technologies; web enablement of applications; enrichment and quality improvement of graphical interfaces through 3D and other capabilities delivered over broadband communications systems [1161]

12.4 Environmental effects of mining

Visual pollution of mine dumps and the scar of open-pit mining left on the surface are some of the results of mining. Erosion, formation of sinkholes, loss of biodiversity and contamination of soil, ground and surface water by chemicals from mining processes are caused by mining [1193–1195]. It results in massive contamination of the area surrounding mines due to the various chemicals used in the mining process as well as the potentially damaging compounds and metals removed from the ground with the ore [1196]. Mining is often accompanied by dissolution and transport of chemicals and heavy metals by run-off and groundwater. Long-term storage of tailings and dust can lead to additional problems, as they can be easily blown off site by wind.

In addition to creating environmental damage, the contamination resulting from the leakage of chemicals affects the health of the local population [1197] because the close physical proximity of industrial mining operations to local towns and villages means that thousands of people are exposed to fumes, dust, noise and effluent water generated by the mines, trucks and processing facilities [1196]. It is obvious that the thousands of trucks that travel to and from the mines and related operations all day and through the night expose residents in the vicinities to heightened air pollution and leaving them rightfully afraid of contracting lung diseases. It is known that chronic exposure to such dust can lead to potentially fatal hard-metal lung disease and may lead to a variety of other pulmonary problems, including asthma, decreased lung function and pneumonia [1198]. There are many diseases that can come from the pollutants that are released into the air and water during the mining process. For example, during smelting operations enormous quantities of air pollutants, such as the suspended particulate matter, SO_x, arsenic particles and cadmium, are emitted. Other metals are also usually emitted into the air as particulates. There are many more occupational health hazards. Most of the miners suffer from various respiratory and skin diseases. Miners working in different types of mines suffer from asbestosis, silicosis, or black lung diseases [1200].

It is unfortunate that some communities also face physical danger when explosions caused by the mine's operations damaged homes and property, for which they may not receive compensation. Clean drinking water is often lost to the communities. In some cases villagers who refuse to be relocated have had their water (and other amenities) cut off without warning and were forced to collect water from the contaminated source. For those villagers who do relocate, this is often enforced rather than agreed, and although they are provided with small houses, the only agricultural land they have access to are the small plots around their houses. The small amount of financial compensation does not replace their ability to support themselves through farming. The combination of forced eviction with inadequate compensation and the subsequent loss of livelihood has serious consequences for people already in fragile economic situations. Evicting and impoverishing rural communities have occurred as a consequence of mining, with thousands of people in poor rural communities losing the means to support their food needs and make a small income selling their agricultural surpluses. Inadequate compensation and insufficient alternative ways of making a living have left people even poorer. This has had a particularly negative impact on women heads of household, who derive independence from their ability to grow food and so maintain adequate nutritional standards for their children and even gain a small income from selling surplus food.

The erosion of exposed hillsides, mining dumps, tailings dams and the resultant siltation of drainages, creeks and rivers [1199] significantly impact the surrounding areas. The ecosystem and habitats are disturbed with the destruction of productive grazing and croplands.

Africa is replete with many cases where a small town is completely dependent economically on a single mine. When the mine closes these small towns are destroyed or become ghost towns. African governments must therefore concern themselves with the balance between a healthy mine, its depletion rate, the town's people and the social costs involved.

12.4.1 Mitigation of environmental impacts

It is necessary to instigate remedial and mitigation techniques such as acid mine drainage (AMD) if a mining project site becomes polluted [1193]. African governments and regulatory authorities must require that mining companies post bonds to be held in escrow accounts until productivity or reclamated land has been satisfactorily and convincingly demonstrated. This will go a long way towards ensuring completion of reclamation, or restoration of mine land for future use. Vegetation and wildlife in previous mining lands can be renewed and even used for farming and ranching. Environmental and rehabilitation codes must be followed to ensure that the areas mined are returned close to their original states. Some damage can be mitigated only by the strictest environmental controls [1193].

12.5 Corporate social responsibility

In the mining industry corporate social responsibility (CSR) [1201] is a very important issue [1202]. It is good for profits for the mining companies to be responsible and to collaborate with stakeholders. Unfortunately, in many African countries the dominant interpretation of CSR by mining companies has been in terms of charitable donations and support for good causes [1203,1204]. Commendable though these activities maybe, they do not alleviate the contribution of mining companies to growing social problems around mines, primarily because they do not impact on core business practices and do not contribute to the necessary cross-sectoral collaboration. The state's legislated transformation programmes premised on state sovereignty over mineral resources must be the key driver of CSR.

Enterprises should demonstrate responsible citizenship besides being profitable tools. They must be responsive to the stakeholders' needs as an obligation to protect, foster, increase and enhance the benefits of stakeholders and society [1205,1206]. They must provide adequate compensation for the inevitable disruptions that their international business brings to the local community. Social responsibility is a moral obligation that dictates that beneficial returns must be shared and sustained over the long term in an equitable manner.

12.5.1 Obstacles in the CSR implementation

The main obstacle to the implementation of CSR programmes is cost. As a reaction to the impact caused by tightened budgets, the charitable behaviours of many corporations are said to be restrained. Some companies that used to support philanthropic work previously may now be targeting and limiting some social projects or coverage from which they get some rebate, e.g. they may now prefer to contribute money on work-training and not on public welfare [1203].

12.6 Investment and financing models

African countries present abundance of opportunities for the world's mining groups [1,2,6,15] who base their investment plans [1207] in respect of deposits and mining development upon anticipated medium- and long-term pricing scenarios. The issue of mineral commodity prices is, therefore, of fundamental importance to them. The anticipated viability of expensive mining projects which may run over many years is determined by the cost of mineral extraction and processing and the commodity prices thought likely to be realised. The mining industry automatically assumes constancy of commodity prices at any time. This assumption is based on the fact that demand for many kinds of minerals is expected to grow, notwithstanding the volatility of mineral prices [1179]. Consequently, there is continuing great interest of mining companies in exploring both established and promising deposits in Africa [1187] with investment focus in countries which combine perspectivity and a mining tradition [1208]. However, there is a negative perception about multinational mining companies whose reputation in Africa is poor because they have been seen as despoilers of the countryside, polluters of water resources, usurpers of ancestral lands, exploiters of cheap labour and accomplices to, if not instigators of, gross human rights abuses and civil wars [1209].

African states, most of whom are dependent on minerals, are exceptionally vulnerable to economic shocks. Few of the African countries are performing well: it is estimated that 12 of the world's most mineral-dependent states are classified as highly indebted poor countries. A strong negative correlation between a country's level of mineral dependence and its Human Development Index (HDI) ranking is observed: worse standard of living is likely the more a state relies on exporting minerals [1209].

12.6.1 Mining costs build-up

The cost built up in mining [100] is contributed by many factors with the *exploration* phase consuming a substantial portion of the mining investment without any income and is highly risky. Another high-cost stage requiring substantial capital for the purchase of equipment, most of which is imported, is the *mine development* phase. The longest and most profitable stage in the mining cycle is the *mineral production* phase when payment to the government usually begins to be generated. Significant rehabilitation costs are incurred and, in some instances, extended liabilities for site management *after mining* ceases.

12.6.2 Hedge funds

Volatility in many commodity markets [1218-1220] may be addressed by *hedging* which consists of locking in a future selling price in order to plan commercial operations with some measure of predictability. Problems arise when companies engage in abusive hedging with related parties. Such abuses happen, for example, when companies use hedging contracts to set artificially low sale prices for their production and therefore record systematic hedging losses, which reduce taxable income in the producing country.

12.6.3 Debt funding and thin capitalisation

When a company is financed through a relatively high level of debt compared to equity a situation of *thin capitalisation* [1221] is created and presents [1222] a particular risk of profit shifting whereby management may choose to finance its investment disproportionally through debt rather than equity. This is a means of avoiding corporate income tax as most countries allow companies to deduct interest expenses in calculating taxable income, including interest paid on debt owed to related parties [1223].

12.7 Ownership patterns

12.7.1 Multinational companies

Africa has been a hive of activity for companies dealing in mining [1,2,6]. These multinational companies have left in their wake trails of breakdown and impoverishment of communities, environmental degradation and industrial injury, sicknesses and deaths across the continent. Some of these multinational companies have been known to pay money to 'murderous armed groups' in order to gain entry into regions for the purpose of exploiting the mineral resources. The rewards of corporate irresponsibility are reaped by overseas share-holders. The mining industry in Africa supplies the world market with substantial amounts of minerals used in a vast number of consumer electronics as well as in industrial applications over the world [4]. The concomitant human rights violations and environmental negligence caused by industrial mining are not only serious, but also structural. These multinational companies commit these with impunity mainly because civil societies in Africa have little voices and are poorly supported. Whilst the companies may not respect human rights, the rule of law and their obligations to communities whose lives are affected by the mines, it is the responsibility of African governments to enforce laws that protect their citizens and natural environments affected by mining operations.

Africa is losing billions of dollars because many of the multinational companies are cheat-ing African governments out of vital revenues by not paying their fair share of taxes. The multinational companies may avoid paying taxes through a practice called mispricing – where a company artificially sets the prices for goods or services sold between its subsid-iaries or parent companies. Companies employ other tax avoidance schemes involving tax havens. They are also known to lobby African governments hard for tax breaks as a reward for basing or retaining their businesses in African countries.

The minerals that are mined in Africa are by and large privately owned and disposed of by multinational mining companies, and the value of the minerals is realised in London, New York, Tokyo, Toronto, Paris etc. where these companies are listed or based. The profits of mining are realised in the First World while Africa sits with the enormous costs of mining which include costs to communities surrounding mines, evictions of communities from their land, destruction of traditional economies and culture, destruction of the natural environ-ment, the consumption and pollution of water resources and the destruction of the health of large populations.

The minerals industry in many African countries operate on a free-enterprise, market-driven basis with mineral rights reverting to the state. However, the bulk of mineral land holdings and production is controlled by few mining investment houses, an arrangement structured during the colonial period. In some instances, the industry has undergone major corporate restructuring, or "unbundling". The aims have been varied, ranging from trying to simplify a complex system of interlocking ownership, to establish separate core-commodity-focussed profit centres, and to create an entry point for the aspirant black entrepreneurs into the mining industry and advance affirmative action, and empowering the historically dis-advantaged. In their effort to create favourable conditions for investment, the role of some African governments has been relegated to policing the public.

12.7.2 Nationalisations of mining companies

Nationalisation [1224] is understood to mean acquisition of privately owned enterprises by a government, with or without compensation. It involves the acquisition of an existing asset and the transfer of its ownership into state hands. The rationale for doing this is varied including economic, financial, social, strategic and nationalistic reasons. For this to make sense these reasons must offset the costs of nationalisation (financial, economic, political and reputational) if the act is to prove beneficial for society as a whole.

There are many reasons why the mining industry, by its nature, is especially prone to nationalisation. Firstly, mining involves the exploitation of a country's natural patrimony. Mining resources are non-renewable and will eventually be depleted. Their exploitation is therefore always an emotive topic if citizens benefit unequally from their depletion both currently and inter-generationally. If the exploiters of these mineral resources are foreigners, emotions run even higher. Nationalisation may become an excuse for expropriation when the true target is the value to be extracted for acquisition. It is also easier to nationalise without compensation when the targets are foreigners with little local political influence. However, international investment treaties may change or influence these dynamics.

Secondly, mines represent a concentrated form of economic "revenue". While collectively other enterprises may have far greater total profits, their dispersion makes acquiring control of such profits more difficult. On the other hand, control of a mine gives control of a concentrated source of value which the acquirers may hope to use for social, economic, political or even personal advantage.

Thirdly, mines are by nature location bound without the option of changing the nature of their business if threatened with nationalisation. In contrast, a local manufacturing or services business can sometimes transform the nature of their business if threatened with expropriation. Alternatively, they may opt to leave the country. Mines cannot change or leave once they are constructed, making them easy targets.

Fourthly, commodity prices are cyclical. When commodity prices are high and mine profitability is high, mines become exceptionally attractive targets for cash-strapped governments.

On the whole, nationalisation of the mines has not been very successful accounting for the many examples of previously nationalised mines which were later privatised. Nationalisation-privatisation cycles do occur in the African mining industry with nationalisation of mining enterprises tending to occur when the price of the corresponding mineral is high.

12.7.3 State mining enterprises

State ownership, or the existence of state-owned enterprises, is not the same thing as nationalisation [1189]. There are a number of reasons for state involvement in some commercial enterprises, including the existence of 'natural monopolies' where private ownership may lead to excessive pricing. States, for strategic or reasons of self-sufficiency, may start new industries which are not economic to private investors. In some instances, the scale of an investment may be too great for individual investors, compelling the state to use its resources to fund the necessary investment, either on its own, or in partnership with private investors [1188].

12.7.4 Privatisation of state mining enterprises

Strategic areas that are designed to promote self-sufficiency in an increasingly politically hostile world have often been the reasons for the establishment of state-owned enterprises. Privatisation is a restructuring of the public sector by reducing government involvement in certain areas [1225] for a variety of reasons. In the mining sector, there may have been growing disappointment with dismal performance of the state-owned companies. Private ownership is perceived to strengthen the incentives for profit maximisation [126] and therefore should help to raise revenue for the government. The fiscal benefits of privatisation are related to the efficiency and welfare advantages of private ownership. However, privatisation may result in an excessive emphasis on profit maximisation at the expense of other socially valuable objectives [1189]. The public debt burden is sometimes responsible for strong privatisation drive with proceeds used to reduce public debt.

12.7.5 Private-Public Partnerships (PPP) models in the African mining sector

In the African mining industry, foreign investment is often an essential springboard for future growth with focus on sustainable relationships between private and public entities. Long-term investments and development are preserved by clear standard operating procedures (SOPs) in individual countries and supranational organisations that ensure community consent and cooperation between corporate and civilian interests.

Given the centrality of the mining sector for many African countries lessons must be drawn from the serious failures in PPP of the past years in the mining sector [1149]. It is imperative that PPPs contribute to sustainable development, good governance and the struggle against corruption. It may be necessary to review the working relationships between governments and their international partners in the following ways [1150–1152].

- Clarifying the mining contracts inherited from the past
- Revising the contracts to determine whether they should be renegotiated, revoked or cancelled
- Setting up an independent mechanism to monitor the implementation of contracts
- Ensuring transparent and fair management of the mining resources.

12.7.6 Black empowerment initiatives

In Africa generally, and Southern Africa particularly, black empowerment activists in the mining industry have been at the forefront in promoting policies aimed at addressing social and economic injustices created by past colonial governments, especially apartheid. The broad objective of the initiatives is to advance the participation of historically disadvantaged Africans in the mining industry [1227]. As such black empowerment or indigenisation initiatives are designed to release the monopoly stranglehold of mining investment [1228] houses and allow entry by the aspirant blacks into the mining industry [1229,1230].

There is a caution to the noble initiative. Instead of benefiting the population as a whole this limited 'nationalisation' may benefit only a small elite who may enter into business alliance with financial and mining interests from industrialised countries. This elite faction may see a more comprehensive nationalisation as a threat to their attempts to accumulate wealth [1231].

12.8 Mineral taxation and royalties

Raising questions of fairness about the exploitation of nations' natural capital [1232] in the taxing of mining activities is an important public policy issue. Geopolitical potential of a site influences investment decisions in mining but is also strongly offset by fiscal and socio-political considerations. Socio-political considerations take into account the stability of the tax system while fiscal considerations include tax rates. The goal of government is to encourage investment in the sector while ensuring the greatest possible benefit for the public. Companies want an adequate return on investment [1233] while the government's objective for the mineral sector is to obtain an appropriate share of income and to foster development. Incentives to invest are likely to be reduced by higher taxation levels whereas raising tax rates will increase government receipts in the short term. However, if an increase in taxation is too high, it will discourage exploration and development, thus reducing the tax revenue generated by the sector over the longer term. Exploration of lower grade ore will be encouraged by an acceptable tax regime and therefore lengthen the life span of some mines. There are inherent risks in mining operations including wide fluctuations in international minerals prices and the difficulties of anticipating all geological, technical, financial and political factors over a mine's lifetime [1234]. These must be recognised together with other considerations such as the geological potential for target minerals, the security of tenure and permitting, ability to repatriate profits, consistency of minerals policies, realistic foreign exchange controls, stability of exploration terms and conditions and ability to pre-determine environmental obligations. An important risk consideration for companies contemplating investment is the perceived stability of a tax regime over time. A profits-based tax system is one among others which tend to distribute mining risks more evenly between company and the state. Because of the complexity of this tax system, there are challenges for countries with limited administrative capacity, such as potential for corruption and tax fraud.

12.9 International trade in mineral products

Concentrates are the first saleable product of mining. These are manufactured from mineral ores. Mineral products are sold subsequently after going through different stages of processing. Minerals from Africa flow to developed countries resulting in substantial mineral trade through which Africa tends to export raw materials and import manufactured goods [1174,1176]. However, with increasing interest in mineral value addition, it is anticipated that African primary ores and concentrates and their metals will be exported simultaneously, the mix depending perhaps on how far each is being transported and how this drives the benefit of value-added processing which requires more capital and skilled labour [1178]. Trade is made complex [1172] by many factors such as international technological differences, country size, distance between trading partners, economies of scale and demand-side (consumption) differences across countries.

12.9.1 Skills and trade

African countries by and large export unprocessed mineral ores and concentrates due to an abundance of the mineral resources and lack of adequate high skills. Raising the skills level of Africa's workers will increase the prospect for mineral value-added processing in Africa. Africa's export of unprocessed mineral ores and concentrates is driven by the abundance of the minerals on the continent, scarcity of capital and professional skills that are required. Countries that are currently importing minerals and metals from Africa, conversely, have a scarcity of mineral resources and an abundance of capital and skilled labour [1238].

12.9.2 Policy and trade

Relative resource abundance in minerals can be, and has been, rendered redundant through public policy that destroys investment [1239]. On the other hand, relative resource scarcity, or impending scarcity as high-grade deposits are mined out, can be overcome through proprietary technological competence and good policy. This is because value-added metal production is capital and skills intensive, and not resource intensive. Initial stages of mineral beneficiation are tied to the location of the resource endowment, but further processing and value addition in other locations is possible and potentially efficient, especially if bulk transportation costs decline. It is important to have a conducive investment policy that is supported by a favourable legislative framework complemented by adequate infrastructure such as roads, ports and ICT. This turns mineral resources into production and exports. Given such conditions, African countries are able to export mineral ores, concentrates and metals to countries with lower endowment in mineral resources but abundant with capital and skilled labour. African countries will be able to import capital-intensive manufactures, machinery and chemicals in return.

African countries as a group should strive to diversify their merchandise export mix away from unprocessed ores and concentrates and towards manufacturing. These manufacturing quantity additions would be in addition to, rather than instead of, quantity increases in mineral and metal exports [1709]

12.9.3 Terms of trade and price volatility

Trade is sensitive to price signals [1173,1175]. Terms of trade may worsen for mineral exporters over time either due to a trend or due to a stationary series with negative random deviations. Income terms of trade is more relevant than price terms of trade, since it is possible that declining export prices have been met with increased export quantities. In addition, it is the full accounting statement, which measures costs of production against revenues from production that matters.

Minerals have high price volatility [1178,1179]. Record high prices are sometimes followed by spectacular collapses. If price volatility is supply driven, decreasing prices will be met with increasing quantities, stabilising revenues. If it is demand driven, the two will reinforce each other. One system that may be applied is to have special reserve funds that deposit windfall earnings into accounts that cannot be raided by treasury. This may mitigate the vagaries created by exporting minerals which expose economies to declining terms of trade and export revenue volatility [1179].

12.9.4 Export concentration

For many mineral-intensive producing African countries, the top three mineral exports make up on average 82% of their total merchandise exports. This makes these countries more susceptible to all of the possible negative impacts of mineral booms [1181,1182]. For them, it more sensible and desirable to have policies that foster and then protect manufacturing activities in mineral-based economies. They are advised to place restrictions on trade for both exports and imports in order to direct industrial policy towards mineral resource-based manufacturing [1215–1217]. Without minerals, the manufacturing and service sectors of the economy cannot survive as natural resources (minerals) provide the foundation for the entire economic system.

12.9.5 Markets for minerals from Africa

Minerals from Africa are sold into competitive markets where prices fluctuate. Given this situation, the African governments are compelled to provide flexibility, such as relief from export duties and VATs, or, in more serious cases, relief from other more substantive taxes. It is estimated that nearly 80% of the strategic minerals that the USA requires are found in Africa, including 90% of the world's cobalt, 90% of the platinum, 40% of the gold, 98% of the chromium, 64% of the manganese and one-third of the uranium. These minerals are needed to make jet engines, cars, missiles, electronic components, iron and steel [1183]. Africa also accounts for 18% of U.S. oil imports, compared to 25% from the Persian Gulf, with Nigeria and Angola respectively being the fifth and ninth largest exporters to the U.S. The Democratic Republic of the Congo (DR Congo) is arguably the richest country in Africa, holding the world's biggest copper, cobalt, and cadmium deposits. South Africa and Ghana are the first and second, respectively, largest gold producers in Africa.

African mineral-producing countries are strategically placed to organise themselves so that they could limit the supply of a specific mineral resource. The oil producers became a very powerful cartel, OPEC, which eventually turned economic power into political power. This is a model that African countries, in concert, can adapt for certain minerals in which some African countries are the main sources individually and/or collectively [1180, 1184–1186] as summarised in Table 1.2. The grouping around each mineral can syndicate themselves for the purposes of negotiating favourable prices and trade terms for that mineral. The Association of African Diamond Producing Countries (ADPA) is already organised according to this model. The effectiveness of these groups would depend on their assertiveness on the global marketplace. These syndicates of African mineral-producing countries can also be the basis for collaborative scientific research in beneficiation/extraction and value addition of the relevant minerals and their products.

Mineral consumption in the world rises as more consumers enter the market for minerals and as the global standard of living increases. As new materials and application are found, markets for mineral commodities expand proportionately. Minerals will retain their dominant role as the basis for products used by society and, therefore, as the basis for the world's manufacturing and agriculture.

12.9.6 Mineral booms, the resource curse, the Dutch disease

The mining industries in Africa are positioned to provide benefits to the local population if they spur the development of related, non-extractive industries. One of the ways is by promoting linkages with upstream industries that supply goods to the industry. Another is through the development of downstream industries that process and add value to the products [3,5,1210]. A third way is for the government to use export revenues to promote other, unrelated sectors of the economy. In practice these linkages tend to be weak. When states undergo resource booms, their currency tends to appreciate at the same time at which the resource sector draws labour and capital away from other sectors of the economy. This phenomenon is known to economists as 'Dutch disease' [1211–1214] which can reduce the international competitiveness of the country's agricultural and industrial exports. This makes it harder for the country to diversify its exports and generate the desirable pro-poor form of growth. A country that is dependent on mineral exports [1181,1182,1215–1217] has difficulties diversifying its economy and promoting other sectors such as agriculture and manufacturing, which provide greater direct benefits to the poor.

Many African economies specialising in mineral resource extraction have suffered from a "resource curse" whereby their incomes per capital are higher than normal but their short-run economic growth is slower than normal. The Dutch disease has been described as "a morbid term that simply denotes the co-existence of booming and lagging sectors in an economy due to a temporary sustained increase in export earnings" [1211–1213].

Mineral booms are of two kinds: (a) an increase in international relative prices of minerals, and (b) an increase in the domestic availability of mineral resources. Type (a) is brought about by sudden variations of the terms of trade of minerals in favour of the production and export of minerals. Type (b) is the result of the increase in the relative endowment of minerals in a country either through exploration efforts or through policy as a substantial increase in the exploitation of mineral discoveries generates an increase in mineral exports and a large external surplus. Whichever way the boom is created it translates into a real change to an economy, because its main impact falls on the level of real income. The trade effects of a mineral book are therefore transmitted into the economy. If there exists some market failure inhibiting an appropriate structural adjustment or if there is some existing distortion in the economy which is intensified by the mineral export boom, the *Dutch disease* becomes a real disease.

The resource depletion as opposed to a market failure associated with the booming minerals sector may explain, in part, the slow growth in Africa's mineral economies. Slower economic growth will be exhibited by those economies with shrinking minerals sector output. Those with increasing mineral output will be characterised by faster economic growth.

Compounding the *Dutch disease*, growth effects in some African countries are consequences of the quality of institutions, the rate of human capital formation or increases in poverty and civil wars [1214]. This is attributable to irresponsible governments who fail to finance development efforts, or enact wealth transfers that further impoverish the poor, or inappropriate governance practices. In these circumstances, any blame for natural resource mismanagement lies with the government.

12.10 Legal issues in Africa's mining sector

It is important to be provided with seamless multi-jurisdictional representation to mining clients, and being offered timely advice on local laws, regulations and procedures across the continent and globe. This is so as there are risks inherent in carrying out mining operations in difficult jurisdictions, including those with political risk. Mining companies may be negatively affected by legislative or regime changes and governmental/regulatory decisions in the countries where they operate or by doing business in internationally contested territories, including exposure risk mitigation [1241].

Mining disputes may be high-risk, complex, sensitive and multi-jurisdictional to be resolved through court proceedings, administrative processes, arbitration, mining convention renegotiations and alternative dispute resolution processes (including statutory adjudication, expert determination and dispute resolution boards). Other matters that may seek court approval include amalgamations, amendments to articles, exchange of securities, compromises with creditors, investment treaties applicable to mining interests, including protections in treaties and customary international law against expropriation without compensation, unfair, inequitable or discriminatory treatment and denial of justice [1186]. The key areas of strength include international arbitration, regulation and investigations, including business ethics and anti-corruption; and transnational litigation, increasingly complex and evolving regulatory, compliance and government enforcement on environmental issues. Regulatory issues and investigations can often span several jurisdictions. A review in one jurisdiction can give rise to consequences in others [1242].

Other issues from which disputes may emanate include prospecting and exploration, staking and surveying of mine claims and delimitation and demarcation of mine boundaries, mining and surface rights (including permits, licenses, concessions and stability agreements and work required to apply for or maintain such rights), project financing, mine construction and operation, including in-situ leaching, mine and mining claims ownership, indigenous rights, environmental issues including closure and rehabilitation plans, resource nationalism, expropriation and deprivation, technical reports, alloy composition and assay results, mineral pricing and proceeds from mining (including royalties), securities law issues, board disputes and directors' duties, contract disputes and employment disputes.

These disputes can be governed by legislation and/or regulation (including foreign investment and permitting laws), public authorizations, investment treaties or mining-stability agreements, joint venture, shareholder, licensing and option agreements, EPC contracts, contracts regarding mine and mine infrastructure construction and delivery or mineral supply contracts.

In addition, there could be disputes concerning ancillary contracts such as insurance (including political risk claims), commodity off-take, storage, processing, commodity purchase, international payments, foreign-market credit, carriage and shipment agreements, off-take agreements, industrial disease class action or labour issues [1242].

12.11 Corruption in the African mining sector

12.11.1 General issues

One of the greatest threats to economic and political development of African nations is corruption. Corruption undermines good governance, fundamentally distorts public policy, leads to the misappropriations of resources, harms the public and private sector development, and particularly hurts the poor. While there are great concerns that Africa is massively prejudiced in the skewed international trading of its minerals, corruption involving locals in a booming economy is not helpful. Mineral booms discussed above (section 12.9.6) may degrade the quality of the institutions due to the mismanagement of the mineral revenues generated during the boom [1240]. This mismanagement manifests itself in various forms. Mineral revenues, generally captured by the government through taxes and royalties, may cater to the ruling elite in the booming country. In this context, a mineral export boom may accentuate income disparities between the poor and the rich, for the poor may be largely excluded from any benefits of the boom. In addition, political control over the mineral revenues may make it profitable for individuals and organisations to spend considerable efforts and resources to appropriate an important share of those revenues causing the emergence of revenue-seeking activities among the social groups associated with the domestic mining industry. Such revenue-seeking activities are totally unproductive, since they are carried out in order to increase the share of the mineral revenues that a particular social group enjoys crowding out productive activities associated with economic growth [1211–1213].

Mineral revenues can be easily appropriated by these groups because minerals are spatially concentrated and can be used to bribe public officers in order to obtain support for mineral activities. Public officers may be bribed in order to authorise exploitation of new deposits, to authorise tax exoneration, to weakly enforce laws that regulate the industry and to reduce the governmental interference in the booming mineral industry such as reductions

of royalties and income taxes, the avoidance of windfall taxes, reductions in the enforcement of environmental laws, etc. Hence, the mineral boom may tend to increase the level of
governmental corruption. The increase in corruption may also tend in turn to erode the credibility and the quality of institutions like the parliament, the judiciary, the police, the local
administrations and so on.

There is also ample evidence of the global nature of graft in the mining sector. Lack of
transparency in public-private agreements appears to be extremely disadvantageous to the
African countries. In a number of cases with total lack of transparency governments negoti-
ate, sign and approve joint venture agreements without an international tendering process.
In some cases, the dimension of the mineral reserves to be transferred exceeds the capacity
of the private partners to exploit them efficiently, which runs the risk that the reserves will
be used speculatively. Sometimes the private companies appear to commit themselves only
to producing a feasibility study; under normal practice it would be appropriate to require
firm commitments for phased investments. These and a myriad of other negative forces continue to torpedo initiatives that are aimed at industrialising the continent thereby militating
against socio-economic developments in the African countries. By omission and/or commission some African governments are accomplices in mortgaging the present and future of the
African people for personal interests.

12.11.2 Some illustrative cases of corruption in the African mining industry

12.11.2.1 Transfer pricing

Transfer pricing is a business practice that consists of setting a price for the purchase of a
good or service between two 'related parties' (e.g., subsidiary companies that are owned or
controlled by the same parent company). Transfer pricing becomes abusive when the related
parties distort the price of a transaction to reduce their taxable income. This is known as
transfer mispricing. Multinational mining companies rely on complex webs of interrelated
subsidiaries. Some of them are domiciled in low-tax and secrecy jurisdictions. These subsidiaries can sell minerals to each other at a discount or purchase goods, services and assets
from each other at inflated prices in order to 'transfer' profits to lower-tax jurisdictions from
higher-tax ones. Thus, the transfer price is the price of a transaction between two entities that
are part of the same group of companies. For example, a South Africa-based company might
sell mining equipment and machinery to its Ghana-based subsidiary. The price agreed is the
'transfer price'. The process for setting it is referred to as 'transfer pricing'. The difficulty in
monitoring and taxing such transactions is that they do not take place on an open market. A
commercial transaction between two independent companies in a competitive market should
reflect the best option for both companies; two affiliated companies are more likely to make
transactions in the best interest of their global parent corporation. It can be in the interest of
the global corporation to make higher profits. This is intended for use in conjunction with
lower-taxed jurisdictions and lower profits in higher-taxed ones, as a means of reducing its
overall tax bill. While the corporations gain from such tax planning, there are winners and
losers among the countries involved.

There are numerous possible transactions between affiliated companies in the mining
industry value chain. They can be broadly grouped into two categories: (i) the sale of minerals and/or mineral rights to related parties; and (ii) the purchase of various goods, services
and assets from related parties. These transactions are common to most mining companies.

The value of these transactions and potential tax revenue leakage vary greatly depending on the size and structure of the operation, commodity type and production processes. Other things being equal, large corporations tend to have more transactions between related parties and more complex financing structures than smaller companies. In each case, the price could be manipulated to reduce taxable income in the country where the mine is located. Transfer pricing can take many forms:

Procurement of goods: A company purchases mining machinery on behalf of its subsidiary. The price charged includes the direct cost, plus a fee for service. *Financing*: The subsidiary receives a loan from its parent, usually to finance its exploration or development costs. This is another way for shareholders to provide capital to a mining project, but its accounting treatment is different from equity.

Support services: The subsidiary pays a fee to a related party in return for a range of administrative, technical and advisory functions.

Mineral sales: Minerals may be sold to a related company – a smelter, for example.

Many governments of countries that have lost tax revenue as a result of transfer mispricing have created rules to regulate the practice. One way that governments can address transfer mispricing is by passing laws that require companies to apply the 'arm's length' principle: related parties price transactions as if they were transactions on an open market [1235]. There are transfer pricing methods that are based on the application of the arm's length principle [1236,1237]. One of the methods that overcomes the challenge of lack of comparable transactions is by requiring taxpayers selling mineral products to benchmark the sale price to the publicly quoted prices of minerals or metals. Transfer pricing rules recommend the application of the arm's length principle when a company engages in a transaction with an affiliated company [1235,1236]. This means that the transaction should reflect the market value of the goods or services exchanged: affiliated companies should trade with each other as if they were not affiliated. If the relevant transaction does not conform to the arm's length principle, transfer pricing rules give governments the legal right to adjust the price in the reported profits of the company. To address transfer mispricing, governments can therefore put laws and regulations in place that define the arm's length principle and detail how it should be implemented.

Transfer pricing can limit income tax receipts, including in low- and middle-income countries. Developing countries depend [1171] twice as much on these receipts as developed economies, and African countries three times as much. As developing countries try to increase funds available for social services and their national development agendas, limiting transfer mispricing is a key element of domestic resource mobilisation.

For revenue authorities in Africa, for instance, applying the arm's length principle can be extremely difficult because there is often a lack of comparable transactions. Parties frequently have to adapt comparable data from other contexts in OECD countries. This can often be time-consuming and expensive and produces results that do not reflect the economic reality of companies operating in Africa. Access to information on related parties based in offshore jurisdictions is a further obstacle for many revenue authorities, preventing them from building a complete picture of global activities of companies.

In light of these implementation challenges, an additional transfer pricing method has emerged called "the sixth method". It is designed specifically to limit the risk of transfer mispricing in commodity transactions. It requires that taxpayers selling commodity products to related parties use the publicly quoted price of the traded goods on the date that the goods are shipped as a reference. This is particularly relevant for resource-rich economies when publicly quoted prices of minerals or metals are widely available (for example, through the London Metals Exchange [1178], Platts and other indexes).

In addition to the sixth method, low-income countries are also beginning to explore other policy and procedural alternatives that try to avoid transfer pricing issues altogether. These alternatives may not follow the arm's length principle, but they can relieve tax administrations of time-consuming transfer pricing monitoring and audits.

For example: limiting the deduction of interest on loans from affiliated companies; separating the operational income of the mine from the income on financial products that set future prices of minerals ("hedging"); advance pricing agreements and "safe harbours," which define an appropriate pricing method for specific related party transactions in advance, for a number of years.

While transfer mispricing is a complex problem, there are different steps that various stakeholders can take to address the issue, such as: Putting in place detailed rules that enable revenue authorities to determine the tax value of intra-company transactions in a rigorous and consistent way, including by spelling out the procedures by which the system is to be administered. Establishing administrative structures that promote a concentration of well-trained, highly skilled officials sufficiently empowered to implement transfer pricing rules. Improving inter-agency coordination on mining revenue collection by clarifying division of audit responsibilities, encouraging joint audits and establishing overarching coordination mechanisms. Equipping revenue authorities with transfer pricing expertise and technical sector knowledge to identify and evaluate transfer pricing risks in the mining sector. Taking proactive steps to narrow the information gap, obtain more regular international and precise information from mining companies and develop automatic development exchange of information with other jurisdictions. Holding the political leadership accountable for implementation of transfer pricing rules in the mining sector. Examining the feasibility of adopting specific tax policy rules to limit the reliance on the arm's length principle and the difficulty of finding comparable transactions.

12.11.2.2 Gold smuggling

As detailed in section 2.1.3, gold is used in jewellery, medicine, investments, electronics, automotive, dental products and for religious artefacts. It is also present in some chemical compounds used in certain semiconductors [498,1259] and as a catalyst in manufacturing processes. The reflective ability of gold makes it the ideal substance for many different uses. Gold is extremely malleable, conducts electricity, doesn't tarnish, alloys well with other metals and is easy to work into wires or sheets. It is unrivalled in its natural brilliant lustre and glossy shine. Because of these unique properties, gold makes its way into almost every sphere of modern life in some way, shape or form.

Tons of gold are routinely smuggled from Africa. There are gold-smuggling syndicates linked to scores of buyers in America, Europe and Asia masquerading as dealers in precious metals. Their roots are in illegal gold mining in African countries. Gold bars worth millions of dollars are often stored in luxury homes, warehouses and farms. From there they are prepared for stamping with official African gold serial numbers designating that the metal had been officially mined and refined in one or other African country. From there the gold is flown to countries in America, Europe and Asia before being smelted, re-refined and distributed. Corrupt customs and mining officials facilitate the metal's passage across borders.

The existing forces in the minerals' thefts and smuggling industry are:

- *Thieves*: the petty thieves that operate as individuals or in small groups; owners of small and medium gold mines who have mining licenses; government partners in the large-scale mines
- *Smugglers*: people who smuggle small amounts in vehicles or flights; medium smugglers of whom some are independent and others who are related to the sharks
- *Sharks*: who, due to corrupted relations, are protected and untouchable.

Illegal miners in Africa supply the syndicates. These illegal miners (sometimes armed with lethal weapons) have tarnished the image of legitimate *small-scale* or *artisanal miners*. With instability within some African countries and corruption by government officials, runners, once they have the gold, take it to the border where, through corrupt officials, it is smuggled across disguised as things such as household products. The gold is taken to farms where illicit refineries smelt and refine it. With the help of African mining officials, gold clearance documentation and special serial and insignia stamps are sourced. Once stamped one would never know the difference. The gold is then distributed through legitimate channels in America, Europe and Asia. Those running the syndicates know what they are doing. They are well-connected and influential businessmen with ties to Africa, Europe, America and Asia. They are linked to the gold powerhouses of the world. These are not 'mickey-mouse' people. They are immensely powerful and extremely well connected to some of the world's top legal firms. They obtain official gold clearance documents, serial stamps and other paperwork with links to mines and importers and exporters. Everything these syndicates do is 'legitimised' through fraudulently obtained documents. They use specially designed reinforced armoured vehicles to carry the gold. They are so well organised that paperwork and everything they have for the vehicles, the properties and the rest of their dealings looks legitimate, but it's far from that. The amount of gold that the syndicates handle is immeasurable.

12.11.2.3 Conflict minerals

The four most commonly mined conflict minerals known as 3TGs are cassiterite, wolframite and coltan, for tin, tungsten and tantalum, respectively, together with gold. These minerals are essential in the manufacture of a variety of devices, including consumer electronics such as mobile phones, laptops, and MP3 players.

Tantalum is extracted from columbite-tantalite (coltan). It is used primarily for the production of tantalum capacitors, particularly for applications requiring high performance, a small compact format and high reliability, from hearing aids and pacemakers, to airbags, GPS, ignition systems and anti-lock braking systems in automobiles, through to laptop computers, mobile phones, video cameras and digital cameras. In its carbide form, tantalum possesses significant hardness and wear-resistance properties. As a result, it is used in jet engines/turbine blades, drill bits, end mills and other tools.

Cassiterite is the chief ore needed to produce tin, essential for the production of tin cans and the solder on the circuit boards of electronic equipment. Tin is also commonly a component of biocides, fungicides and as tetrabutyl tin/tetraoctyl tin, and intermediate in polyvinyl chloride (PVC) and high-performance paint manufacturing.

Wolframite is an important source of the element tungsten. Tungsten is a very dense metal and is frequently used for this property, such as in fishing weights, dart tips and golf club heads. Like tantalum carbide, tungsten carbide possesses hardness and wear resistance properties and is frequently used in applications like metalworking tools, drill bits and milling.

Smaller amounts are used to substitute lead in "green ammunition". Minimal amounts are used in electronic devices, including the vibration mechanism of cell phones.

As in gold, these three minerals are extracted and passed through a variety of intermediaries before being purchased. They are extracted in conflict zones and sold to perpetuate the fighting. Belligerent accessibility to these precious commodities prolongs conflicts. Various armies, rebel groups and outside actors profit from mining while contributing to violence and exploitation during wars in the areas where mining takes place. Armed groups smuggle the materials and unprocessed minerals out of the affected countries. As the mines produce ore, the militia, rebels and other armed forces in the area smuggle them out of the countries. The groups sell the ore to refineries. The refineries then sell refined metals to smelting corporations, most of which are in Asia. The militia uses the money from selling ore to purchase weapons and other supplies. The more politically motivated groups may use the money to influence government officials. Visitors to the involved countries are obliged to pay arbitrary fees and bribes to avoid harassment. Most of the money from the mining operations goes to the armed groups in control of the area. Very little makes its way back to the actual people mining the ore. Meanwhile, the smelting companies refine the ore into the metals used in electronics. The smelting companies sell the metals to component manufacturers. These companies then sell components to big companies like microchip processors and electronics companies. Ultimately, many of these products end up in the hands of consumers around the world, who are unaware of the conflict behind the source of the minerals needed to produce them.

12.11.2.4 Blood diamonds

Blood diamonds are diamonds mined in a war zone and sold to finance an insurgency, an invading army's war efforts, or a warlord's activity. The term refers to analogous situations involving conflict minerals discussed prior and is used to highlight the negative consequences of the diamond trade in certain areas, or to label an individual diamond as having come from such an area. The diamonds produced are sold for illegal and unethical purposes which are specifically *conflict* in nature.

There are links between diamonds and conflicts in certain countries. Smugglers from some African countries sell blood diamonds through channels less sophisticated such as social media posts. For example, a rebel group in one country would produce these diamonds and trade them to another country to get them a certificate of naturalisation to then be sold as legitimate. The country would become a route for exporting diamonds from war-torn country. The rough diamonds would therefore be exported out of the country to neighbouring states and international trading centres through controlled section of the country where rebel groups would use the funds to re-arm. Often fake Kimberley certificates would accompany the gems. Some countries with no official diamond mining industry would be exporting large quantities of diamonds, the origin of which they are not able to detail. They would be falsifying certificates of origin.

The huge consequences of blood diamonds include people being abused by the security forces, including rape and the use of excessive force on detainees, including teenagers. Child abuse and child labour are other serious issues. As they need a huge number of workers, the security forces start kidnapping and forcing young adults to be their slaves; children are forced to join their army as soldiers, and women are raped. In some cases, entire villages are burned down. Thousands of men, women and children are used as slaves to collect diamonds, and they are forced to use their bare hands to dig in mud along riverbanks instead of digging with tools.

The funds from these diamonds create opportunities for tax evasion and financial support of crime. The money obtained from selling these diamonds does not reach the public and it does not provide benefit to anyone in the communities which continue to have no basic facilities, like electricity and repairing of roads.

The Kimberley Process Certification Scheme (KPCS) is meant to strengthen the diamond industry's ability to block sales of conflict diamonds. It is an international certification system on the export and import of diamonds, with legislation in all countries to accept only officially sealed packages of diamonds, for countries to impose criminal charges on anyone trafficking in conflict diamonds, and to institute a ban on any individual found trading in conflict diamonds from the diamond bourses of the World Federation of Diamond Bourses. This system tracks diamonds from the mine to the market and regulates the policing surrounding the export, manufacture and sale of the products. Kimberley members are not allowed to trade with non-members. Before a gemstone is allowed through the airports to other countries, the Kimberley Certification must be presented by the gem's owner or obtained from a renowned attorney. The certificate should also be requested by the customer when the gems have reached a retail store to ensure its precedence.

The KPCS attempts to curtail the flow of conflict diamonds, help stabilise fragile countries and support their development. As the KPCS makes life harder for criminals, it brings large volumes of diamonds onto the legal market that would not otherwise make it there. This has increased the revenues of poor governments and helped them to address their countries' development challenges.

However, the KPCS has ultimately failed to stem the flow of blood diamonds. There is no guarantee that diamonds with a Kimberley Process Certification are in fact conflict-free. This is due to the nature of the corrupt government officials in the leading diamond-producing countries. It is common for these officials to be bribed in exchange for paperwork declaring that blood diamonds are Kimberley Process Certified.

The KPCS attempted to increase governments' transparency by forcing them to keep records of the diamonds they are exporting and importing and how much they are worth. In theory, this would show governments their finances so that they can be held accountable for how much they are spending for the benefit of the country's population. However, non-compliance by countries has led to the failure of accountability.

12.11.2.5 Oil and gas scams

In some African countries the oil and gas industry has been marred by political and economic strife largely due to a long history of corrupt regimes and complicity of multinational corporations. Corruption in the oil and gas industry takes place in the context of corruption in the other aspects of the society's economic, social, political, religious, academic institutions and in the services. Oil and gas are very big businesses but very small fractions of transaction values in the sector can equate to very large sums of money, representing a very serious temptation to corruption.

At the production level, the oil and gas revenue flows to the government tend to be concentrated, coming from relatively few taxpayers, mostly foreign rather than domestic. In this sort of environment, the accountability to the government agencies in receipt of the revenue flows may be limited. The oil industry is technically and structurally complex, and the legal and fiscal arrangements governing revenue flows are typically even more complex. This makes it relatively easy for those who manipulate revenue flows for political gains to conceal their activities.

The sheer scale of the oil and gas industry and its supporting infrastructure often result in natural monopolies in areas such as pipeline transport, terminating and port facilities. Monopoly control creates opportunities for corrupt abuse through discretionary control of access and through the setting of fees or traffic for use.

Oil is almost universally regarded as being of strategic significance. From the standpoint of producing-country government, oil is one of the "commanding heights" of the economy, an argument that is used in support of wide-ranging government involvement in the sector. Government intervention ranges from ownership of the resource through policy formulation and legislation, control of access to infrastructure and regulation of operations to the establishment of national oil companies. Each of these areas of government involvement may spawn innumerable opportunities for corruption.

Corruption may come from the government officials in selecting a contractor to build the sector infrastructure, through the process of selecting a less competent contractor because of the desire to make personal gain by the government official. Corruption in the industry may include overpricing, inventory recycling, syndicated bidding, connivance, espionage, collusion and fraud. Thus, corruption can take place at policy, administrative and commercial level. It may also take the form of direct theft of massive amounts of money through diversion of production, products or revenues.

Because of the demand and supply relationship between the consuming-country governments (mostly the developed nations) and the African producing-country governments the former, in an attempt to maintain security of oil supply from the producers, tend to support the corrupt practices in the producing African nations. Consuming countries may overlook corruption in the producing African countries as long as they get their constant and continuous supply of what they want, using their economic, political and military leverage to influence the corrupt behaviour of the African nations and their governments.

Increasing global demand is driving new oil and gas discoveries in Africa. Yet many African countries rich in oil and gas are home to some of the world's poorest people. How can this happen? Too often, wealth stays in the hands of politicians and industry insiders. Revenues don't get published. Payments made to governments to exploit resources remain secret. Bribery and embezzlement go unchecked.

Many oil and gas companies protect the identities of their equity holders and subsidiaries. This allows corrupt leaders to hide stolen funds unnoticed. Inadequate financial statements make it easy to disguise corrupt deals, and impossible to monitor them. Many oil and gas companies don't publish information country by country. This allows them to hide the royalties, taxes and fees they pay. But without this information, it is difficult to hold governments to account for the money they receive. Stolen oil and gas income have terrible consequences. It benefits an elite few. But for everyone else, it fuels conflict over resources. And it traps people in poverty they'd otherwise avoid.

12.11.2.6 Sea piracy

Piracy is a serious problem and it poses a real threat not only to the safety of vessels and their crews, but also to the economies of affected countries. In Africa, while piracy in Somalia's Gulf of Aden is currently on the decline, it has spread to West Africa. In both cases it was initially a threat to international fishing vessels but has expanded to international shipping. International organisations have expressed concern over the piracy due to its high cost to global trade and the incentive to profiteer by insurance companies and others.

Weapons and tactics are used such as taking over a foreign ship until their owners paid a ransom. Due to the profitability of ransom payments, some financiers and former militiamen fund pirate activities, splitting the profits evenly with the pirates. The frequency and sophistication of the attacks as well as the size of vessels being targeted vary. Large cargo ships, oil and chemical tankers on international voyages become the targets of choice for hijackers. Attacks are directed against smaller, more vulnerable vessels carrying trade or employed in the coastal trade. Pirates may hijack ships well off shore, attacking strategically important waterways for international trade carrying mineral ores or oil tankers. Major commodities routes come under threat from gangs wanting to snatch cargoes and crews. Some pirates steal goods, particularly oil. Many attacks end up with crew members injured or killed. Pirate attacks do not only result in killings and injuries, they also damage the economy as the losses affect international insurance rates and other trade-related costs by increasing the cost of imports and decreasing the competitiveness of exports. The decrease also affects the livelihoods of the country's citizens.

Analysis of the methods used in a typical pirate attack have been made. Apparently most attacks occur during the day involving two or more skiffs that can reach speeds of up to 25 knots. With the help of motherships that include captured fishing and merchant vessels, the operating range of the skiffs can be increased far off shore. An attacked vessel is approached from quarter or stern. Small arms are used to intimidate the operator to slow down and allow boarding. Light ladders are brought along to climb aboard. Pirates then try to get control of the bridge to take operational control of the vessel. Pirates often jettison their equipment in the sea before arrest, as this lowers the likelihood of a successful prosecution.

The funding of piracy operations is now structured in a stock exchange, with investors buying and selling shares in upcoming attacks in a bourse. Pirates say ransom money is paid in large denomination US$ bills. It is delivered to them in burlap sacks which are either dropped from helicopters or cased in waterproof suitcases loaded onto tiny skiffs. Ransom money may also be delivered to pirates via parachute, perhaps to secure the release of ship and crew. To authenticate the banknotes, pirates use currency-counting machines, the same technology used at foreign exchange bureaus worldwide. These machines are, in turn, purchased from business connections in various areas. Hostages seized by the pirates usually have to wait 45 days or more for the ships' owners to pay the ransom and secure their release.

Weapons dealers in the surrounding cities receive a deposit from a dealer on behalf of the pirates and the weapons are then driven to a location where the pirates pay the balance. Other armed groups may extort the pirates, demanding protection money from them and forcing seized pirate gang leaders to hand over a percentage of future ransom proceeds. It is inconceivable to intelligence agencies that international criminal organisations would not be getting some financial reward from the successful hijackings. The funding links are able to keep the group satisfied as piracy gains more publicity and higher ransoms.

The indirect costs of piracy are much higher as they include insurance, naval support, legal proceedings, re-routing of slower ships and individual protective steps taken by ship owners. A veritable industry of profiteers has also risen around the piracy. Insurance companies, in particular, have profited from the pirate attacks, as insurance premiums have increased significantly. In order to keep premiums high, insurance firms may not demand that ship owners take security precautions that would make hijackings more difficult. For their part, shipping companies often do not comply with naval guidelines on how best to prevent pirate attacks in order to cut down on costs. Security contractors and the arms industry have profited from the phenomenon.

As is often the case, corruption, weak law enforcement and poverty are the main causes of piracy. Piracy tends to be conducted or supported by marginalised communities that have not been participating in economic development. Some pirate attacks have been driven mainly by corruption in the oil sector. Pirates have an incentive to steal oil, since they know that they will be able to sell it on the black market. Illegal bunkering (filling ships with fuel) is enormously profitable. The damage caused by thieves has forced some oil companies to shut down pipelines.

12.11.3 Fight against corruption in the African mining industry

From the foregoing, it is evident that fighting corruption is a formidable task that needs to match the determination of the varied corruption schemes perpetrated in Africa's mining industry. Although on the surface, it would appear as if corruption in the African mining sector is a local phenomenon, this is an incomplete picture. At the heart of corruption in the African mining sector is the global hunger for minerals from Africa to meet the demand created for consumer high-tech goods including military and nuclear arsenals. *Without international market for Africa's minerals local corruption in the sector would be very minimal.* Corruption in its varied manifestations is globally funded by very powerful and well-organized global syndicates who will not stop at anything to get access to the minerals in Africa. So, the fight against corruption in the African mining sector is ultimately a fight against global interests in minerals from Africa.

At the local level, the fight against corruption is a fight against, with a few exceptions, multinational mining officials on the ground and against government officers spanning the entire hierarchy of government right to the top. It is a fight against the system with entrenched personal interests protected by institutions that are supposed to serve the interests of the citizens. With a few exceptions, the personnel in institutions like the media, the police, the judiciary, parliament, etc. are often all compromised. The measure of success in fighting the corruption which progressively becomes more endemic with the passage of time is related to the support given from the highest office of the land. Without that support, the fight is almost suicidal for anyone who attempts it. The fight against corruption in the mining sector is a fight against the media, especially state media, which is roped in to underplay the role of some government officials in corruption and to discredit fighters against corruption.

12.12 Concluding remarks: Africa is poor by choice!

This book wades into the continent's paradox; the paradox of a continent that is richly endowed with resources, specifically mineral resources, that have catalysed industrialisation in other regions of the world, yet its citizens are wallowing in abject poverty made manifest in the form of underdevelopment. The book progressively moves in sequence from information about mineral resources in Africa, through the scientific and technological means necessary to exploit them, to policy considerations that should spur bold decisions that are imperative for industrialisation of the African countries. It has been the author's objective and ambition to appeal, in a single volume, to Africa's constituencies in the economic, scientific and political spheres to consolidate their initiatives and strengthen every endeavour for industrialisation in a continent which hungers for good quality of life for its citizens that is commensurate with the abundant natural resources.

This book attempts to demonstrate that the materials that are required for the continent's industrialisation are plentiful in the continent, the science and technologies required to exploit the mineral resources are available, and the African human capacity in the continent and in the diaspora for that is potentially adequate. What is required is the determined and concerted political will to act especially on twin issues to make the industrialisation of the continent happen: (i) a purposeful promotion of local mineral beneficiation and value addition and (ii) a serious fight against corruption in the sector without fear or favour. This process will transform the MINERALS IN AFRICA into MINERALS OF AFRICA. Without this double pronged approach, *Africa is poor by choice!*

References

1. Yager, T.R., Bermúdez-Lugo, O., Mobbs, P.M., Newman, H.R., Taib, M., Wallace, G.J. & Wilburn, D.R. (2012) The mineral industries of Africa (PDF). 2012 Minerals Yearbook. U.S. Geological Survey, PD license (U.S. government source, public domain), August.
2. Mining in Africa Towards 2020 kpmgafrica.com.
3. The International Study Group Report on Africa's Mineral Regimes: "Minerals and Africa's Development", Economic Commission for Africa African Union, 2011.
4. Curtis, M. (2016) *The New Colonialism: Britain's scramble for Africa's energy and mineral resources*. Available from: www.waronwant.org@waronwant.
5. African Union, Africa Mining Vision, February 2009.
6. Brown, T.J., Hetherington, L.E., Hannis, S.D., Bide, T., Benham, A.J., Idoine, N.E. & Lusty, P.A.J. (2009) World Mineral Production 2003–07 (PDF). British Geological Survey > World Mineral Statistics. p. 82. ISBN 978-0-85272-638-9.
7. Rodney, W. (1972) *How Europe Underdeveloped Africa*. Bogle-L'Ouverture Publications, London, UK, in partnership with the Tanzania Publishing House, Dar es Salaam ISBN 0-9501546-4-4.
8. Lagos Plan of Action for the Economic Development of Africa (1980–2000).
9. Johannesburg Political Declaration and Plan of Implementation [chapter 46 and paragraphs (f and g) of chapter 62 (Sustainable development for Africa)] of the World Summit on Sustainable Development, 2002.
10. Yaoundé Vision on Artisanal and Small-Scale Mining, 2002.
11. Africa Mining Partnership's Sustainable Development Charter and Mining Policy Framework.
12. SADC Framework and Implementation Plan for Harmonisation of Mining Policies, Standards, Legislative and Regulatory Frameworks.
13. UEMOA's Common Mining Policy and "Code Miniere Communautaire".
14. Big Table on "Managing Africa's Natural Resources for Growth and Poverty Reduction" 2007 jointly organized by ECA and the AfDB; Cawood, F.T., Macfarlane, K.S.A. and Minnitt, R. (2001) *Mining, Minerals and Economic Development and the Transition to Sustainable Development in Southern Africa*. University of the Witwatersrand, Witwatersrand.
15. Roy, R. in The effects of mining in Africa. *Mining Africa*, 26 January 2017.
16. Hitzer, E. & Perwass, C. (2006) The hidden beauty of gold" (PDF). *Proceedings of the International Symposium on Advanced Mechanical and Power Engineering 2007 (ISAMPE 2007) between Pukyong National University (Korea), University of Fukui (Japan) and University of Shanghai for Science and Technology (China), November 22–25, 2006*, hosted by the University of Fukui (Japan), 22 November. pp. 157–167. (Figs 15, 16, 17, 23 revised.).
17. Khapoya, V.B. (1998) [1994] *The African Experience*, 2nd edition. Prentice Hall, Upper Saddle River, NJ. ISBN 0137458525.
18. Lovejoy, P.E. (2012) *Transformations of Slavery: A History of Slavery in Africa*, 3rd edition. Cambridge University Press, London. ISBN 9780521176187.
19. Ferguson, N. (2003) *Empire: How Britain Made the Modern World*. Allen Lane, London. ISBN 0-7139-9615-3.

20. Mbembe, A. (1992) Provisional notes on the postcolony. *Africa: Journal of the International African Institute*, 62, 3–37.

21. *UK/Africa High Level Prosperity Partnership*, 19 November 2013. Available from: www.gov.uk/government/news/ukafrica-highlevel-prosperity-partnership.

22. Smith, C.S. (1981) *A Search for Structure*. MIT Press, Cambridge, MA. ISBN 0262191911.

23. Roy, R. (1979) Interdisciplinary science on campus. In: Kockelmans, J.J. (ed.) *Interdisciplinarity and Higher Education*. Pennsylvania State University Press. pp. 161–196.

24. Hemminger, J.C. (2010) Science for Energy Technology: Strengthening the Link between Basic Research and Industry (Report). United States Department of Energy, Basic Energy Sciences Advisory Committee, August.

25. Alivisatos, P. & Buchanan, M. (2010) Basic Research Needs for Carbon Capture: Beyond 2020 (Report). United States Department of Energy, Basic Energy Sciences Advisory Committee, March.

26. Tanja, G., Parkin, S.S.P. & Felser, C. (2011). Simple rules for the understanding of heusler compounds. *Progress in Solid State Chemistry*, 39(1), 1–50.

27. Jowi, J.O., Obamba, M., Sehoole, C., Alabi, G., Oanda, O. & Barifaijo, M. *Programme on Innovation, Higher Education and Research for Development (IHERD)*. OECD.

28. Clancy, P. & Dill, D.D. (2009) The research mission of the university: An introduction. In: Clancy, P. & Dill, D.D. (eds.). Volume 1. Sense Publishers, Rotterdam.

29. Mulumba, B.M., Obaje, A., Kobedi, K. & Kishun, R. (2008) International student mobility in and out of Africa: Challenges and opportunities. In: Teferra, D. & Knight, J. (eds.). Association of African Universities, Accra, Ghana.

30. Teferra, D. & Knight, J. (2008) *Higher Education in Africa: The International Dimension*. Centre for International Higher Education, Boston College and Association of African Universities, Chestnut Hill, Association of African Universities, Accra, Ghana.

31. World Bank Task Force on Higher Education and Society (2000) *Higher Education in Developing Countries: Peril and Promise*. World Bank, Washington, DC.

32. Government-Sponsored Research and Development in Mining Technology & National Research Council (2002) *Evolutionary and Revolutionary Technologies for Mining*. The National Academies Press, Washington, DC. doi:10.17226/10318.

33. SAIMM, Mining research and development reborn: The mining precinct. *Journal of South African Institute of Mining and Metallurgy*, 117(12), Johannesburg, December 2017.

34. Chi, G., Fuerstenau, M.C. & Marsden, J.O. (1997) A study of Merrill-Crowe processing. Part I: Solubility of zinc in alkaline cyanide solution. *International Journal of Mineral Processing*, 49, 171–183.

35. Kesler, S.E. *Mineral Supply and Demand into the 21st Century*. Available from: https://pubs.usgs.gov/circ/2007/1294/reports/paper9.pd.

36. Yepes, T., Pierce, J. & Foster, V. (2009) Making sense of Africa's infrastructure endowment: A benchmarking approach. Policy Research Working Paper 4912, World Bank.

37. Fauconnier, C.J. (2004) Bulk mining and associated infrastructure development can support Africa's economic growth. *Investing in African Mining Conference*, Cape Town, South Africa.

38. Foster, V. (2008) Overhauling the engine of growth: Infrastructure in Africa, Executive Summary report on Africa infrastructure country diagnostic. *World Bank*. Available from: http://siteresources.worldbank.org/INTAFRICA/Resources/AICD_exec_summ_9-30-08a.pdf.

39. Adams, M.D. (2016) *Gold Ore Processing: Project Development and Operations*, 2nd edition. Elsevier, Singapore.

40. Ashby, M. & Jones, D.R.H. (1992) *Engineering Materials 2*. Pergamon Press, Oxford. ISBN 0-08-032532-7.

41. The demand for gold by industry (PDF). Gold bulletin. 2009.

42. Krech III, S., Merchant, C. & McNeill, J. R. (eds.) (2004) *Uses of Gold*. ISBN 978-0-442-23898-8.

43. Shepard, K., McNeill, R.J. & Merchant, C. (2004) *Encyclopedia of World Environmental History*, Volume 3. Routledge, Abingdon, UK. p. 597. ISBN 0-415-93734-5.

44. General electric contact materials. Electrical Contact Catalog (Material Catalog). Tanaka Precious Metals (2005)

45. Fulay, P. & Lee, J.-K. (2016) *Electronic, Magnetic, and Optical Materials*, 2nd edition. CRC Press, 18 November. ISBN 978-1-4987-0173-0.

46. Peckham, J. (2016) Japan wants citizens to donate their old phone to make 2020 Olympics medals. *TechRadar*, 23 August.

47. Kean, W.F. & Kean, I.R.L. (2008) Clinical pharmacology of gold. *Inflammopharmacology*, 16(3), 112–125.

48. Moir, D.M. (1831) *Outlines of the Ancient History of Medicine*. Available from: https://en.wikipedia.org/wiki/David_Macbeth_Moir.

49. Mortier, T. (2006) *An Experimental Study on the Preparation of Gold Nanoparticles and Their Properties*. PhD thesis, University of Leuven, Belgium, May.

50. Richards, D.G., McMillin, D.L., Mein, E.A. & Nelson, C.D. (2002) Gold and its relationship to neurological/glandular conditions. *The International Journal of Neuroscience*, 112(1), January, 31–53.

51. Harris, D.C. & Cabri, L.J. (1991) Nomenclature of platinum-group-element alloys: Review and revision. *The Canadian Mineralogist*, 29(2), 231–237.

52. Rollinson, H. (1993) *Using Geochemical Data: Evaluation, Presentation, Interpretation*. Longman Scientific and Technical. ISBN 0-582-06701-4.

53. Hunt, L.B. & Lever, F.M. (1969) Platinum metals: A survey of productive resources to industrial Uses (PDF). *Platinum Metals Review*, 13(4), 126–138.

54. Bernardis, F.L., Grant, R.A. & Sherrington, D.C. (2005) A review of methods of separation of the platinum-group metals through their chloro-complexes. *Reactive and Functional Polymers*, 65, 205–217.

55. Emsley, J. (2003) *"Iridium": Nature's Building Blocks: An A–Z Guide to the Elements*. Oxford University Press, Oxford, England, UK. pp. 201–204. ISBN 0-19-850340-7.

56. Emsley, J. (2011) *Nature's Building Blocks: An A-Z Guide to the Elements*, New edition. Oxford University Press, New York, NY. ISBN 978-0-19-960563-7.

57. Perry, D.L. (1995) *Handbook of Inorganic Compounds*. CRC Press, Boca Raton, FL. pp. 203–204. ISBN 1439814619.

58. Lagowski, J.J. (ed.) (2004) *Chemistry Foundations and Applications*, Volume 2. Thomson Gale, University of Texas, Austin. pp. 250–251. ISBN 0028657233.

59. Greenwood, N.N. & Earnshaw, A. (1997) *Chemistry of the Elements*, 2nd edition. Butterworth-Heinemann, Oxford. pp. 1113–1143, 1294. ISBN 0-7506-3365-4.

60. Hunt, L.B. (1987) A history of iridium (PDF). *Platinum Metals Review*, 31(1), 32–41.

61. Kittel, C. (2004) *Introduction to Solid State Physics*, 7th edition. Wiley-India. ISBN 81-265-1045-5.

62. George, M.W. (2008) Platinum-group metals (PDF). *U.S. Geological Survey Mineral Commodity Summaries* (USGS Mineral Resources Program), 126–127

63. Xiao, Z. & Laplante, A.R. (2004) Characterizing and recovering the platinum group minerals: A review. *Minerals Engineering*, 17(9–10), 961–979.

64. Platinum-Group Metals. U.S. Geological Survey Mineral Commodity Summaries.

65. Renner, H., Schlamp, G., Kleinwächter, I., Drost, E., Lüschow, H.M., Tews, P., Panster, P., Diehl, M., et al. (2002) Platinum group metals and compounds. In: *Ullmann's Encyclopedia of Industrial Chemistry*. Wiley, New York.

66. Seymour, R.J. & O'Farrelly, J.I. (2001) Platinum-group metals. In: *Kirk Othmer Encyclopedia of Chemical Technology*. Wiley, New York.

67. Gilchrist, R. (1943) The platinum metals. *Chemical Reviews*, 32(3), 277–372.

68. Ohriner, E.K. (2008) Processing of iridium and iridium alloys. *Platinum Metals Review*, 52(3), 186–197.

69. Hunt, L.B. & Lever, F.M. (1969) Platinum metals: A survey of productive resources to industrial uses (PDF). *Platinum Metals Review*, 13(4), 126–138.

70. Handley, J.R. (1986) Increasing applications for iridium (PDF). *Platinum Metals Review*, 30(1), 12–13.

71. Stallforth, H. & Revell, P.A. (2000) *Euromat 99*. Wiley-VCH, Weinheim, Germany. ISBN 978-3-527-30124-9.

72. Jollie, D. (2008) *Platinum 2008* (PDF). Johnson Matthey. ISSN 0268-7305.

73. Egorova, R.V., Korotkov, B.V., Yaroshchuk, E.G., Mirkus, K.A., Dorofeev, N.A. & Serkov, A.T. (1979) Spinnerets for viscose rayon cord yarn. *Fibre Chemistry*, 10(4), 377–378.

74. Emsley, J. (2005) Iridium (PDF). *Visual Elements Periodic Table: Royal Society of Chemistry*, January 18.

75. Crookes, W. (1908) On the use of iridium crucibles in chemical operations. *Proceedings of the Royal Society of London: Series A, Containing Papers of a Mathematical and Physical Character*, 80(541), 535–536.

76. Cheung, H., Tanke, R.S. & Torrence, G.P. (2000) Acetic acid. In: *Ullmann's Encyclopedia of Industrial Chemistry*. Wiley, New York.

77. Ziegler, E., Hignette, O., Morawe, C. & Tucoulou, R. (2001) High-efficiency tunable X-ray focusing optics using mirrors and laterally-graded multilayers. *Nuclear Instruments and Methods in Physics Research Section A: Accelerators, Spectrometers, Detectors and Associated Equipment*, 467–468, 954–957.

78. Arblaster, J.W. (1995) Osmium, the densest metal known. *Platinum Metals Review*, 39(4), 164.

79. Hammond "Osmium", C. R. In: Lide, D.R. (ed.) (2005) *CRC Handbook of Chemistry and Physics*, 86th edition. CRC Press, Boca Raton, FL. pp. 4–25. ISBN 0-8493-0486-5.

80. Weinberger, M., Tolbert, S. & Kavner, A. (2008) Osmium metal studied under high pressure and nonhydrostatic stress. *Physics Review Letters*, 100(4), 045506.

81. Cynn, H., Klepeis, J.E., Yeo, C.S. & Young, D.A. (2002) Osmium has the lowest experimentally determined compressibility. *Physical Review Letters*, 88(13), 135701.

82. Sahu, B.R. & Kleinman, L. (2005) Osmium is not harder than diamond. *Physical Review B*, 72(11), 113106.

83. Xiao, Z. & Laplante, A.R. (2004) Characterizing and recovering the platinum group minerals: A review. *Minerals Engineering*, 17(9–10), 961–979.

84. Holleman, A.F. & Wiberg, N. (2007) *Lehrbuch der Anorganischem Chemie (102nd ed)*. De Gruyter ISBN 978-3-11-017770-1.

85. George, M.W. (2006) Minerals yearbook: Platinum-group metals (PDF). United States Geological Survey USGS.

86. Renner, H., Schlamp, G., Kleinwächter, I. & Drost, E. et al. (2002) Platinum group metals and compounds. In: *Ullmann's Encyclopedia of Industrial Chemistry*. Wiley, New York.

87. Gilchrist, R. (1943) The platinum metals. *Chemical Reviews*, 32(3), 277–372.

88. Hunt, L.B. & Lever, F.M. (1969) Platinum metals: A survey of productive resources to industrial uses (PDF). *Platinum Metals Review*, 13(4), 126–138.

89. Lassner, E. & Schubert, W.-D. (1999) Low temperature brittleness. In: *Tungsten: Properties, Chemistry, Technology of the Element, Alloys, and Chemical Compounds*. Springer, London. p. 2021. ISBN 978-0-306-45053-2.

90. MacDonell, H.L. (1960) The use of hydrogen fluoride in the development of latent fingerprints found on glass surfaces. *The Journal of Criminal Law, Criminology, and Police Science*, 51(4), 465–470.

91. Bozzola, J.J. & Russell, L.D. (1999) Specimen preparation for transmission electron microscopy. In: *Electron Microscopy: Principles and Techniques for Biologists*. Jones and Bartlett, Sudbury, MA, pp. 21–31. ISBN 978-0-7637-0192-5.

92. Chadwick, D. (2002) *Role of the Sarcoplasmic Reticulum in Smooth Muscle*. John Wiley and Sons, Hoboken, NJ. pp. 259–264. ISBN 0-470-84479-5.

93. Kielhorn, J., Melber, C., Keller, D. & Mangelsdorf, I. (2002) Palladium: A review of exposure and effects to human health. *International Journal of Hygiene and Environmental Health*, 205(6), 417–432.

94. Fajarnes, P. (2011) Palladium. *United Nations Conference on Trade and Development*, UNCTD.

95. Hesse, R.W. (2007) Palladium. In: *Jewelry-Making through History: An Encyclopedia*. Greenwood Publishing Group. p. 146. ISBN 978-0-313-33507-5.

96. Rushforth, R. (2004) Palladium in restorative dentistry: Superior physical properties make palladium an ideal dental metal. *Platinum Metals Review*, 48(1).

97. Colon, P., Pradelle-Plasse, N. & Galland, J. (2003) Evaluation of the long-term corrosion behavior of dental amalgams: Influence of palladium addition and particle morphology. *Dental Materials*, 19(3), 232–239.

98. Tsuji, J. (2004) *Palladium Reagents and Catalysts: New Perspectives for the 21st Century*. John Wiley and Sons, Hoboken, NJ. p. 90. ISBN 0-470-85032-9.

99. Drahl, C. (2008) Palladium's hidden talent. *Chemical & Engineering News*, 86(35), 53–56.

100. Gary Poxleitner, SRK Consulting (Canada) Inc. Operating costs for miners: Reducing mining costs and value optimization.

101. Gupta, D.C., Langer, P.H. & ASTM Committee F-1 on Electronics (1987) Emerging semiconductor technology: A symposium. *ASTM International*. pp. 273. ISBN 978-0-8031-0459-4.

102. Hindsen, M., Spiren, A. & Bruze, M. (2005) Cross-reactivity between nickel and palladium demonstrated by systemic administration of nickel. *Contact Dermatitis*, 53(1), 2–8.

103. Hilliard, H.E. *Daily Metal Prices: September 2001*. Johnson Matthey. London.

104. Holmes, E. (2007) Palladium, platinum's cheaper sister, makes a bid for love. *Wall Street Journal* (Eastern edition), 13 February, B.1.

105. Ware, M. (2005) Book review of photography in platinum and palladium. *Platinum Metals Review*, 49(4), 190–195.

106. Platinum Archived 22 December 2011 at the Wayback Machine. Available from: mysite.du.edu.

107. Loferski, P.J. (2011) 2010 minerals yearbook: Platinum-group metals (PDF). *USGS Mineral Resources Program*, October.

108. Heiserman, D.L. (1992) *Exploring Chemical Elements and Their Compounds*. TAB Books, McGraw-Hill. pp. 272–274. ISBN 0-8306-3018-X.

109. Tabet, N. & Salim, M. (1998) "KRXPS study of the oxidation of GeO(001) surface". *Applied Surface Chemistry Science*, 134(1–4), 275–282.

110. Kauffman, G.B., Teter, L.A. & Rhoda, R.N. (1963) Recovery of platinum from laboratory residues. *Inorganic Syntheses*., 7, 232–236.

111. Jha, M.K., Lee, J.-C.; Kimi, M.-S., Jeong, J., Kim, B-S. & Kumar, V. (2013) Hydrometallurgical recovery/recycling of platinum by the leaching of spent catalyst: A review. *Hydrometallurgy*, 133, 23–32.

112. Krebs, R.E. (1998) Platinum. In: *The History and Use of our Earth's Chemical Elements*. Greenwood Press. pp. 124–127. ISBN 0-313-30123-9.

113. Petrucci, R.H. (2007) *General Chemistry: Principles & Modern Applications*, 9th edition. Prentice Hall. p. 606. ISBN 0-13-149330-2.

114. Laramie, J. & Dicks, A. (2003) *Fuel Cell System Explained*. John Wiley and Sons, Hoboken, NJ. ISBN 0-470–84857-X.

115. Wang, C., Daimon, H., Onodera, T., Koda, T. & Sun, S. (2008) A general approach to the size- and shape-controlled synthesis of platinum nanoparticles and their catalytic reduction of oxygen. *Angewandte Chemie International Edition*, 47(19), 3588–3591.

116. Feltham, A.M. & Spiro, M. (1971) Platinized platinum electrodes. *Chemical Reviews*, 71(2), 177–193.

117. Cramer, S.D., Covino, Jr. & Bernard, S. (eds.) (1990) *ASM Handbook*. ASM International, Materials Park, OH. pp. 393–396. ISBN 0-87170-707-1.

118. Emsley, J. (2001) *Nature's Building Blocks*, Hardcover, 1st edition. Oxford University Press, Oxford. p. 363. ISBN 0-19-850340-7.

119. Shelef, M. & Graham, G.W. (1994) Why rhodium in automotive three-way catalysts? *Catalysis Reviews*, 36(3), 433–457.

120. Lide, D.R. (2004) *CRC Handbook of Chemistry and Physics: A Ready-Reference Book of Chemical and Physical Data*. CRC Press, Boca Raton. pp. 4–26. ISBN 0-8493-0485-7.

121. Griffith, W.P. (2003) Bicentenary of four platinum group metals: Osmium and iridium-events surrounding their discoveries. *Platinum Metals Review*, 47(4), 175–183.

122. Loferski, P.J. (2013) Commodity report: Platinum-group metals (PDF). United States Geological Survey.

123. Amatayakul, W. & Ramnäs, O. (2001) Life cycle assessment of a catalytic converter for passenger cars. *Journal of Cleaner Production*, 9(5), 395.

124. Heck, R. & Farrauto, R.J. (2001) Automobile exhaust catalysts. *Applied Catalysis A: General*, 221, 443.

125. Heck, R., Gulati, S. & Farrauto, R.J. (2001) The application of monoliths for gas phase catalytic reactions. *Chemical Engineering Journal*, 82, 149.

126. Roth, J.F. (1975) Rhodium catalysed carbonylation of methanol" (PDF). *Platinum Metals Review*, 19, 1 January, 12–14.

127. Heidingsfeldova, M. & Capka, M. (2003) Rhodium complexes as catalysts for hydrosilylation crosslinking of silicone rubber. *Journal of Applied Polymer Science*, 30(5), 1837.

128. Halligudi, S.B. et al. (1992) Hydrogenation of benzene to cyclohexane catalyzed by rhodium(I) complex supported on montmorillonite clay. *Reaction Kinetics and Catalysis Letters*, 48(2), 547.

129. Fischer, T., Fregert, S., Gruvberger, B. & Rystedt, I. (1984) Contact sensitivity to nickel in white gold. *Contact Dermatitis*, 10(1), 23–24.

130. Lassner, E. & Schubert, W.-D. (1999) Low temperature brittleness. In: *Tungsten: Properties, Chemistry, Technology of the Element, Alloys, and Chemical Compounds*. Springer, London. pp. 20–21. ISBN 978-0-306-45053-2.

131. Weisberg, A.M. (1999) Rhodium plating. *Metal Finishing*, 97(1), 296–299.

132. Smith, W.J. (2007) Reflectors. In: *Modern Optical Engineering: The Design of Optical Systems*. McGraw-Hill. pp. 247–248. ISBN 978-0-07-147687-4.

133. McDonagh, C.P. et al. (1984) Optimum x-ray spectra for mammography: Choice of K-edge filters for tungsten anode tubes. *Physics of Medicine and Biology*, 29(3), 249.

134. Sokolov, A.P., Pochivalin, G.P., Shipovskikh, Y.M., Garusov, Y.V., Chernikov, O.G. & Shevchenko, V.G. (1993) Rhodium self-powered detector for monitoring neutron fluence, energy production, and isotopic composition of fuel. *Atomic Energy*, 74(5), 365–367.

135. Landolt, R.R., Berk, H.W. & Russell, HT. (1972) Studies on the toxicity of rhodium trichloride in rats and rabbits. *Toxicology and Applied Pharmacology*, 21(4), 589–590.

136. Leikin, J.B. & Paloucek, F.P. (2008) *Poisoning and Toxicology Handbook*. Informal Health Care. p. 846. ISBN 978-1-4200-4479-9.

137. Summary. Ruthenium. platinum.matthey.com. p. 9 (2009).

138. Hartman, H.L. & Britton, S.G. (eds.) (1992) *SME Mining Engineering Handbook*. Society for Mining, Metallurgy, and Exploration, Littleton, CO. p. 69. ISBN 978-0-87335-100-3.

139. Harris, D.C. & Cabri, L.J. (1973) The nomenclature of the natural alloys of osmium, iridium and ruthenium based on new compositional data of alloys from world-wide occurrences. *The Canadian Mineralogist*, 12(2), 104–112.

140. George, M.W. (2006) Minerals yearbook: Platinum-group metals (PDF). United States Geological Survey USGS. [accessed 16 September 2008].

141. Commodity report: Platinum-group metals (PDF). United States Geological Survey USGS.

142. Renner, H., Schlamp, G., Kleinwächter, I., Drost, E., Lüschow, H.M., Tews, P., Panster, P., Diehl, M., Lang, J., Kreuzer, T., Knödler, A., Starz, K.A., Dermann, K., Rothaut, J. & Drieselman, R. (2002) Platinum group metals and compounds. In: *Ullmann's Encyclopedia of Industrial Chemistry*. Wiley, New York.

143. Seymour, R.J. & O'Farrelly, J.I. (2001) Platinum-group metals. In: *Kirk Othmer Encyclopedia of Chemical Technology*. Wiley, New York.

144. Gilchrist, R. (1943) The platinum metals. *Chemical Reviews*, 32(3), 277–372.

145. Cotton, S. (1997) *Chemistry of Precious Metals*. Springer-Verlag, New York, LLC. pp. 1–20. ISBN 0-7514-0413-6.

146. Hunt, L.B. & Lever, F.M. (1969) Platinum metals: A survey of productive resources to industrial uses (PDF). *Platinum Metals Review*, 13(4), 126–138.

147. Kolarik, Z. & Renard, E.V. (2005) Potential applications of fission platinoids in industry (PDF). *Platinum Metals Review*, 49(2), 79.

148. Kolarik, Z. & Renard, E.V. (2003) Recovery of value fission platinoids from spent nuclear fuel, Part I: General considerations and basic chemistry (PDF). *Platinum Metals Review*, 47(2), 74–87.

149. Kolarik, Z. & Renard, E.V. (2003) Recovery of value fission platinoids from spent nuclear fuel, Part II: Separation process (PDF). *Platinum Metals Review*, 47(2), 123–131.

150. Brown, G.M. & Butler, J.H. (1997) New method for the characterization of domain morphology of polymer blends using ruthenium tetroxide staining and low voltage scanning electron microscopy (LVSEM). *Polymer*, 38(15), 3937–3945.

151. Greenwood, N.N. & Earnshaw, A. (1997) *Chemistry of the Elements*, 2nd edition. Butterworth-Heinemann, Oxford. ISBN 0-7506-3365-4.

152. Hunt, L.B. & Lever, F.M. (1969) Platinum metals: A survey of productive resources to industrial uses (PDF). *Platinum Metals Review*, 13(4), 126–138.

153. Rao, C. & Trivedi, D. (2005) Chemical and electrochemical depositions of platinum group metals and their applications. *Coordination Chemistry Reviews*, 249(5–6), 613.

154. Weisberg, A. (1999) Ruthenium plating. *Metal Finishing*, 97, 297.

155. Prepared under the direction of the ASM International Handbook Committee; Minges, M.L. (technical chairman) (1989) *Electronic Materials Handbook*. ASM International, Materials Park, OH. p. 184. ISBN 978-0-87170-285-2.

156. Emsley, J. (2003) Ruthenium. In: *Nature's Building Blocks: An A-Z Guide to the Elements*. Oxford University Press, Oxford, England, UK. pp. 368–370. ISBN 0-19-850340-7.

157. Busana, M.G., Prudenziati, M. & Hormadaly, J. (2006) Microstructure development and electrical properties of RuO_2-based lead-free thick film resistors. *Journal of Materials Science: Materials in Electronics*, 17(11), 951.

158. Rane, S., Prudenziati, M. & Morten, B. (2007) Environment friendly perovskite ruthenate based thick film resistors. *Materials Letters*, 61(2), 595.

159. Slade, P.G. (ed.) (1999) *Electrical Contacts: Principles and Applications*. Dekker, New York, NY. pp. 184, 345. ISBN 978-0-8247-1934-0.

160. Schutz, R.W. (1996) Ruthenium enhanced titanium alloys (PDF). *Platinum Metals Review*, 40(2), 54–61.

161. Cardarelli, F. (2008) Dimensionally Stable Anodes (DSA) for chlorine evolution. In: *Materials Handbook: A Concise Desktop Reference*. Springer, London. pp. 581–582. ISBN 978-1-84628-668-1.

162. Varney, M.S. (2000) Oxygen microoptode. In: *Chemical Sensors in Oceanography*. Gordon & Breach, Amsterdam. p. 150. ISBN 978-90-5699-255-2.

163. Hayat, M.A. (1993) Ruthenium red. In: *Stains and Cytochemical Methods*. Plenum Press, New York, NY. pp. 305–310. ISBN 978-0-306-44294-0.

164. Wiegel, T. (1997) *Radiotherapy of Ocular Disease, Ausgabe 13020*. Karger, Basel, Freiburg. ISBN 978-3-8055-6392-5.

165. Gudyanga, F.P., Shonhiwa, C.S. & Chiguvare, Z. (2009) The role of Jatropha in biodiesel production and sustainable development in Africa. In: Muhongo, S.M., Gudyanga, F.P., Enow, A.A. & Nyanganyura, D. (eds.) *Science, Technology, and Innovation for Socio-Economic Development: Success Stories from Africa*. ICSU Regional Office for Africa, Pretoria, South Africa. ISBN 978 0 620 45741 5.

166. Fürstner, A. (2000) Olefin metathesis and beyond. *Angewandte Chemie International Edition*, 39(17), 3012–3043.

167. Perry, R., Kitagawa, K., Grigera, S., Borzi, R., MacKenzie, A., Ishida, K. & Maeno, Y. (2004) Multiple first-order metamagnetic transitions and quantum oscillations in ultrapure $Sr_3Ru_2O_7$. *Physical Review Letters*, 92(16), 66602.

168. Maeno, Y., Rice, T.M. & Sigrist, M. (2001) The intriguing superconductivity of strontium ruthenate (PDF). *Physics Today*, 54, 42.

169. Shlyk, L., Kryukov, S., Schüpp-Niewa, B., Niewa, R. & De Long, L.E. (2008) High-temperature ferromagnetism and tunable semiconductivity of $(Ba, Sr)M_{2\pm x}Ru_{4\mp x}O_{11}$ (M = Fe, Co): A new paradigm for spintronics. *Advanced Materials*, 20(7), 1315.

170. Austin, A. (2007) *The Craft of Silversmithing: Techniques, Projects, Inspiration*. Sterling Publishing Company, Inc. p. 43. ISBN 1600591310.

171. Edwards, H.W. & Petersen, R.P. (1936) Reflectivity of evaporated silver films. *Physics Review*, 9(9), 871.

172. Silver vs. Aluminium. Gemini Observatory. [accessed 1 August 2014].

173. Ronen, Y. (2010) Some remarks on the fusible isotopes. *Annals of Nuclear Energy*, 37(12), 1783–1784.

174. Gold Jewellery Alloys > Utilise Gold. Scientific, industrial and medical applications, products, suppliers from the World Gold Council.

175. Sykora, A. (2010) Rising solar-panel generation means increasing industrial demand for silver. *Kitco News*.

176. Jaworske, D.A. (1997) Reflectivity of silver and silver-coated substrates from 25°C to 800°C (for solar collectors). *Energy Conversion Engineering Conference, 1997: IECEC-97, Proceedings of the 32nd Intersociety 1*. p. 407. ISBN 0-7803-4515-0.

177. Silver in windows and glass: The silver institute, 20 July 2014.

178. Nikitin, P.V., Lam, S. & Rao, K.V.S. (2005) Low cost silver ink RFID tag antennas. *2005 IEEE Antennas and Propagation Society International Symposium (PDF) 2B*. p. 353. ISBN 0-7803-8883-6.

179. Wilson, R.N. (2004) *Reflecting Telescope Optics: Basic Design Theory and Its Historical Development*. Springer. pp. 15, 241. ISBN 3-540-40106-7.

180. Gemini mirror is first with silver lining. Gemini Observatory.

181. Wilson, T. (2007) *Reflecting Telescope Optics I: Basic Design Theory and Its Historical Development*. Springer Science & Business Media. Berlin/Heidelberg, Germany.

182. Rossing, T.D. (1998) *The Physics of Musical Instruments*. Springer. pp. 728–732. ISBN 0-387-98374-0.

183. Meyers, A. (2004) *Musical Instruments: History, Technology, and Performance of Instruments of Western Music*. Oxford University Press, Oxford. p. 132. ISBN 0-19-816504-8.

184. Oliver, C. (1994) Use of immunogold with silver enhancement. *Immunocytochemical Methods and Protocols: Methods in Molecular Biology*, 34, 211–216.

185. Beattie, M. & Taylor, J. (2011) Silver alloy vs. Uncoated urinary catheters: A systematic review of the literature. *Journal of Clinical Nursing*, 20(15–16), 2098–2108.

186. Bouadma, L., Wolff, M. & Lucet, J.C. (2012) Ventilator-associated pneumonia and its prevention. *Current Opinion in Infectious Diseases*, 25(4), August, 395–404.

187. Maillard, J.-Y. & Hartemann, P. (2012) Silver as an antimicrobial: Facts and gaps in knowledge. *Critical Reviews in Microbiology*, 1.

188. Lansdown, A.B.G. (2010) *Silver in Healthcare: Its Antimicrobial Efficacy and Safety in Use*. Royal Society of Chemistry. Burlington House, London, UK. p. 159.

189. Duquesne, S. et al. (2007) *Multifunctional Barriers for Flexible Structure: Textile, Leather, and Paper*. p. 26. ISBN 3-540-71917-2.

190. Shakhashiri, B.Z. (2008) Chemical of the week: Aluminium (PDF). SciFun.org. University of Wisconsin, 17 March.

191. Bauxite and Alumina (PDF). U.S. Geological Survey. p. 2.

192. Dickin, A.P. (2005) In situ cosmogenic isotopes. In: *Radiogenic Isotope Geology*. Cambridge University Press, Cambridge. ISBN 978-0-521-53017-0.

193. Frank, W.B. (2009) Aluminium. In: *Ullmann's Encyclopedia of Industrial Chemistry*. Wiley-VCH.

194. Totten, G.E. & Mackenzie, D.S. (2003) *Handbook of Aluminium*. Marcel Dekker. p. 40. ISBN 978-0-8247-4843-2.

195. Singh, B.J. (2014) *RSM: A Key to Optimize Machining: Multi-Response Optimization of CNC Turning with Al-7020 Alloy*. Anchor Academic Publishing (aap_verlag). ISBN 9783954892099.

196. Aluminium. Encyclopædia Britannica.

197. Millberg, L.S. Aluminium foil. How Products are Made.

198. Lyle, J.P., Granger, D.A. & Sanders, R.E. (2005) Aluminium alloys. In: *Ullmann's Encyclopedia of Industrial Chemistry*. Wiley-VCH.

199. Selke, S. (1994) *Packaging and the Environment: Alternatives, Trends and Solutions*. CRC Press, 21 April. ISBN 9781566761048.

200. Sustainability of aluminium in buildings (PDF). *European Aluminium Association*.

201. Materials in watchmaking: From traditional to exotic. Watches. Infoniac.com.

202. Brown, R.E. (2008) *Electric Power Distribution Reliability*, 2nd edition. CRC Press, 9 September. ISBN 9780849375682.

203. Tokushichi mishima MK magnetic steel. Japan Patent Office, 7 October 2002.

204. Hellweg, P. (1987) *The Insomniac's Dictionary*. Facts on File Publications. p. 115. ISBN 0-8160-1364-0.

205. Wang, H.Z., Leung, D.Y.C., Leung, M.K.H. & Ni, M. (2009) A review on hydrogen production using Aluminium and Aluminium alloys. *Renewable and Sustainable Energy Reviews*, 13(4), 845–854.

206. Ghali, E. (2010) *Corrosion Resistance of Aluminium and Magnesium Alloys: Understanding, Performance, and Testing*. John Wiley and Sons, Hoboken, NJ, 5 May. ISBN 9780470531761.

207. Roe, Jay and Marieli. World's coinage uses 24 chemical elements, Part 1. *World Coin News*, 17 February 1992.

208. Roe, Jay and Marieli Roe. World's coinage uses 24 chemical elements, Part 2. *World Coin News*, 2 March 1992.

209. What is the difference between paper-cone and Aluminium-cone woofers in your bass guitar speaker cabinets? MUSIC Group: All rights reserved. *musicgroup-prod.mindtouch.us*.

210. Skachkov, V.M., Pasechnik, L.A. & Yatsenko, S.P. (2014) Introduction of scandium, zirconium and hafnium into Aluminium alloys: Dispersion hardening of intermetallic compounds with nanodimensional particles (PDF). *Nanosystems: Physics, Chemistry, Mathematics*, 5(4).

211. Ryan, R.S.M. & World Health Organization (2004) *WHO Model Formulary, 2004*. World Health Organization. ISBN 9789241546317.

212. Stevenson C.J. (1989) Occupational skin disease. *Postgraduate Medical Journal*, 65, 374–380.

213. Galbraith, A., Bullock, S., Manias, E., Hunt, B. & Richards, A. (1999) *Fundamentals of Pharmacology: A Text for Nurses and Health Professionals*. Pearson, Harlow. p. 482.

214. Papich, M.G. (2007) Aluminium hydroxide and aluminium carbonate. In: *Saunders Handbook of Veterinary Drugs*, 2nd edition. Saunders, Elsevier, St. Louis, MO. pp. 15–16. ISBN 9781416028888.

215. Brown, H.C. (1951) Reductions by Lithium Aluminium hydride. *Organic Reactions*, 6, 469.

216. Gerrans, G.C. & Hartmann-Petersen, P. (2007) Lithium Aluminium hydride. In: *Sasol Encyclopaedia of Science and Technology*. New Africa Books. p. 143. ISBN 1-86928-384-8.

217. Witt, M. & Roesky, H.W. (2000) Organo Aluminium chemistry at the forefront of research and development (PDF). *Current Science*, 78(4), 410. Archived from the original (PDF) on 6 October 2014.

218. Andresen, A., Cordes, H.G., Herwig, J., Kaminsky, W., Merck, A., Mottweiler, R., Pein, J., Sinn, H. & Vollmer, H.J. (1976) Halogen-free soluble ziegler-catalysts for the polymerization of ethylene. *Angewandte Chemie International Edition*, 15(10), 630.

219. Butterman, C. & Carlin, J.F., Jr. (2003) *Mineral Commodity Profiles: Antimony*. United States Geological Survey.

220. Grund, S.C., Hanusch, K., Breunig, H.J. & Wolf, H.U. (2006) Antimony and antimony compounds. In: *Ullmann's Encyclopedia of Industrial Chemistry*. Wiley-VCH, Weinheim.

221. Norman, N.C. (1998) *Chemistry of Arsenic, Antimony, and Bismuth*. Springer, The Netherlands. p. 45. ISBN 978-0-7514-0389-3.222. Wilson, N.J., Craw, D. & Hunter, K. (2004) Antimony distribution and environmental mobility at an historic antimony smelter site, New Zealand. *Environmental Pollution*, 129(2), 257–266.

223. Weil, E.D. & Levchik, S.V. (2009) Antimony trioxide and related compounds. In: *Flame Retardants for Plastics and Textiles: Practical Applications*, 4 June. ISBN 978-3-446-41652-9. *Polymer International* 57, 431–448. https: doi.org/10.3139/9783446430655.fm.

224. Hastie, J.W. (1973) Mass spectrometric studies of flame inhibition: Analysis of antimony trihalides in flames. *Combustion and Flame*, 21, 49.

225. Kiehne, H.A. (2003) Types of alloys. In: *Battery Technology Handbook*. CRC Press. pp. 60–61. ISBN 978-0-8247-4249-2.

226. De Jong, B.H.W.S., Beerkens, R.G.C. & Van Nijnatten, P.A. (2000) Ullmann's encyclopedia of industrial chemistry. ISBN 3-527-30673-0.

227. O'Mara, W.C., Herring, R.B. & Hunt, L.P. (1990) *Handbook of Semiconductor Silicon Technology*. William Andrew. p. 473. ISBN 978-0-8155-1237-0.

228. Committee on New Sensor Technologies: Materials And Applications, National Research Council (U.S.) (1995) Expanding the vision of sensor materials. p. 68. ISBN 978-0-309-05175-0.

229. Kinch, M.A. (2007) Fundamentals of infrared detector materials. p. 35. ISBN 978-0-8194-6731-7.

230. Willardson, R.K. & Beer, A.C. (1970) Infrared detectors. p. 15. ISBN 978-0-12-752105-3.

231. Jang, H. and Kim, S. (2000) The effects of antimony trisulfide Sb S and zirconium silicate in the automotive brake friction material on friction. *Journal of Wear*.

232. Randich, E., Duerfeldt, W., McLendon, W. & Tobin, W. (2002) A metallurgical review of the interpretation of bullet lead compositional analysis. *Forensic Science International*, 127(3), 174–191.

233. Whelan, J.M., Struthers, J.D. & Ditzenberger, J.A. (1960) Separation of sulfur, selenium, and tellurium from arsenic. *Journal of the Electrochemical Society*, 107(12), 982–985.

234. Grund, S.C., Hanusch, K. & Wolf, H.U. (2005) Arsenic and arsenic compounds. In: *Ullmann's Encyclopedia of Industrial Chemistry*. Wiley-VCH, Weinheim. doi:10.1002/14356007.a03_113.pub2.

235. Bagshaw, N.E. (1995) Lead alloys: Past, present and future. *Journal of Power Sources*, 53, 25–30. Bibcode:1995JPS . . . . 53 . . . 25B. doi:10.1016/0378-7753(94)01973-Y.

236. Nachman, K.E., Graham, J.P., Price, L.B. & Silbergeld, E.K. (2005) Arsenic: A roadblock to potential animal waste management solutions. *Environmental Health Perspectives*, 113(9), 1123–1124.

237. Arsenic (PDF). Agency for Toxic Substances and Disease Registry. Section 5.3, p. 310.

238. Jones, F.T. (2007) A broad view of arsenic. *Poultry Science*, 86(1), 2–14.

239. Gray, T. (2012) Arsenic. In: Gray, T. & Mann, N. (eds.) *Elements: A Visual Exploration of Every Known Atom in the Universe*. Hachette Books, 3 April. ISBN 1579128955.

240. Jakubke, H.-D. & Jeschkeit, H. (eds.) (1994) *Concise Encyclopedia Chemistry* M. Eagleson (trans. rev.). Walter de Gruyter, Berlin.

241. Weast, R. (1984) *CRC, Handbook of Chemistry and Physics*. Chemical Rubber Company Publishing, Boca Raton, FL. p. E110. ISBN 0-8493-0464-4.

242. Beryllium. In: *Periodic Table of Elements*. Los Alamos National Laboratory, Los Alamos. 2010.

243. Veness, R., Ramos, D., Lepeule, P., Rossi, A., Schneider, G. & Blanchard, S. Installation and commissioning of vacuum systems for the LHC particle detectors (PDF). CERN.

244. Wieman, H., Bieser, F., Kleinfelder, S., Matis, H.S., Nevski, P., Rai, G. & Smirnov, N. (2001) A new inner vertex detector for STAR (PDF). *Nuclear Instruments and Methods in Physics Research Section A*, 473, 295.

245. Davis, J.R. (1998) Beryllium. In: *Metals Handbook*. ASM International. pp. 690–691. ISBN 978-0-87170-654-6.

246. Walsh, K.A. (2009) Sources of Beryllium. In: *Beryllium Chemistry and Processing*. ASM International Materials, Ohio. pp. 20–25. ISBN 978-0-87170-721-5.

247. McGraw-Hill contributors (2004) *Concise Encyclopedia of Chemistry* Geller, E. (ed.). McGraw-Hill, New York City. ISBN 0-07-143953-6.

248. Schwartz, M.M. (2002) *Encyclopedia of Materials, Parts, and Finishes*. CRC Press. p. 62. ISBN 1-56676-661-3.

249. Museum of mountain bike art & technology: American bicycle manufacturing.

250. Ward, W. Aluminium-Beryllium. Ret-Monitor.

251. Collantine, K. Banned! Beryllium.

252. Hausner, H.H. (1965) Nuclear properties. In: *Beryllium Its Metallurgy and Properties*. University of California Press, Berkeley. p. 239.

253. Barnaby, F. (1993) *How Nuclear Weapons Spread*. Routledge, Abingdon, UK. p. 35. ISBN 0-415-07674-9.

254. Byrne, J. (2011) *Neutrons, Nuclei, and Matter*. Dover Publications, Mineola, NY. pp. 32–33. ISBN 0486482383.

255. Clark, R.E.H. & Reiter, D. (2005) *Nuclear Fusion Research*. Springer. p. 15. ISBN 3-540-23038-6.

256. Petti, D., Smolik, G., Simpson, M., Sharpe, J., Anderl, R., Fukada, S., Hatano, Y., Hara, M. et al. (2006) JUPITER-II molten salt Flibe research: An update on tritium, mobilization and redox chemistry experiments. *Fusion Engineering and Design*, 81(8–14), 1439.

257. Purdue engineers create safer, more efficient nuclear fuel, model its performance. Purdue University, 27 September 2005.

258. Holleman, A.F., Wiberg, E. & Wiberg, N. (1985) Cadmium. In: *Lehrbuch der Anorganischen Chemie*, Walter de Gruyter & Co. Berlin/NY. 91–100.

259. Case Studies in Environmental Medicine (CSEM) cadmium. Agency for Toxic Substances and Disease Registry.

260. Taki, M. (2013) Chapter 5: Imaging and sensing of cadmium in cells. In: Sigel, A., Sigel, H., Sigel, R.K.O. (eds.) *Cadmium: From Toxicology to Essentiality: Metal Ions in Life Sciences*, Volume 11. Springer. p. 99115. ISBN 978-94-007-5179-8 5.

261. Ayres, R.U., Ayres, L. & Råde, I. (2003) *The Life Cycle of Copper, Its Co-Products and Byproducts*. Springer. pp. 135–141. ISBN 978-1-4020-1552-6.

262. Plachy, J. (1998) Annual average cadmium price (PDF). U.S. Geological Survey. pp. 17–19.

263. Fthenakis, V.M. (2004) Life cycle impact analysis of cadmium in CdTe PV production. *Renewable and Sustainable Energy Reviews*, 8(4), 303.

264. Jiao, Y., Grant, C.A. & Bailey, L.D. (2004) Effects of phosphorus and zinc fertilizer on cadmium uptake and distribution in flax and durum wheat. *Journal of the Science of Food and Agriculture*, 84(8), 777–785.

265. Bettinelli, M., Baroni, U. & Pastorelli, N. (1988) Determination of arsenic, cadmium, lead, antimony, selenium and thallium in coal fly ash using the stabilised temperature platform furnace and Zeeman-effect background correction. *Journal of Analytical Atomic Spectrometry*, 3(7), 1005–1011.

266. Scoullos, M.J. (2001) *Mercury, Cadmium, Lead: Handbook for Sustainable Heavy Metals Policy and Regulation*. Springer. pp. 104–116. ISBN 978-1-4020-0224-3.

267. Buxbaum, G. & Pfaff, G. (2005) Cadmium pigments. In: *Industrial Inorganic Pigments*. Wiley-VCH. pp. 121–123.

268. Smith, C.J.E., Higgs, M.S. & Baldwin, K.R. (1999) Advances to protective coatings and their application to ageing aircraft (PDF). RTO MP-25, 20 April.

269. Scoullos, M.J., Vonkeman, G.H., Thornton, I. & Makuch, Z. (2001) *Mercury, Cadmium, Lead: Handbook for Sustainable Heavy Metals Policy and Regulation*. Springer. ISBN 978-1-4020-0224-3.

270. Lee, C.-H. & Hsi, C.S. (2002) Recycling of scrap cathode ray tubes. *Environmental Science & Technology*, 36(1), 69–75.

271. Miller, L.S. & Mullin, J.B. (1991) Crystalline cadmium sulfide. In: *Electronic Materials: From Silicon to Organics*. Springer. p. 273. ISBN 978-0-306-43655-0.

272. Nambiar, K.R. (2006) Helium-cadmium Laser. In: *Lasers: Principles, Types and Applications*. Newage Publishers. ISBN 978-81-224-1492-9.

273. Cadmium selenium testing for microbial contaminants. NASA, 10 June 2002.

274. Dye, J.L. (1979) Compounds of alkali metal anions. *Angewandte Chemie International Edition*, 18(8), 587–598.

275. Butterman, W.C., Brooks, W.E. & Reese, R.G., Jr. (2004) Mineral commodity profile: Caesium (PDF). United States Geological Survey.

276. Kaner, R. (2003) C&EN: It's elemental: The periodic table-caesium. American Chemical Society.

277. Chemical data: Caesium-Cs. Royal Society of Chemistry.

278. Clark, J. (2005) Flame tests. Chemguide.

279. Burt, R.O. (1993) Caesium and caesium compounds. In: *Kirk-Othmer Encyclopedia of Chemical Technology*, Volume 5, 4th edition. John Wiley and Sons, New York. pp. 749–764. ISBN 978-0-471-48494-3.

280. Benton, W. & Turner, J. (2000) Caesium formate fluid succeeds in North Sea HPHT field trials (PDF). *Drilling Contractor*, May/June, 38–41.

281. Eagleson, M. (ed.) (1994) *Concise Encyclopedia Chemistry* M. Eagleson (ed.). de Gruyter, Berlin. p. 198. ISBN 978-3-11-011451-5.

282. Essen, L. & Parry, J.V.L. (1955) An atomic standard of frequency and time interval: A caesium resonator. *Nature*, 176(4475), 280–282.

283. Reel, M. (2003) Where timing truly is everything. *The Washington Post*, –22 July, p. B1.

284. Haven, M.C., Tetrault, G.A. & Schenken, J.R. (1994) Internal standards. In: *Laboratory Instrumentation*. John Wiley and Sons, New York. p. 108. ISBN 978-0-471-28572-4.

285. Fawcett, E. (1988) Spin-density-wave antiferromagnetism in chromium. *Reviews of Modern Physics*, 60, 209.

286. Wallwork, G.R. (1976) The oxidation of alloys. *Reports on the Progress Physics*, 39(5), 401–485.

287. Holleman, A.F., Wiberg, E. & Wiberg, N. (1985) Chromium. In: *Lehrbuch der Anorganischen Chemie*, 91–100 edition. Walter de Gruyter. pp. 1081–1095. ISBN 3-11-007511-3.

288. National Research Council (U.S.) & Committee on Coatings (1970) High-temperature oxidation-resistant coatings: Coatings for protection from oxidation of superalloys, refractory metals, and graphite. National Academy of Sciences. ISBN 0-309-01769-6.

289. National Research Council (U.S.) & Committee on Biologic Effects of Atmospheric Pollutants (1974) Chromium. National Academy of Sciences. p. 155. ISBN 978-0-309-02217-0.

290. Papp, J.F. Commodity summary 2009: Chromium (PDF). United States Geological Survey.

291. Kotaś, J. & Stasicka, Z. (2000) Chromium occurrence in the environment and methods of its speciation. *Environmental Pollution*, 107(3), 263–283.

292. Kelsall, G.H., House, C.I. & Gudyanga, F.P. (1988) Chemical and electrochemical equilibria and kinetics in aqueous Cr(III)/Cr(II) chloride solutions. *Journal of Electroanalytical Chemistry*, 244(1988), 179–201.

293. Gonzalez, A.R., Ndung'u, K. & Flegal, A.R. (2005) Natural occurrence of hexavalent chromium in the aromas red sands aquifer, California. *Environmental Science and Technology*, 39(15), 5505–5511.

294. Papp, J.F. & Lipin, B.R. (2006) Chromite. In: *Industrial Minerals & Rocks: Commodities, Markets, and Uses*, 7th edition. SME. ISBN 978-0-87335-233-8.

295. Bhadeshia, H.K.D.H. (2003) *Nickel-Based Superalloys*. Cambridge University Press, Cambridge, UK.

296. Edwards, J. (1997) *Coating and Surface Treatment Systems for Metals*. Finishing Publications Ltd. and ASMy International. pp. 66–71. ISBN 0-904477-16-9.

297. Gawne, D.T. & Gudyanga, T.F.P. (1984) Durability of chromium plating under dry sliding conditions. *TRIBOLOGY International*, 17, 123–128.

298. Zhao, J., Xia, L., Sehgal, A., Lu, D., McCreery, R.L. & Frankel, G.S. (2001) Effects of chromate and chromate conversion coatings on corrosion of Aluminium alloy 2024-T3 (PDF). *Surface and Coatings Technology*, 140(1), 51–57.

299. Gawne, D.T. & Gudyanga, T.F.P. (1984) Wear behavior of chromium electrodeposits. In: Strafford, K.N., Datta, P.K. & Googan, C.G. (eds.) *Coatings and Surface Treatment for Corrosion and Wear Resistance*. Ellis Horwood, Chichester, England, Chapter 2. pp. 28–45.

300. Sprague, J.A. & Smidt, F.A. (1994) *ASM Handbook: Surface Engineering*. ASM International. ISBN 978-0-87170-384-2.

301. Dennis, J.K. & Such, T.E. (1993) History of chromium plating. In: *Nickel and Chromium Plating*. Woodhead Publishing. pp. 9–12. ISBN 978-1-85573-081-6.

302. Worobec, M.D. & Hogue, C. (1992) *Toxic substances controls guide: Federal regulation of chemicals in the environment*. BNA Books, Washington, DC. p. 13. ISBN 978-0-87179-752-0.

303. Gettens, R.J. (1966) Chrome yellow. In: *Painting Materials: A Short Encyclopaedia*. Courier Dover Publications. pp. 105–106. ISBN 978-0-486-21597-6.

304. Anger, G. et al. (2005) Chromium compounds. In: *Ullmann's Encyclopedia of Industrial Chemistry*. Wiley-VCH, Weinheim.

305. Marrion, A. (2004) *The Chemistry and Physics of Coatings*. Royal Society of Chemistry. p. 287. ISBN 978-0-85404-604-1.

306. Moss, S.C. & Newnham, R.E. (1964) The chromium position in ruby (PDF). *Zeitschrift fur Kristallographie*, 120(4–5), 359–363.

307. Hingston, J. et al. (2001) Leaching of chromated copper arsenate wood preservatives: A review. *Environmental Pollution*, 11(1), 53–66.

308. Brown, E.M. (1997) A conformational study of collagen as affected by tanning procedures. *Journal of the American Leather Chemists Association*, 92, 225–233.

309. Sreeram, K. & Ramasami, T. (2003) Sustaining tanning process through conservation, recovery and better utilization of chromium. *Resources, Conservation and Recycling*, 38(3), 185–212.

310. Weckhuysen, B.M. & Schoonheydt, R.A. (1999) Olefin polymerization over supported chromium oxide catalysts. *Catalysis Today*, 51(2), 215–221.

311. Twigg, M.V.E. (1989) *The Water-Gas Shift Reaction*. Catalyst Handbook. ISBN 978-0-7234-0857-4.

312. Rhodes, C., Hutchings, G.J. & Ward, A.M. (1995) Water-gas shift reaction: Finding the mechanistic boundary. *Catalysis Today*, 23, 43–58.

313. Lazier, W.A. & Arnold, H.R. (1939) Copper chromite catalyst. *Organic Syntheses*, 19, 31; *Coll.*, 2, 142.

314. Mallinson, J.C. (1993) Chromium dioxide. In: *The Foundations of Magnetic Recording*. Academic Press. ISBN 978-0-12-466626-9.

315. Garverick, L. (1994) *Corrosion in the Petrochemical Industry*. ASM International. ISBN 978-0-87170-505-1.

316. Biggs, T., Taylor, S.S. & Van Der Lingen, E. (2005) The hardening of platinum alloys for potential jewellery application. *Platinum Metals Review*, 49, 2–15.

317. Luborsky, F.E., Mendelsohn, L.I. & Paine, T.O. (1957) Reproducing the properties of alnico permanent magnet alloys with elongated single-domain cobalt-iron particles. *Journal of Applied Physics*, 28(344), 344.

318. Davis, J.R. (2000) *ASM Specialty Handbook: Nickel, Cobalt, and Their Alloys*. ASM International. p. 347. ISBN 0-87170-685-7.

319. Holleman, A.F., Wiberg, E. & Wiberg, N. (2007) Cobalt. In: *Lehrbuch der Anorganischen Chemie*, 102nd edition. de Gruyter. pp. 1146–1152. ISBN 978-3-11-017770-1.

320. Shedd, K.B. Mineral yearbook 2006: Cobalt (PDF). United States Geological Survey.

321. Shedd, K.B. Commodity report 2008: Cobalt (PDF). US Geological Survey.

322. Hawkins, M. (2001) Why we need cobalt. *Applied Earth Science: Transactions of the Institution of Mining & Metallurgy, Section B*, 110(2), 66–71.

323. Zhang, P., Yokoyama, T., Itabashi, O., Wakui, Y., Suzuki, T.M. & Inoue, K. (1999) Recovery of metal values from spent nickel-metal hydride rechargeable batteries. *Journal of Power Sources*, 77(2), 116–122.

324. Armstrong, R.D., Briggs, G.W.D. & Charles, E.A. (1988) Some effects of the addition of cobalt to the nickel hydroxide electrode. *Journal of Applied Electrochemistry*, 18(2), 215–219.

325. Khodakov, A.Y., Chu, W. & Fongarland, P. (2007) Advances in the development of novel cobalt fischer-tropsch catalysts for synthesis of long-chain hydrocarbons and clean fuels. *Chemical Reviews*, 107(5), 1692–1744.

326. Hebrard, F. & Kalck, P. (2009) Cobalt-catalyzed hydroformylation of alkenes: Generation and recycling of the carbonyl species, and catalytic cycle. *Chemical Reviews*, 109(9), 4272–4282.

327. Overman, F. (1852) *A Treatise on Metallurgy*. D. Appleton & Company. pp. 631–635.

328. Muhlethaler, B., Thissen, J. & Muhlethaler, B. (1969) Smalt. *Studies in Conservation*, 14(2), 47–61.

329. Gehlen, A.F. (1803) Ueber die Bereitung einer blauen Farbe aus Kobalt, die eben so schön ist wie Ultramarin. Vom Bürger Thenard. *Neues Allgemeines Journal der Chemie*, 2. H. Frölich. (German translation from L.J. Thénard; Journal des Mines; Brumaire 12 1802; pp. 128–136).

330. Witteveen, H.J. & Farnau, E.F. (1921) Colors developed by cobalt oxides. *Industrial & Engineering Chemistry*, 13(11), 1061–1066.

331. Venetskii, S. (1970) The charge of the guns of peace. *Metallurgist*, 14(5), 334–336.

332. Payne, L.R. (1977) The hazards of cobalt. *Occupational Medicine*, 27(1), 20–25.

333. Committee on Technological Alternatives for Cobalt Conservation. National Research Council (U.S.) (1983) Ground: Coat frit. In: *Cobalt Conservation through Technological Alternatives*. The National Academies Press, Washington, DC. p. 129.

334. Randazzo, R. (2011) A new method to harvest copper. *Azcentral.com*, 19 June.

335. Gordon, R.B., Bertram, M. & Graedel, T.E. (2006) Metal stocks and sustainability. *Proceedings of the National Academy of Sciences*, 103(5), 1209–1214.

336. Greenwood, N.N. & Earnshaw, A. (1997) *Chemistry of the Elements*, 2nd edition. Butterworth-Heinemann. ISBN 0-08-037941-9.

337. Watling, H.R. (2006) The bioleaching of sulphide minerals with emphasis on copper sulphides: A review (PDF). *Hydrometallurgy*, 84(1), 81–108.

338. Bahadir, A.M. & Duca, G. (2009) *The Role of Ecological Chemistry in Pollution Research and Sustainable Development*. Springer, 3 August. ISBN 9789048129034.

339. Emsley, J. (2003) *Nature's Building Blocks: An A-Z Guide to the Elements*. Oxford University Press, Oxford, 11 August. pp. 121–125. ISBN 978-0-19-850340-8.

340. Plimmer, J.R., Ragsdale, N.N. & Gammon, D. (2007) Nonsystematic (contact) fungicides. In: *Ullmann's Agrochemicals*, Wiley-VCH. 2 April. p. 623. ISBN 978-3-527-31604-5.

341. Joseph, G. (1999) *Copper: Its Trade, Manufacture, Use, and Environmental Status* K.J.A. Kundig (ed.). ASM International. pp. 141–192, 331–375.

342. Joseph, G. (1999) *Copper: Its Trade, Manufacture, Use, and Environmental Status* K.J.A. Kundig (ed.). ASM International. p. 348.

343. Energy-efficiency policy opportunities for electric motor-driven systems, International Energy Agency, 2011 Working Paper in the Energy Efficiency Series, by Paul Waide and Conrad U. Brunner, OECD/IEA 2011.

344. Fuchsloch, J. & Brush, E.F. (2007) Systematic design approach for a new series of Ultra-NEMA premium copper rotor motors. *EEMODS 2007 Conference Proceedings, 10–15 June, Beijing*.

345. Kramer, D.A. Mineral commodity summary 2006: Gallium (PDF). United States Geological Survey.

346. Frenzel, M., Ketris, M.P., Seifert, T. & Gutzmer, J. (2016) On the current and future availability of gallium. *Resources Policy*, 47, March, 38–50.

347. Tsai, W.L., Hwu, Y., Chen, C.H., Chang, L.W., Je, J.H., Lin, H.M. & Margaritondo, G. (2003) Grain boundary imaging, gallium diffusion and the fracture behavior of Al-Zn Alloy: An in situ study. *Nuclear Instruments and Methods in Physics Research Section B*, 199, 457–463.

348. Xiao-quan, S., Wen, W. & Bei, W. (1992) Determination of gallium in coal and coal fly ash by electrothermal atomic absorption spectrometry using slurry sampling and nickel chemical modification. *Journal of Analytical Atomic Spectrometry*, 7(5), 761.

349. Gallium in West Virginia coals. *West Virginia Geological and Economic Survey*, 2 March 2002.

350. Font, O., Querol, X., Juan, R., Casado, R., Ruiz, C.R., López-Soler, Á., Coca, P. & Peña, F.G. (2007) Recovery of gallium and vanadium from gasification fly ash. *Journal of Hazardous Materials*, 139(3), 413–423.

351. Headlee, A.J.W. & Hunter, R.G. (1953) Elements in coal ash and their industrial significance. *Industrial and Engineering Chemistry*, 45(3), 548–551.

352. Preston-Thomas, H. (1990) The international temperature scale of 1990 (ITS-90) (PDF). *Metrologia*, 27, 3–10.

353. Magnum, B.W. & Furukawa, G.T. (1990) *Guidelines for Realizing the International Temperature Scale of 1990 (ITS-90)* (PDF). National Institute of Standards and Technology, August. NIST TN 1265.

354. Strouse, G.F. (1999) NIST realization of the gallium triple point. *National Institute of Standards and Technology. Proc. TEMPMEKO*, 1, 147–152.

355. Frenzel, M., Hirsch, T. & Gutzmer, J. (2016) Gallium, germanium, indium, and other trace and minor elements in sphalerite as a function of deposit type: A meta-analysis. *Ore Geology Reviews*, 76, July, 52–78.

356. Frenzel, M., Tolosana-Delgado, R. & Gutzmer, J. (2015) Assessing the supply potential of high-tech metals: A general method. *Resources Policy*, 46, 45–58.

357. Frenzel, M., Mikolajczak, C., Reuter, M.A. & Gutzmer, J. (2017) Quantifying the relative availability of high-tech by-product metals: The cases of gallium, germanium and indium. *Resources Policy*, 52, June, 327–335.

358. Greber, J.F. (2012) Gallium and Gallium compounds. In: *Ullmann's Encyclopedia of Industrial Chemistry*. Wiley-VCH, Weinheim.

359. Vigilante, G.N., Trolano, E. & Mossey, C. (1999) *Liquid Metal Embrittlement of ASTM A723 Gun Steel by Indium and Gallium*. Defense Technical Information Center, June.

360. Emsley, J. (2001) *Nature's Building Blocks*. Oxford University Press, Oxford. pp. 506–510. ISBN 0-19-850341-5.

361. Naumov, A.V. (2007) World market of germanium and its prospects. *Russian Journal of Non-Ferrous Metals*, 48(4), 265–272.

362. Moskalyk, R.R. (2004) Review of germanium processing worldwide. *Minerals Engineering*, 17(3), 393–402.

363. U.S. Geological Survey (2008) Germanium: Statistics and information. U.S. Geological Survey, Mineral Commodity Summaries.

364. Rieke, G.H. (2007) Infrared detector arrays for astronomy. *Annual Review of Astronomy and Astrophysics*, 45(1), 77–115.

365. Brown, R.D., Jr. (2000) Germanium (PDF). U.S. Geological Survey.

366. Ellis, A. & Sorokina, M. *Chapter III: Optical Fiber For Communications* (PDF). Stanford Research Institute, Jenny Stanford Publishing.

367. Thiele, U.K. (2001) The current status of catalysis and catalyst development for the industrial process of poly(ethylene terephthalate) polycondensation. *International Journal of Polymeric Materials*, 50(3), 387–394.

368. Lettington, A.H. (1998) Applications of diamond-like carbon thin films. *Carbon*, 36(5–6), 555–560.

369. Gardos, M.N., Soriano, B.L. & Propst, S.H. (1990) Feldman, A. & Holly, S. (eds.) Study on correlating rain erosion resistance with sliding abrasion resistance of DLC on germanium. *SPIE Conference Proceedings*, 1325 (Mechanical Properties), 99.

370. Bailey, S.G., Raffaelle, R. & Emery, K. (2002) Space and terrestrial photovoltaics: Synergy and diversity. *Progress in Photovoltaics: Research and Applications*, 10(6), 399–406.

371. Crisp, D., Pathare, A. & Ewell, R.C. (2004) The performance of gallium arsenide/germanium solar cells at the Martian surface. *Acta Astronautica*, 54(2), 83–101.

372. Schemel, J.H. (1977) *ASTM Manual on Zirconium and Hafnium*. ASTM International. pp. 1–5. ISBN 978-0-8031-0505-8.

373. Larsen, E., Fernelius, W.C. & Quill, L. (1943) Concentration of hafnium: Preparation of hafnium-free zirconia. *Industrial and Engineering Chemistry Analytical Edition*, 15(8), 512–515.

374. van Arkel, A.E. & de Boer, J.H. (1924) Die Trennung des Zirkoniums von anderen Metallen, einschließlich Hafnium, durch fraktionierte Distillation (The separation of zirconium and hafnium by fractionated distillation). *Zeitschrift für anorganische und allgemeine Chemie* (in German), 141, 289–296.

375. Hedrick, J.B. Hafnium (PDF). United States Geological Survey.

376. Griffith, R.F. (1952) Zirconium and hafnium. Minerals yearbook metals and minerals (except fuels). The first production plants Bureau of Mines. pp. 1162–1171.

377. Gilbert, H.L. & Barr, M.M. (1955) Preliminary investigation of hafnium metal by the kroll process. *Journal of the Electrochemical Society*, 102(5), 243.

378. Holleman, A.F., Wiberg, E. & Wiberg, N. (1985) *Lehrbuch der Anorganischen Chemie*, 91–100 edition. Walter de Gruyter. pp. 1056–1057. ISBN 3-11-007511-3.

379. van Arkel, A.E. & de Boer, J.H. (1925) Darstellung von reinem Titanium-, Zirkonium-, Hafnium- und Thoriummetall (Production of pure titanium, zirconium, hafnium and Thorium metal). *Zeitschrift für anorganische und allgemeine Chemie*, 148, 345–350.

380. Hebda, J. (2001) Niobium alloys and high temperature applications (PDF). CBMM.

381. Rankin, William John (2011) *Minerals, Metals and Sustainability: Meeting Future Material Needs. 1st Edition*. CRC Press. ISBN 9780415684590.

382. Forsberg, C.W., Takase, K. & Nakatsuka, N. (2011) Water reactor. In: Yan, X.L. & Hino, R. (eds.) *Nuclear Hydrogen Production Handbook*. CRC Press. p. 192. ISBN 978-1-4398-1084-2.

383. Maslenkov, S.B., Burova, N.N. & Khangulov, V.V. (1980) Effect of hafnium on the structure and properties of nickel alloys. *Metal Science and Heat Treatment* 22(4), 283–285.

384. Beglov, V.M., Pisarev, B.K. & Reznikova, G.G. (1992) Effect of boron and hafnium on the corrosion resistance of high-temperature nickel alloys. *Metal Science and Heat Treatment*, 34(4), 251–254.

385. Voitovich, R.F. & Golovko, É.I. (1975) Oxidation of hafnium alloys with nickel. *Metal Science and Heat Treatment*, 17(3), 207–209.

386. Patchett, P.J. (1983) Importance of the Lu-Hf isotopic system in studies of planetary chronology and chemical evolution. *Geochimica et Cosmochimica Acta*, 47(1), 81–91.

387. Söderlund, U., Patchett, P.J., Vervoort, J.D. & Isachsen, C.E. (2004) The 176Lu decay constant determined by Lu-Hf and U-Pb isotope systematics of Precambrian mafic intrusions. *Earth and Planetary Science Letters*, 219(3–4), March, 311–324.

388. Blichert-Toft, J. & Albarède, F. (1997) The Lu-Hf isotope geochemistry of chondrites and the evolution of the mantle-crust system. *Earth and Planetary Science Letters*, 148(1–2), April, 243–258.

389. Patchett, P.J. & Tatsumoto, M. (1980) Lu-Hf total-rock isochron for the eucrite meteorites. *Nature*, 288(5791), 11 December, 571–574.

390. Kinny, P.D. (2003) Lu-Hf and Sm-Nd isotope systems in zircon. *Reviews in Mineralogy and Geochemistry*, 53(1), 1 January, 327–341.

391. Albarède, F., Duchêne, S., Blichert-Toft, J., Luais, B., Télouk, P. & Lardeaux, J.-M. (1997) The Lu-Hf dating of garnets and the ages of the Alpine high-pressure metamorphism. *Nature*, 387(6633), 5 June, 586–589.

392. Haynes, W.M. (2010) *CRC Handbook of Chemistry and Physics: A Ready-Reference Book of Chemical and Physical Data*. Lide, D.R. (ed.). CRC Press. ISBN 978-1-4398-2077-3.

393. Alfantazi, A.M. & Moskalyk, R.R. (2003) Processing of indium: A review. *Minerals Engineering*, 16(8), 687–694.
394. Schwarz-Schampera, U. & Herzig, P.M. (2002) *Indium: Geology, Mineralogy, and Economics*. Springer. ISBN 978-3-540-43135-0.
395. Indium price supported by LCD demand and new uses for the metal (PDF). *Geology.com*.
396. Bachmann, K.J. (1981) Properties, preparation, and device applications of indium phosphide. *Annual Review of Materials Science*, 11, 441–484.
397. Bhuiyan, G., Hashimoto, A. & Yamamoto, A. (2003) Indium nitride (InN): A review on growth, characterization, and properties. *Journal of Applied Physics*, 94(5), 2779.
398. Powalla, M. & Dimmler, B. (2000) Scaling up issues of CIGS solar cells. *Thin Solid Films*, 361–362, 540–546.
399. Schubert, E.F. (2003) *Light-Emitting Diodes*. Cambridge University Press, Cambridge. p. 16. ISBN 0-521-53351-1.
400. Shenai, D.V., Timmons, M.L., DiCarlo, R.L., Jr. & Marsman, C.J. (2004) Correlation of film properties and reduced impurity concentrations in sources for III/V-MOVPE using high-purity trimethylindium and tertiarybutylphosphine. *Journal of Crystal Growth*, 272(1–4), 603–608.
401. Geological Survey (U.S.) (2010) *Minerals Yearbook, 2008, V. 1, Metals and Minerals*. Government Printing Office. pp. 80–81. ISBN 978-1-4113-3015-3.
402. Rabilloud, G. (1997) *High-Performance Polymers: Conductive Adhesives*. Editions TECHNIP. p. 263. ISBN 2-7108-0716-5.
403. Weissler, G.L. (ed.) (1990) *Vacuum Physics and Technology*. Acad. Press, San Diego. p. 296. ISBN 978-0-12-475914-5.
404. Reading, M., Hourston, D.J. (2006) *Modulated Temperature Differential Scanning Calorimetry*. Springer. p. 245. ISBN 978-1-4020-3749-8.
405. Surmann, P. & Zeyat, H. (2005) Voltammetric analysis using a self-renewable non-mercury electrode. *Analytical and Bioanalytical Chemistry*, 383(6), November, 1009–1013.
406. Downs, A.J. (1993) *Chemistry of Aluminium, Gallium, Indium, and Thallium*. Springer. pp. 89 and 106. ISBN 978-0-7514-0103-5.
407. Preston-Thomas, H. (1990) Procès-Verbaux du Comité International des Poids et Mesures. *Metrologia*, 27(1), 3–10.
408. Scoullos, M.J. (2001) Other types of cadmium alloys. *Mercury, cadmium, Lead: Handbook for Sustainable Heavy Metals Policy and Regulation*, 31 December. p. 222. ISBN 978-1-4020-0224-3.
409. Berger, H., National Bureau Of Standards, United States & Committee E-7 On Nondestructive Testing, American Society for Testing and Materials (1976) Image detectors for other neutron energies. *Practical Applications of Neutron Radiography and Haging: A Symposium*. pp. 50–51.
410. Tong, X.C. (2011) *Advanced Materials for Thermal Management of Electronic Packaging*. Springer. p. 323. ISBN 978-1-4419-7759-5.
411. Van Nostrand, D., Abreu, S.H., Callaghan, J.J., Atkins, F.B., Stoops, H.C. & Savory, C.G. (1988) In-111-labeled white blood cell uptake in noninfected closed fracture in humans: Prospective study. *Radiology* (Radiological Society of North America, Inc.), 167(2, May), 495–498.
412. Verhoeven, J.D. (1975) *Fundamentals of Physical Metallurgy*. Wiley, New York. p. 326.
413. Bramfitt, B.L. & Benscoter, A.O. (2002) The Iron Carbon phase diagram. *Metallographer's Guide: Practice and Procedures for Irons and Steels*. ASM International. pp. 24–28. ISBN 978-0-87170-748-2.
414. Dlouhy, A.C. & Outten, C.E. (2013) *Metallomics and the Cell: Metal Ions in Life Sciences*, Volume 12, Banci, L. (ed.). Springer, London.
415. Yee, G.M. & Tolman, W.B. (2015) *Sustaining Life on Planet Earth: Metalloenzymes Mastering Dioxygen and Other Chewy Gases: Metal Ions in Life Sciences*, Volume 15, Kroneck, P.M.H. & Torres, M.E.S. (eds.). Springer, London. pp. 131–204.
416. Kolasinski, K.W. (2002) Where are heterogenous reactions important. In: *Surface Science: Foundations of Catalysis and Nanoscience*. John Wiley and Sons, Hoboken, NJ. pp. 15–16. ISBN 978-0-471-49244-3.

417. McKetta, J.J. (1989) Nitrobenzene and nitrotoluene. In: *Encyclopedia of Chemical Processing and Design, Volume 31: Natural Gas Liquids and Natural Gasoline to Offshore Process Piping: High Performance Alloys*. CRC Press. pp. 166–167. ISBN 978-0-8247-2481-8.

418. Wildermuth, E., Stark, H., Friedrich, G., Ebenhöch, F.L., Kühborth, B., Silver, J. & Rituper, R. (2000) Iron compounds. *Ullmann's Encyclopedia of Industrial Chemistry*. ISBN 3-527-30673-0.

419. Fleet, M.E. & Mumin, A.H. (1997) Gold-bearing arsenian pyrite and marcasite and arsenopyrite from Carlin Trend gold deposits and laboratory synthesis (PDF). *American Mineralogist*, 82, 182–193.

420. Terkel Rosenqvist (2004) *Principles of Extractive Metallurgy*, 2nd edition. Papir Academic Press. p. 52. ISBN 82-519-1922-3.

421. Cylindrical Primary Lithium [battery]. *Lithium-Iron Disulfide (Li-FeS_2) (PDF). Handbook and Application Manual. Energizer Corporation*, 19 September 2017.

422. Ellmer, K. & Tributsch, H. (2000) Iron disulfide (Pyrite) as photovoltaic material: Problems and opportunities. *Proceedings of the 12th Workshop on Quantum Solar Energy Conversion – (QUANTSOL 2000)*, 11 March.

423. The Principles Underlying Radio Communication. *U.S. Army Signal Corps. Radio Pamphlet. 40*. 10 December 1918. Section 179. pp. 302–305.

424. Lee, T.H. (2004) *The design of radio frequency integrated circuits*, 2nd edition. Cambridge University Press, Cambridge, UK. pp. 4–6.

425. Wadia, C., Alivisatos, A.P. & Kammen, D.M. (2009) Materials availability expands the opportunity for large-scale photovoltaics deployment. *Environmental Science & Technology*, 43(6), 7.

426. Sanders, R. (2009) *Cheaper Materials Could Be Key to Low-Cost Solar Cells*. University of California, Berkeley, CA, 17 February.

427. Hesse, R.W. (2007) *Jewelrymaking through History: An Encyclopedia*. Greenwood Publishing Group. p. 15. ISBN 0-313-33507-9.

428. Young, C.A., Taylor, P.R. & Anderson, C.G. (2008) Hydrometallurgy 2008. *Proceedings of the Sixth International Symposium*. SME. ISBN 9780873352666.

429. Da Silva, G. (2004) Kinetics and mechanism of the bacterial and ferric sulphate oxidation of galena. *Hydrometallurgy*, 75, 99.

430. Metropolitan Museum of Art. (2005) *The Art of Medicine in Ancient Egypt*. The Museum, New York. p. 10. ISBN 1-58839-170-1.

431. Lithium (PDF) (2016) US Geological Survey (USGS).

432. Elena Brandaleze, Gustavo Di Gresia, Leandro Santini, Alejandro Martin and Edgardo Benavidez (2012); "Mould fluxes in the steel continuous casting process". Doi: 10.5772/50874.

433. Lu, Y.Q., Zhang, G.D., Jiang, M.F., Liu, H.X. & Li, T. (2011) Effects of Li_2CO_3 on properties of mould flux for high speed continuous casting. *Materials Science Forum*, 675–677, 877–880.

434. Haupin, W. (1987) Chemical and physical properties of the hall-héroult electrolyte. In: Mamantov, G. & Marassi, R. (eds.) *Molten Salt Chemistry: An Introduction and Selected Applications*. Springer. p. 449.

435. Davis, J.R. ASM International. Handbook Committee (1993) Aluminium and Aluminium alloys. *ASM International*. p. 121. ISBN 978-0-87170-496-2.

436. Koch, E.-C. (2004) Special materials in pyrotechnics: III. Application of lithium and its compounds in energetic systems. *Propellants, Explosives, Pyrotechnics*, 29(2), 67–80.

437. Hammond, C.R. (2000) *The Elements, in Handbook of Chemistry and Physics*, 81st edition. CRC Press. ISBN 0-8493-0481-4.

438. Wiberg, E., Wiberg, N. & Holleman, A.F. (2001) *Inorganic Chemistry Archived 18 June 2016 at the Wayback Machine*. Academic Press. p. 1089. ISBN 0-12-352651-5.

439. Mulloth, L.M. & Finn, J.E. (2005) Air quality systems for related enclosed spaces: Spacecraft air. *The Handbook of Environmental Chemistry*, 4H, 383–404.

440. Markowitz, M.M., Boryta, D.A. & Stewart, H. (1964) Lithium perchlorate oxygen candle: Pyrochemical source of pure oxygen. *Industrial & Engineering Chemistry Product Research and Development*, 3(4), 321–330.

441. Point Defects in Lithium Fluoride Films Induced by Gamma Irradiation. *Proceedings of the 7th International Conference on Advanced Technology & Particle Physics: (ICATPP-7): Villa Olmo, Como, Italy.* 2001. World Scientific. 2002. p. 819. ISBN 981-238-180-5.

442. Sinton, W.M. (1962) Infrared spectroscopy of planets and stars. *Applied Optics*, 1(2), 105.

443. Yurkovetskii, A.V., Kofman, V.L. & Makovetskii, K.L. (2005) Polymerization of 1,2-dimethylenecyclobutane by organolithium initiators. *Russian Chemical Bulletin*, 37(9), 1782–1784.

444. Quirk, R.P. & Cheng, P.L. (1986) Functionalization of polymeric organolithium compounds: Amination of poly(styryl)lithium. *Macromolecules*, 19(5), 1291–1294.

445. Stone, F.G.A. & West, R. (1980) *Advances in Organometallic Chemistry.* Academic Press. p. 55. ISBN 0-12-031118-6.

446. Bansal, R.K. (1996) *Synthetic Approaches in Organic Chemistry*, 1st edition. Jones and Bartlett Learning. p. 192. ISBN 0-7637-0665-5.

447. Hughes, T.G., Smith, R.B. & Kiely, D.H. (1983) Stored chemical energy propulsion system for underwater applications. *Journal of Energy*, 7(2), 128–133.

448. Agarwal, A. (2008) *Nobel Prize Winners in Physics.* APH Publishing. p. 139. ISBN 81-7648-743-0.

449. National Research Council (U.S.). Committee on Separations Technology and Transmutation Systems (1996) *Nuclear Wastes: Technologies for Separations and Transmutation.* National Academies Press. p. 278. ISBN 0-309-05226-2.

450. Afenya, P.M. (1991) Treatment of carbonaceous refractory gold ores. *Minerals Engineering*, 4(7–11), 1043–1055.

451. Allan, G.C. & Woodcock, J.T. (2001) A review of the flotation of native gold and electrum. *Minerals Engineering*, 14(9), 931–962.

452. de Klerk, A. (2013) Fischer-Tropsch process. In: *Kirk-Othmer Encyclopedia of Chemical Technology.* Wiley-CH, Weinheim. ISBN 978-0471238966.

453. Bulatovic, S.M. (1997) Flotation behavior of gold during processing of porphyry copper-gold ores and refractory gold bearing sulphides. Miner. Eng., 10(9), 895–908.

454. Celep, O., Alp, L., Deveci, H. & Vicil, M. (2009) Characterization of refractory behavior of complex gold/silver ore by diagnostic leaching. Transactions of Nonferrous Metals Society of China, 19, 707–713.

455. Cybulski, A. & Moulijin, J.A. (eds.) (2005). Structured Catalysts and Reactors, 2nd edition. CRC Press. p. 35. ISBN 978-0-8247-2343-9.

456. Mackiw, V.N., Benz, T.W. & Evans, D.J. (1966) A review of recent developments in pressure metallurgy. *Metallurgical Review*, 11(109), 143–158.

457. Kim, S.H., Kim, S.J. & Oh, S.M. (1999) *Chemistry of Material.*, 11, 557.

458. Chadov, S., Qi, X.-L., Kubler, J., Fecher, G.H., Felser, C. & Zhang, S.-C. (2010) Tunable multifunctional topological insulators in ternary heusler compounds. Nature Materials, 9, 541.

459. Gathje, J.J., Oberg, K.C. & Simmons, G. (1995) Pressure oxidation process development: Beware of lab results. Minerals Engineering, 47(6), 520–523.

460. Ohzuku, T. & Ueda, A. (1995) Journal of Electrochemical Society, 142, 1431.

461. Zhou, X., Weston, J., Chalkova, E., Hofmann, M.A., Ambler, C.M., Allcock, H.R. & Lvov, S.N. (2003) High temperature transport properties of polyphosphazene membranes for direct methanol fuel cells. Electrochemica Acta, 48, 2173–2180.

462. Abotsi, G.M.K. & Osseo-Asare, K. (1986) Surface chemistry of carbonaceous gold ores: I. Characterization of the carbonaceous matter and absorptive behavior in aurocyanide solution. International Journal of Mineral Processing, 18, 217–236.

463. Thackeray, M.M., Kang, S.-H., Johnson, C.S., Vaughey, J.T. & Hackney, S.A. (2006) *Electrochemical Communications*, 8, 1531.

464. Beale, G. (2016) Water management in gold ore processing. In Adams, M.D. (ed.) *Gold Ore Processing: Project Development and Operations*, 2nd edition. Elsevier. ISBN 978-0-444-63658-4.

465. Malherbe, J.B., Friedland, E. & van der Berg, N.G. (2008) "Ion beam analysis of materials in the PBMR reactor." *Nuclear Instruments & Methods in Physics* B266(8) 1373–1377.

466. Mahlangu, T., Gudyanga, F.P. & Simbi, D.J. (2006) Reductive leaching of stibnite (Sb_2S_3) flotation concentrates using metallic iron in a hydrochloric acid medium. I: Thermodynamics. *Hydrometallurgy*, 84, 192–203.

467. Fernandez, R.R., Sohn, H.Y. & LeVier, K.M. (2000) Process for treating refractory gold ores by roasting under oxidizing conditions. *Mineral and Metallurgical Processing*, 17(1), 1–6.

468. Diat, O. & Gebel, G. (2008) "Proton channels" *Nature Materials*, 7, 13–14.

469. Schulz, H. (1999) Short history and present trends of Fischer-Tropsch synthesis. *Applied Catalysis A: General*, 186, 3–12.

470. Kusema, B.T., Campo, B.C., Maki-Arvela, P., Salmi, T. & Yu, M.D. (2010) Selective catalytic oxidation of arabinose – A comparison of gold and palladium catalyst, *Applied. Catalysis A: General; General*, 386, 101–108.

471. Wolff, I.M. & Hill, P.J. (2000) Platinum metals-based intermetallics for high temperature service, *Platinum Metals Review*, 44(4), 158–166.

472. Gudyanga, F.P., Mahlangu, T., Chifamba, J. & Simbi, D.J. (1998) Reductive-oxidative pretreatment of a stibnite flotation concentrate: Thermodynamic and kinetic considerations. *Minerals Engineering*, 11(6), 563–580.

473. Cornish, L.A., Suss, R., Douglas, A., Chown, L.H. & Glaner, L. (2009) *Plat. Metals Rev.*, 53(1), 2–10.

474. Roy, R., Agrawal, D.K. & McKinstry, H.A. (1989) *Annu. Rev. Mater. Sci.*, 19, 58–59.

475. Luna-Sanchez, R.M. & Lapidus, G.T. (2000) Cyanidation kinetics of silver sulphide. *Hydrometallurgy*, 56, 171–188.

476. Murr, L.E. & Barry, V.K. (1979) Observations of a natural thermophilic micro-organism in the leaching of a large experimental copper-bearing waste body. *Metallurgical and Materials Transactions B*, 10B, 523–531.

477. Yang, S. & Langmuir, R.P. (2011) Nanoparticle flotation collectors II. *The Role of Nanoparticle Hydrophobicity*, 27(18), 11409–11415; Copyright © 2011 American Chemical Society.

478. Flemming, C.A. (2016) Cyanide recovery in Mike D. Adams. In: *Gold Ore Processing: Project Development and Operations*, 2nd edition. Elsevier. ISBN 978-0-444-63658-4.

479. Shuro, I., Umemoto, M., Todaka, Y. & Yokoyama, S. (2010) Phase transformation and allealing behavior of SUS 304 austenitic stainless steel deformed by high temperature torsion. *Materuaks Science Forum*, 654–656, 334–337.

480. Gallagher, N.P., Hentrix, J.L., Molisavljevic, E.B. & Nelson, J.H. (1989) Affinity of carbon for gold complexes: Dissolution of finely disseminated gold using a flow electrochemical cell. *Journal of Electrochemical Society*, 136, 2546–2551.

481. Kreuer, K.D. (1996) Proton conductivity: Materials and applications. *Chemistry of Materials*, 8, 610–641.

482. Bell, J.G., et al. (1977) *Journal of Applied Physics*, 10, 1379–1387.

483. Celep, O., Alp, I., Paktung, D. & Thibault, Y. (2011) Implementation of sodium hydroxide pretreatment for refractory antimonial gold and silver ores. *Hydrometallurgy*, 108, 109–114.

484. Douglas, A., Hill, P.J., Cornish, L.A. & Suss, R. (2009) Platinum-based alloys and coatings materials for the future. *Platinum Materials Review*, 53(2), 69–77.

485. Arehart, G.B., Chryssoulis, S.I. & Kesler, S.E. (1993). Gold and arsenic in iron sulfides from sediment-hosted disseminated gold deposits: implications for depositional process. *Economic Geology*, 88, 177–185.

486. Aricoa, A.S., Srinivasan, S. & Antonuccia, V. (2001). DMFCs: From fundamental aspects to technology development. *Fuel Cells*, 1, 133.

487. Dai, X. & Jeffry, M.I. (2006) The effect of sulfide minerals on the leaching of gold in aerated cyanide solution. *Hydrometallurgy*, 82, 118–125.

488. Ren, X., Zelenay, P., Thomas, S., Davey, J. & Gottesfeld, S. (2010) Recent advances in direct methanol fuel cell at Los Almos National Laboratory. *Journal of Power Sources*, 86, 111.

489. Feng, D. & van Deventer, J.S.J. (2001) Preg-robbing phenomena in the thiosulfate leaching of gold ores. *Minerals Engineering*, 14(11), 1387–1402.

490. Prasad, P.N. (2004) *Nanophotonics*, John Wiley, New Jersey.

491. Chyrssoulis, S.L. & Cabri, L.J. (1990) Significance of gold mineralogical balance in mineral processing. *Transactions of the Institute of Mining and Metallurgy (Section C. Mineral Processing and Extractive. Metallurgy)*, 99, C1–C9.

492. Wang, S., Jarrett, B.R., Kauzlarich, S.M. & Louie, A.Y. (2007) Core/shell quantum dots with high relaxivity and photoluminescence for multimodality imaging. *Journal of American Chemical Society*, 129, 3848–3856.

493. Gudyanga, F.P., Mahlangu, T., Chifamba, J. & Simbi, D.J. (1999) Reductive decomposition of galena (PbS) using Cr (II) ionic species in an aqueous chloride medium for silver (Ag) recovery. *Minerals Engineering*, 12(7), 787–797.

494. Hao, Y., Lai, Q., Xu, Z., Liu, X. & Ji, X. (2005) Synthesis by TEA sol-gel method and electrochemical properties of $Li_4Ti_5O_{12}$ anode material for lithium-ion battery. *Solid State Ionics*, 176, 1201–1206.

495. Cornish, L.A., Suss, R., Chown, L.H. & Glaner, L. (2009). New Pt-based alloys for high temperature applicatioin aggressive environments: The next stage. *Platinum Metals Review*, 53(3), 155–163.

496. Shao, G., Yin, G., Wang, Z. & Gao, Y. (2007) Proton exchange membrane fuel cell from low temperature to high temperature: materials challenge. *Journal of Power Sources*, 167, 235.

497. Zako, T., et al. (2010) Development of near infrared-fluorescent nanoparticles and applications for cancer diagnosis and therapy. *Journal of Nanomaterials* 2010, 1–8.

498. Talapin, D.V., Gaponik, N., Borchert, H., Rogach, A.L., Haase, M. & Weller, H. (2002) Etching of colloidal InP nanocrystals with fluorides: photochemical nature of the process resulting in high photoluminescence efficiency. *Journal of Physics and Chemistry B*, 106, 12659–12663.

499. Kelsall, G.H. and Gudyanga, F.P. (1992) Electrochemical reactor design considerations for simultaneous tin electrowinning and Cr2+ electrogeneration. *DECHEMA Monographs*, 23, 167–185.

500. Barnaby, F. (1993) *How Nuclear Weapons Spread: Nuclear-Weapon Proliferation in the 1990s*. Routledge, Abingdon, UK. p. 39. ISBN 0-415-07674-9.

501. Baesjr, C. (1974) The chemistry and thermodynamics of molten salt reactor fuels. *Journal of Nuclear Materials*, 51, 149–162.

502. Pal, U.B. & Powell, A.C. (2007) The use of solid-oxide-membrane technology for electrometallurgy. *JOM*, 59(5), 44–49.

503. Birbilis, N., Williams, G., Gusieva, K., Samaniego, A., Gibson, M.A. & McMurray, H.N. (2013) Poisoning the corrosion of magnesium. *Electrochemistry Communications*, 34, 295.

504. Baker, H.D.R. & Avedesian, M. (1999) *Magnesium and Magnesium Alloys*. Materials Information Society, Materials Park, OH. p. 4. ISBN 0-87170-657-1.

505. Amundsen, K., Aune, T.K., Bakke, P., Eklund, H.R., Haagensen, J.O., Nicolas, C. et al. (2002) Magnesium. In: *Ullmann's Encyclopedia of Industrial Chemistry*. Wiley-VCH. ISBN 3527306730.

506. Aghion, E. & Bronfin, B. (2000) Magnesium alloys development towards the 21st century. *Materials Science Forum*, 350–351, 19–30.

507. Bronfin, B. et al. (2007) Elektron 21 specification. In: Kainer, K. (ed.) *Magnesium: Proceedings of the 7th International Conference on Magnesium Alloys and Their Applications*. Wiley, Weinheim, Germany. p. 23. ISBN 978-3-527-31764-6.

508. AAHS, American aviation historical society (1999) *AAHS Journal*, 44–45.

509. Dreizin, E.L., Berman, C.H. & Vicenzi, E.P. (2000) Condensed-phase modifications in magnesium particle combustion in air. *Scripta Materialia*, 122, 30–42.

510. Magnesium (Powder). *International Programme on Chemical Safety (IPCS), IPCS INCHEM, April 2000*.

511. Linsley, T. (2011) Properties of conductors and insulators. In: *Basic Electrical Installation Work*. CRC Press, Routledge. p. 362. ISBN 978-0-08-096628-1.

512. Lide, D.R. (2004) Magnetic susceptibility of the elements and inorganic compounds. In *Handbook of Chemistry and Physics (PDF)*. CRC Press. ISBN 0-8493-0485-7.

513. Corathers, L.A. & Machamer, J.F. (2006) Manganese. In: *Industrial Minerals & Rocks: Commodities, Markets, and Uses*, 7th edition. SME. pp. 631–636. ISBN 978-0-87335-233-8.

514. Zhang, W. & Cheng, C.Y. (2007) Manganese metallurgy review. Part I: Leaching of ores/secondary materials and recovery of electrolytic/chemical manganese dioxide. *Hydrometallurgy*, 89(3–4), 137–159.

515. Corathers, L.A. (2008) *2006 Minerals Yearbook: Manganese* (PDF), June. United States Geological Survey, Washington, DC.

516. Verhoeven, J.D. (2007) *Steel Metallurgy for the Non-Metallurgist*. ASM International, Materials Park, OH. pp. 56–57. ISBN 978-0-87170-858-8.

517. Dastur, Y.N. & Leslie, W.C. (1981) Mechanism of work hardening in Hadfield manganese steel. *Metallurgical Transactions A*, 12(5), 749.

518. Corathers, L.A. (2009) *Mineral Commodity Summaries 2009: Manganese* (PDF). United States Geological Survey.

519. Stansbie, J.H. (2007) *Iron and Steel*. Read Books. pp. 351–352. ISBN 978-1-4086-2616-0.

520. Bradt, G.S., Clauser, H.R. & Vaccari, J.A. (2002) *Materials Handbook: An Encyclopedia for Managers, Technical Professionals, Purchasing and Production Managers, Technicians, and Supervisors*. McGraw-Hill, New York, NY. pp. 585–587. ISBN 978-0-07-136076-0.

521. Tweedale, G. (1985) Sir Robert Abbott Hadfield F.R.S. (1858–1940), and the discovery of manganese steel geoffrey tweedale. *Notes and Records of the Royal Society of London*, 40(1), 63–74.

522. Kaufman, J.G. (2000) Applications for aluminium alloys and tempers. In: *Introduction to Aluminium Alloys and Tempers*. ASM International. pp. 93–94. ISBN 978-0-87170-689-8.

523. Shepard, A.O. (1956) Manganese and iron: Manganese paints. In: *Ceramics for the Archaeologist*. Carnegie Institution of Washington. pp. 40–42. ISBN 978-0-87279-620-1.

524. Mccray, W.P. (1998) Glassmaking in renaissance Italy: The innovation of venetian cristallo. *Journal of the Minerals, Metals and Materials Society*, 50(5), 14.

525. Graham, L.A., Fout, A.R., Kuehne, K.R., White, J.L., Mookherji, B., Marks, F.M., Yap, G.P.A., Zakharov, L.N., Rheingold, A.L. & Rabinovich, D. (2005) Manganese(I) poly(mercaptoimidazolyl) borate complexes: Spectroscopic and structural characterization of MnH-B interactions in solution and in the solid state. *Dalton Transactions*, (1), 171–180.

526. Dell, R.M. (2000) Batteries fifty years of materials development. *Solid State Ionics*, 134, 139–158.

527. Chow, N., Nacu, A., Warkentin, D., Aksenov, I. & Teh, H. (2010) The recovery of manganese from low grade resources: Bench scale metallurgical test program completed (PDF). *Kemetco Research Inc.*

528. Preisler, E. (1980) Moderne Verfahren der Großchemie: Braunstein. *Chemie in unserer Zeit*, 14(5), 137–148.

529. Kuwahara, R.T., Skinner III, R.B. & Skinner Jr., R.B. (2001) Nickel coinage in the United States. *Western Journal of Medicine*, 175(2), 112–114.

530. The CRB Commodity Yearbook (annual). (2000) p. 173. ISSN 1076-2906.

531. Leopold, B.R. (2002) Chapter 3: Manufacturing processes involving mercury: Use and release of mercury in the United States. (PDF). *National Risk Management Research Laboratory, Office of Research and Development, U.S. Environmental Protection Agency, Cincinnati, Ohio*.

532. Gibson, B.K. (1991) Liquid mirror telescopes: History. *Journal of the Royal Astronomical Society of Canada*, 85, 158.

533. Brans, Y.W. & Hay, W.W. (1995) *Physiological Monitoring and Instrument Diagnosis in Perinatal and Neonatal Medicine*. CUP Archive. p. 175. ISBN 0-521-41951-4.

534. Hammond, C.R. (2005) The elements. In: Lide, D.R. (ed.) *CRC Handbook of Chemistry and Physics*, 86th edition. CRC Press, Boca Raton, FL. ISBN 0-8493-0486-5.

535. Zoski, C.G. (2007) *Handbook of Electrochemistry*. Elsevier Science, 7 February. ISBN 0-444-51958-0.

536. Kissinger, P. & Heineman, W.R. (1996) *Laboratory Techniques in Electroanalytical Chemistry*, 2nd edition, Revised and Expanded, 2nd edition. CRC Press, 23 January. ISBN 0-8247-9445-1.

537. Hopkinson, G.R., Goodman, T.M. & Prince, S.R. (2004) *A Guide to the Use and Calibration of Detector Array Equipment.* SPIE Press. p. 125. ISBN 0-8194-5532-6.

538. Howatson, A.H. (1965) Chapter 8. In: *An Introduction to Gas Discharges.* Pergamon Press, Oxford. ISBN 0-08-020575-5.

539. Milo, G.E. & Casto, B.C. (1990) *Transformation of Human Diploid Fibroblasts.* CRC Press. p. 104. ISBN 0-8493-4956-7.

540. Shionoya, S. (1999) *Phosphor Handbook.* CRC Press. p. 363. ISBN 0-8493-7560-6.

541. Lide, D.R. (ed.) (1994) Molybdenum. In: *CRC Handbook of Chemistry and Physics 4.* Chemical Rubber Publishing Company. p. 18. ISBN 0-8493-0474-1.

542. Emsley, J. (2001) *Nature's Building Blocks.* Oxford University Press, Oxford. pp. 262–266. ISBN 0-19-850341-5.

543. Shpak, A.P., Kotrechko, S.O., Mazilova, T.I. & Mikhailovskij, I.M. (2009) Inherent tensile strength of molybdenum nanocrystals. *Science and Technology of Advanced Materials,* 10(4), 045004.

544. Holleman, A.F., Wiberg, E. & Wiberg, N. (1985) *Lehrbuch der Anorganischen Chemie,* 91–100 edition. Walter de Gruyter. pp. 1096–1104. ISBN 3-11-007511-3.

545. Gupta, C.K. (1992) *Extractive Metallurgy of Molybdenum.* CRC Press. pp. 1–2. ISBN 978-0-8493-4758-0.

546. Considine, G.D. (ed.) (2005) Molybdenum. In: *Van Nostrand's Encyclopedia of Chemistry.* Wiley-Interscience, New York. pp. 1038–1040. ISBN 978-0-471-61525-5.

547. Smallwood, R.E. (1984) TZM moly alloy. In: *ASTM Special Technical Publication 849: Refractory Metals and Their Industrial Applications: A Symposium.* ASTM International. p. 9. ISBN 9780803102033.

548. Cubberly, W.H. & Bakerjian, R. (1989) *Tool and Manufacturing Engineers Handbook.* Society of Manufacturing Engineers. p. 421. ISBN 978-0-87263-351-3.

549. Lal, S. & Patil, R.S. (2001) Monitoring of atmospheric behaviour of NO_x from vehicular traffic. *Environmental Monitoring and Assessment,* 68(1), 37–50.

550. Lancaster, J.L. Ch. 4: Physical determinants of contrast. Physics of Medical X-Ray Imaging (PDF). University of Texas Health Science Center.

551. Gray, T. The Elements. pp. 105–107.

552. Winer, W. (1967) Molybdenum disulfide as a lubricant: A review of the fundamental knowledge. *Wear,* 10(6), 422.

553. Topsøe, H., Clausen, B.S. & Massoth, F.E. (1996) *Hydrotreating Catalysis, Science and Technology.* Springer-Verlag, Berlin.

554. Moulson, A.J. & Herbert, J.M. (2003) *Electroceramics: Materials, Properties, Applications.* John Wiley and Sons, Hoboken, NJ. p. 141. ISBN 0-471-49748-7.

555. Gottschalk, A. (1969) Technetium-99m in clinical nuclear medicine. *Annual Review of Medicine,* 20(1), 131–140.

556. Stixrude, L., Waserman, E. & Cohen, R. (1997) Composition and temperature of Earth's inner core. *Journal of Geophysical Research: American Geophysical Union,* 102(B11, November), 24729–24740.

557. Coey, J.M.D., Skumryev, V. & Gallagher, K. (1999) Rare-earth metals: Is gadolinium really ferromagnetic?. *Nature,* 401(6748), 35–36.

558. Kittel, C. (1996) *Introduction to Solid State Physics.* Wiley. p. 449. ISBN 0-471-14286-7.

559. Mond, L., Langer, K. & Quincke, F. (1890) Action of carbon monoxide on nickel. *Journal of the Chemical Society,* 57, 749–753.

560. Kerfoot, D.G.E. (2005) Nickel. In: *Ullmann's Encyclopedia of Industrial Chemistry.* Wiley-VCH, Weinheim.

561. Neikov, O.D., Naboychenko, S., Gopienko, V.G. & Frishberg, I.V. (2009) *Handbook of Non-Ferrous Metal Powders: Technologies and Applications.* Elsevier, January 15. p. 371. ISBN 978-1-85617-422-0.

562. Engineer, Engineering Record, Building Record, and Sanitary (1896). American plumbing practice: From the engineering record (Prior to 1887 the sanitary engineer): A selected reprint of articles describing notable plumbing installations in the United States, and questions and answers on problems arising in plumbing and house draining. *With Five Hundred and Thirty-six Illustrations*, 01 January. Engineering Record. p. 119.

563. Davis, J.R. (2000) Uses of nickel. In: *ASM Specialty Handbook: Nickel, Cobalt, and Their Alloys*. ASM International. pp. 7–13. ISBN 978-0-87170-685-0.

564. Kharton, V.V. (2011) *Solid State Electrochemistry II: Electrodes, Interfaces and Ceramic Membranes*. Wiley-VCH. p. 166. ISBN 978-3-527-32638-9.

565. Bidault, F., Brett, D.J.L., Middleton, P.H. & Brandon, N.P. A new cathode design for alkaline fuel cells(AFCs) (PDF). Imperial College London.

566. Angara, R. (2009) *High Frequency High Amplitude Magnetic Field Driving System for Magnetostrictive Actuators*. University of Maryland, Baltimore City. p. 5. ISBN 9781109187533.

567. Cheburaeva, R.F., Chaporova, I.N. & Krasina, T.I. (1992) Structure and properties of tungsten carbide hard alloys with an alloyed nickel binder. *Soviet Powder Metallurgy and Metal Ceramics*, 31(5), 423–425.

568. Patel, Z. & Khul'ka, K. (2001) Niobium for steelmaking. *Metallurgist*, 45(11–12), 477–480.

569. Heisterkamp, F. & Carneiro, T. (2001) Niobium: Future possibilities: Technology and the market place (PDF). In: Minerals, M. & Materials Society, Metals and Materials Society Minerals (eds.) *Niobium Science & Technology: Proceedings of the International Symposium Niobium 2001*. Niobium 2001 Ltd, Orlando, FL, USA, 2002. ISBN 978-0-9712068-0-9.

570. Eggert, P., Priem, J. & Wettig, E. (1982) Niobium: A steel additive with a future. *Economic Bulletin*, 19(9), 8–11.

571. Hillenbrand, H.-G., Gräf, M. & Kalwa, C. (2001) Development and production of high strength pipeline steels (PDF). *Niobium Science & Technology: Proceedings of the International Symposium Niobium 2001 (Orlando, FL, USA), 2 May*. Europipe.

572. Donachie, M.J. (2002) *Superalloys: A Technical Guide*. ASM International. pp. 29–30. ISBN 978-0-87170-749-9.

573. Hebda, J. (2001) Niobium alloys and high temperature applications (PDF). *Niobium Science & Technology: Proceedings of the International Symposium Niobium 2001 (Orlando, FL, USA), 2 May*. Companhia Brasileira de Metalurgia e Mineração.

574. Dinardi, A., Capozzoli, P. & Shotwell, G. (2008) Low-cost Launch opportunities provided by the Falcon family of launch vehicles (PDF). *Fourth Asian Space Conference, Taipei*.

575. Lindenhovius, J.L.H., Hornsveld, E.M., Den Ouden, A., Wessel, W.A.J. et al. (2000) Powder-in-tube (PIT) Nb/sub 3/Sn conductors for high-field magnets. *IEEE Transactions on Applied Superconductivity*, 10, 975–978.

576. Glowacki, B.A., Yan, X. -Y., Fray, D., Chen, G., Majoros, M. & Shi, Y. (2002) Niobium based intermetallics as a source of high-current/high magnetic field superconductors. *Physica C: Superconductivity*, 372–376(3), 1315–1320.

577. Grunblatt, G., Mocaer, P., Verwaerde, C. & Kohler, C. (2005) A success story: LHC cable production at ALSTOM-MSA. *Fusion Engineering and Design (Proceedings of the 23rd Symposium of Fusion Technology)*, 75–79, 1–5.

578. Lilje, L., Kako, E., Kostin, D., Matheisen, A., et al. (2004) Achievement of 35 MV/m in the superconducting nine-cell cavities for TESLA. *Nuclear Instruments and Methods in Physics Research Section A: Accelerators, Spectrometers, Detectors and Associated Equipment*, 524(1–3), 1–12.

579. Geballe, T.H. (1993) Superconductivity: From physics to technology. *Physics Today*, 46(10, October), 52–56.

580. Connelly, N.G. & Damhus, T. (eds.) (2005) *Nomenclature of Inorganic Chemistry: IUPAC Recommendations 2005 (PDF)*. With Hartshorn, R.M. & Hutton, A.T. RSC Publishing, Cambridge. ISBN 0-85404-438-8.

581. Haxel, G., Hedrick, J. & Orris, J. (2002) Rare earth elements: Critical resources for high technology (PDF). In: Stauffer, P.H. & Hendley, J.W., II (eds.) *Graphic Design by Gordon B. Haxel, Sara Boore, and Susan Mayfield: United States Geological Survey.* Reston, Virginia, USA. USGS Fact Sheet: 087-02.

582. Zepf, V. (2013) *Rare Earth Elements: A New Approach to the Nexus of Supply, Demand and Use: Exemplified Along the Use of Neodymium in Permanent Magnets.* Springer, Berlin and London. ISBN 9783642354588.

583. Rollinson, H.R. (1993) *Using Geochemical Data: Evaluation, Presentation, Interpretation.* Longman Scientific & Technical, Harlow, Essex, England. ISBN 9780582067011.

584. Brownlow, A.H. (1996) *Geochemistry.* Prentice Hall, Upper Saddle River, NJ. ISBN 0133982726.

585. Wolfgang Stoll, M. (2012) Thorium and thorium compounds. In: *Ullmann's Encyclopedia of Industrial Chemistry.* Wiley-VCH, Weinheim.

586. McGill, M.I. (2005) Rare earth elements. In: *Ullmann's encyclopedia of industrial chemistry.* Wiley-VCH, Weinheim.

587. Gupta, M.C.K. & Mukherjee, T.K. (1990) *Hydrometallurgy in Extraction Processes.* CRC Press, Boca Raton, FL.

588. Gupta, M.C.K. & Krishnamurthy, N. (2005) *Extraction Metallurgy of Rare Earths.* CRC Press, Boca Raton, FL.

589. Beatty, B.R. (2007) *The Lanthanides.* Published by Marshall Cavendish. Singapore.

590. Gupta, B.C.K. (2004) *Extractive Metallurgy of Rare Earths.* CRC Press. ISBN 0-415-33340-7.

591. Vallina, B.B., Rodriguez-Blanco, J.D., Blanco, J.A. & Benning, L.G. (2014) The effect of heating on the morphology of crystalline neodymium hydroxycarbonate, $NdCO_3OH$. *Mineralogical Magazine,* 78, 1391–1397.

592. Long, B., Keith, R., Van Gosen, B.S., Foley, N.K. & Cordier, D. (2010) Scientific investigations report 2010–5220. *The Principal Rare Earth Elements Deposits of the United States: A Summary of Domestic Deposits and a Global Perspective.* USGS.

593. Ohly, J. (1910) Rubidium. In: *Analysis, Detection and Commercial Value of the Rare Metals.* Mining Science Pub. Co.

594. Holleman, A.F., Wiberg, E. & Wiberg, N. (1985) Vergleichende Übersicht über die Gruppe der Alkalimetalle. In: *Lehrbuch der Anorganischen Chemie (in German),* 91–100 edition. Walter de Gruyter. pp. 953–955. ISBN 3-11-007511-3.

595. Butterman, W.C., Brooks, W.E. & Reese, Jr., R.G. (2003) Mineral commodity profile: Rubidium (PDF). *United States Geological Survey.*

596. Wise, M.A. (1995) Trace element chemistry of lithium-rich micas from rare-element granitic pegmatites. *Mineralogy and Petrology,* 55(13), 203–215.

597. Norton, J.J. (1973) Lithium, caesium, and rubidium: The rare alkali metals. In Brobst, D.A. & Pratt, W.P., *United States Mineral Resources. Paper 820. U.S. Geological Survey Professional.* pp. 365–378.

598. Bolter, E., Turekian, K. & Schutz, D. (1964) The distribution of rubidium, caesium and barium in the oceans. *Geochimica et Cosmochimica Acta,* 28(9), 1459.

599. Bulletin 585. (1995) United States. Bureau of Mines.

600. Caesium and rubidium hit market (1959) *Chemical & Engineering News,* 37(22), 50.

601. Boikess, R.S. & Edelson, E. (1981) *Chemical Principles.* Harper & Row, New York. p. 193. ISBN 978-0-06-040808-4.

602. Cornell, E. et al. (1996) Bose-einstein condensation (all 20 articles). *Journal of Research of the National Institute of Standards and Technology,* 101(4), 419–618.

603. Martin, J.L., McKenzie, C.R., Thomas, N.R., Sharpe, J.C., Warrington, D.M., Manson, P.J., Sandle, W.J. & Wilson, A.C. (1999) Output coupling of a Bose-Einstein condensate formed in a TOP trap. *Journal of Physics B: Atomic, Molecular and Optical Physics,* 32(12), 3065.

604. Gentile, T.R., Chen, W.C., Jones, G.L., Babcock, E. & Walker, T.G. (2005) Polarized ^{3}He spin filters for slow neutron physics (PDF). *Journal of Research of the National Institute of Standards and Technology,* 110(3), 299–304.

605. Neutron spin filters based on polarized helium-3. NIST Center for Neutron Research 2002 Annual Report.

606. Eidson, J.C. (2006) GPS. *Measurement, Control, and Communication Using IEEE 1588, 11 April.* p. 32. ISBN 978-1-84628-250-8.

607. King, T. & Newson, D. (1999) Rubidium and crystal oscillators. In: *Data Network Engineering, 31 July.* p. 300. ISBN 978-0-7923-8594-3.

608. Marton, L. (1977) Rubidium vapor cell. *Advances in Electronics and Electron Physics, 01 January.* ISBN 978-0-12-014644-4.

609. Mittal (2009) *Introduction to Nuclear And Particle Physics.* p. 274. ISBN 978-81-203-3610-0.

610. Li, Z., Wakai, R.T. & Walker, T.G. (2006) Parametric modulation of an atomic magnetometer. *Applied Physics Letters,* 89(13), 134105.

611. Jadvar, H. & Anthony Parker, J. (2005) Rubidium-82. *Clinical PET and PET/CT.* p. 59. ISBN 978-1-85233-838-1.

612. Yen, C.K., Yano, Y., Budinger, T.F., Friedland, R.P., Derenzo, S.E., Huesman, R.H. & O'Brien, H.A. (1982) Brain tumor evaluation using Rb-82 and positron emission tomography. *Journal of Nuclear Medicine,* 23(6), 532–537.

613. Paschalis, C., Jenner, F.A. & Lee, C.R. (1978) Effects of rubidium chloride on the course of manic-depressive illness. *Journal of the Royal Society of Medicine,* 71(9), 343–352.

614. Malekahmadi, P. & Williams, J.A. (1984) Rubidium in psychiatry: Research implications. *Pharmacology Biochemistry and Behavior,* 21, 49.

615. Canavese, C., Decostanzi, E., Branciforte, L., Caropreso, A., Nonnato, A. & Sabbioni, E. (2001) Depression in dialysis patients: Rubidium supplementation before other drugs and encouragement?. *Kidney International,* 60(3), 1201–1201.

616. Lake, J.A. (2006) *Textbook of Integrative Mental Health Care.* Thieme Medical Publishers, New York. pp. 164–165. ISBN 1-58890-299-4.

617. Torta, R., Ala, G., Borio, R., Cicolin, A., Costamagna, S., Fiori, L. & Ravizza, L. (1993) Rubidium chloride in the treatment of major depression. *Minerva psichiatrica,* 34(2), 101–110.

618. House, J.E. (2008) *Inorganic Chemistry.* Academic Press. p. 524. ISBN 0-12-356786-6.

619. Greenwood, N.N. & Earnshaw, A. (1997) *Chemistry of the Elements,* 2nd edition. Butterworth-Heinemann. pp. 751–752. ISBN 0080379419.

620. Kabata-Pendias, A. (1998) Geochemistry of selenium. *Journal of Environmental Pathology, Toxicology and Oncology: Official Organ of the International Society for Environmental Toxicology and Cancer,* 17(3–4), 173–177.

621. Fordyce, F. (2007) Selenium geochemistry and health. *AMBIO: A Journal of the Human Environment,* 36, 94.

622. Amouroux, D., Liss, P.S., Tessier, E. et al. (2001) Role of oceans as biogenic sources of selenium. *Earth and Planetary Science Letters,* 189(3–4), 277.

623. Naumov, A.V. (2010) Selenium and tellurium: State of the markets, the crisis, and its consequences. *Metallurgist,* 54(3–4), 197.

624. Hoffmann, J.E. (1989) Recovering selenium and tellurium from copper refinery slimes. *JOM,* 41(7), 33.

625. Hyvärinen, O., Lindroos, L. & Yllö, E. (1989) Recovering selenium from copper refinery slimes. *JOM,* 41(7), 42.

626. Sun, Y. Tian, X., He, B., et al. (2011) Studies of the reduction mechanism of selenium dioxide and its impact on the microstructure of manganese electrodeposit. *Electrochimica Acta,* 56(24), 8305.

627. Langner, B.E. (2005) Selenium and selenium compounds. In *Ullmann's Encyclopedia of Industrial Chemistry.* Wiley-VCH, Weinheim.

628. Davis, J.R. (2001) *Copper and Copper Alloys.* ASM International. p. 91. ISBN 978-0-87170-726-0.

629. Isakov, E. (2008) Cutting Data for Turning of Steel, 31 October. p. 67. ISBN 978-0-8311-3314-6.

630. Gol'Dshtein, Y.E., Mushtakova, T.L. & Komissarova, T.A. (1979) Effect of selenium on the structure and properties of structural steel. *Metal Science and Heat Treatment,* 21(10), 741.

631. Davis, J.R. (2001) *Copper and Copper Alloys*. ASM International. p. 278. ISBN 978-0-87170-726-0.

632. Deutsche Gesellschaft für Sonnenenergie (2008) Copper indium diselenide (CIS) cell. In: *Planning and Installing Photovoltaic Systems: A Guide for Installers, Architects and Engineers*. Earthscan. pp. 43–44. ISBN 978-1-84407-442-6.

633. Springett, B.E. (1988) Application of selenium-tellurium photoconductors to the xerographic copying and printing processes. *Phosphorus and Sulfur and the Related Elements*, 38(3–4), 341.

634. Williams, R. (2006) *Computer Systems Architecture: A Networking Approach*. Prentice Hall. pp. 547–548. ISBN 978-0-321-34079-5.

635. Diels, J.-C. & Arissian, L. (2011) The laser printer. In *Lasers*. Wiley-VCH. pp. 81–83. ISBN 978-3-527-64005-8.

636. Meller, G. & Grasser, T. (2009) *Organic Electronics*. Springer. pp. 3–5. ISBN 978-3-642-04537-0.

637. Kasap, S.; Frey, J.B., Belev, G. et al. (2009) Amorphous selenium and its alloys from early xeroradiography to high resolution X-ray image detectors and ultrasensitive imaging tubes. *Physica Status Solidi (b)*, 246(8), 1794.

638. Hai-Fu, F., Woolfson, M.M. & Jia-Xing, Y. (1993) New techniques of applying multi-wavelength anomalous scattering data. *Proceedings of the Royal Society A: Mathematical, Physical and Engineering Sciences*, 442(1914), 13.

639. MacLean, M.E. (1937) A project for general chemistry students: Color toning of photographic prints. *Journal of Chemical Education*, 14, 31.

640. Penichon, S. (1999) Differences in image tonality produced by different toning protocols for matte collodion photographs. *Journal of the American Institute for Conservation*, 38(2), 124–143.

641. McKenzie, J. (2003) *Exploring Basic Black & White Photography*. Delmar. p. 176. ISBN 978-1-4018-1556-1.

642. Moseley, P.T. & Seabrook, C.J. (1973) The crystal structure of β-tantalum. *Acta Crystallographica Section B Structural Crystallography and Crystal Chemistry*, 29(5), 1170–1171.

643. Ronen, Y. (2006) A rule for determining fissile isotopes. *Nuclear Science and Engineering*, 152(3), 334–335. [1].

644. Zhu, Z. & Cheng, C.Y. (2011) Solvent extraction technology for the separation and purification of niobium and tantalum: A review. *Hydrometallurgy*, 107, 1–12.

645. Agulyanski, A. (2004) *Chemistry of Tantalum and Niobium Fluoride Compounds*, 1st editions. Elsevier, Burlington. ISBN 9780080529028.

646. Okabe, T.H. & Sadoway, D.R. (1998) Metallothermic reduction as an electronically mediated reaction. *Journal of Materials Research*, 13(12), 3372–3377.

647. Cohen, R., Della Valle, C.J. & Jacobs, J.J. (2006) Applications of porous tantalum in total hip arthroplasty. *Journal of the American Academy of Orthopaedic Surgeons*, 14(12), 646–655.

648. Walters, W., Cooch, W., Burkins, M. & Burkins, M. (2001) The penetration resistance of a titanium alloy against jets from tantalum shaped charge liners. *International Journal of Impact Engineering*, 26, 823.

649. Russell, A.M., Lee, K.L. (2005) *Structure-Property Relations in Nonferrous Metals*. Wiley-Interscience, Hoboken, NJ. p. 218. ISBN 978-0-471-64952-6.

650. Craig, B.D., Anderson, D.S. & International, A.S.M. (1995) *Handbook of Corrosion Data*, January. ASTM International. p. 126. ISBN 978-0-87170-518-1.

651. Schwartz, M. (2002) Tin and alloys, properties. In: *Encyclopedia of Materials, Parts and Finishes*, 2nd edition. CRC Press. ISBN 1-56676-661-3.

652. Greenwood, N.N. & Earnshaw, A. (1997) *Chemistry of the Elements*, 2nd edition. Butterworth-Heinemann. ISBN 0-08-037941-9.

653. Holleman, A.F. & Wiberg, E. (2001) *Inorganic Chemistry*, Wiberg, N. (ed.), Eagleson, M. & Brewer, W. (trans.). Academic Press and De Gruyter, San Diego and Berlin, ISBN 0-12-352651-5.

654. Sutphin, D.M., Reed, D.M., Sutphin, A.E., Sabin, B.L., Sabin, A.E. & Reed, B.L. (1992) *Tin: International Strategic Minerals Inventory Summary Report*, 01 June. p. 9. ISBN 978-0-941375-62-7.

655. Black, H. (2005) Getting the lead out of electronics. *Environmental Health Perspectives*, 113(10), A682–685.

656. Childs, P. (1995) The tin-man's tale. *Education in Chemistry*, 32(4, July). Royal Society of Chemistry. p. 92.

657. Control, T.U. (1945) Tin under control. pp. 10–15. ISBN 978-0-8047-2136-3.

658. Panel On Tin, National Research Council (U.S.). Committee on Technical Aspects of Critical and Strategic Materials (1970) Trends in the use of tin. pp. 10–22.

659. Williams, R.S. (2007) *Principles of Metallography*. Read Books. pp. 46–47. ISBN 978-1-4067-4671-6.

660. Geballe, T.H. (1993) Superconductivity: From physics to technology. *Physics Today*, 46(10, October), 52–56.

661. Campbell, F.C. (2008) Zirconium. *Elements of Metallurgy and Engineering Alloys*. p. 597. ISBN 978-0-87170-867-0.

662. Audsley, G.A. (1988) Metal pipes: And the materials used in their construction. In: *The Art of Organ Building Audsley, George Ashdown*. Courier Dover Publications. p. 501. ISBN 978-0-486-21315-6.

663. Pilkington, L.A.B. (1969) Review lecture: The float glass process. *Proceedings of the Royal Society of London. Series A, Mathematical and Physical Sciences*, 314(1516), 1–25.

664. Lucas, I.T., Syzdek, J. & Kostecki, R. (2011) Interfacial processes at single-crystal β-Sn electrodes in organic carbonate electrolytes. *Electrochemistry Communications*, 13(11), 1271–1275.

665. Hattab, F. (1989) The state of fluorides in toothpastes. *Journal of Dentistry*, 17(2, April), 47–54.

666. Perlich, M.A., Bacca, L.A., Bollmer, B.W., Lanzalaco, A.C., McClanahan, S.F., Sewak, L.K., Beiswanger, B.B., Eichold, W.A., Hull, J.R. et al. (1995) The clinical effect of a stabilized stannous fluoride dentifrice on plaque formation, gingivitis and gingival bleeding: A six-month study. *The Journal of Clinical Dentistry*, 6(Special Issue), 54–58.

667. Atkins, P., Shriver, D.F., Overton, T. & Rourke, J. (2006) *Inorganic Chemistry*, 4th edition. W.H. Freeman. pp. 343, 345. ISBN 0-7167-4878-9.

668. Wilkes, C.E., Summers, J.W., Daniels, C.A. & Berard, M.T. (2005) PVC handbook, August. p. 108. ISBN 978-1-56990-379-7.

669. Lucas, I. & Syzdek, J. (2011) Electrochemistry communications. *Electrochemistry Communications*, 13(11), 1271.

670. Chen, G.Z., Fray, D.J. & Farthing, T.W. (2000) Direct electrochemical reduction of titanium dioxide to titanium in molten calcium chloride. *Nature*, 407(6802), 361–364.

671. ASTM International (2006) *Annual Book of ASTM Standards*, Volume 02.04: Non-ferrous Metals. ASTM International, West Conshohocken, PA. section 2. ISBN 0-8031-4086-X. ASTM International (1998) *Annual Book of ASTM Standards*, Volume 13.01: Medical Devices; Emergency Medical Services. ASTM International, West Conshohocken, PA. sections 2 & 13. ISBN 0-8031-2452-X.

672. Titanium (2000–2006) In: *Columbia Encyclopedia*, 6th edition. Columbia University Press, New York. ISBN 0-7876-5015-3.

673. Krebs, R.E. (2006) *The History and Use of Our Earth's Chemical Elements: A Reference Guide*, 2nd edition. Greenwood Press, Westport, CT. ISBN 0-313-33438-2.

674. Stwertka, A. (1998) Titanium. In: *Guide to the Elements*, Revised edition. Oxford University Press, Oxford. pp. 81–82. ISBN 0-19-508083-1.

675. Moiseyev, V.N. (2006) *Titanium Alloys: Russian Aircraft and Aerospace Applications*. Taylor and Francis, LLC. p. 196. ISBN 978-0-8493-3273-9; Titanium. *Encyclopædia Britannica*. 2006.

676. Kleefisch, E.W. (ed.) (1981) *Industrial Application of Titanium and Zirconium*. ASTM International, West Conshohocken, PA. ISBN 0-8031-0745-5.

677. Bunshah, R.F. (ed.) (2001) Ch. 8. In: *Handbook of Hard Coatings*. William Andrew Inc, Norwich, NY. ISBN 0-8155-1438-7.

678. Bell, T. et al. (2001) Heat treating. *Proceedings of the 20th Conference, 9–12 October 2000*. ASM International. p. 141. ISBN 0-87170-727-6.

679. Davis, J.R. (1998) *Metals Handbook*. ASM International. p. 584. ISBN 0-87170-654-7.
680. Lütjering, G. & Williams, J.C. (2007) Appearance related applications. *Titanium*, 12 June. ISBN 978-3-540-71397-5.
681. Gafner, G. (1989) The development of 990 Gold-Titanium: Its production, use and properties (PDF). *Gold Bulletin*, 22(4), 112–122.
682. Alwitt, R.S. (2002) Electrochemistry Encyclopedia.
683. Lide, D.R. (ed.) (2005) *CRC Handbook of Chemistry and Physics*, 86th edition. CRC Press, Boca Raton, FL. ISBN 0-8493-0486-5.
684. Smook, G.A. (2002) *Handbook for Pulp & Paper Technologists*, 3rd edition. Angus Wilde Publications. p. 223. ISBN 0-9694628-5-9.
685. Sibum, H., Günther, V., Roidl, O., Habashi, F. & Wolf, H.U. (2005) Titanium, titanium alloys, and titanium compounds. In *Ullmann's Encyclopedia of Industrial Chemistry*. Wiley-VCH, Weinheim.
686. Mineral Commodity Summary: Titanium Mineral Concentrates (PDF). USGS: 175. 2012.
687. Banfield, J.F., Veblen, D.R., Smith, D.J. (1991) Photo-induced hydrophobicity of brookite TiO_2 prepared by hydrothermal conversion from Mg_2TiO_4. *American Mineralogist*, 76, 343.
688. Daintith, J. (2005) *Facts on File Dictionary of Chemistry*, 4th edition. Checkmark Books, New York. ISBN 0-8160-5649-8.
689. deLaubenfels, B., Weber, C. & Bamberg, K. (2009) *Knack Planning Your Wedding: A Step-by-Step Guide to Creating Your Perfect Day*. Globe Pequot. p. 35, 8 December. ISBN 978-1-59921-397-2.
690. Schultz, K. (2009) *Ken Schultz's Essentials of Fishing: The Only Guide You Need to Catch Freshwater and Saltwater Fish*. John Wiley and Sons, Hoboken, NJ, 18 November. p. 138. ISBN 978-0-470-44431-3.
691. Ramakrishnan, P. (2007) Powder metallurgy for aerospace applications. In: *Powder Metallurgy: Processing for Automotive, Electrical/Electronic and Engineering Industry*. New Age International. p. 38. ISBN 81-224-2030-3.
692. Stwertka, A. (2002) *A Guide to the Elements*, 2nd edition. Oxford University Press, New York. ISBN 0-19-515026-0.
693. Hesse, R.W. (2007) Tungsten. In: *Jewelrymaking through History: An Encyclopedia*. Greenwood Press, Westport, CT. pp. 190–192. ISBN 978-0-313-33507-5.
694. Voort, G.F.V. (1984) *Metallography, Principles and Practice*. ASM International. p. 137. ISBN 978-0-87170-672-0.
695. Cardarelli, F. (2008) *Materials Handbook: A Concise Desktop Reference*. Springer. p. 338. ISBN 978-1-84628-668-1.
696. Holleman, A.F., Wiberg, E. & Wiberg, N. (1985) Vanadium. In: *Lehrbuch der Anorganischen Chemie (in German)*, 91–100 edition. Walter de Gruyter. pp. 1071–1075. ISBN 3-11-007511-3.
697. Moskalyk, R.R. & Alfantazi, A.M. (2003) Processing of vanadium: A review. *Minerals Engineering*, 16(9), 793.
698. Carlson, O.N. & Owen, C.V. (1961) Preparation of high-purity vanadium metals by the iodide refining process. *Journal of the Electrochemical Society*, 108, 88.
699. Bauer, G., Güther, V., Hess, H., Otto, A., Roidl, O., Roller, H. & Sattelberger, S. (2005) Vanadium and vanadium compounds. In *Ullmann's Encyclopedia of Industrial Chemistry*. Wiley-VCH, Weinheim.
700. Chandler, H. (1998) *Metallurgy for the Non-Metallurgist*. ASM International. pp. 6–7. ISBN 978-0-87170-652-2.
701. Davis, J.R. (1995) *Tool Materials: Tool Materials*. ASM International. ISBN 978-0-87170-545-7.
702. Neikov, O.D., Naboychenko, S., Mourachova, I.B., Gopienko, V.G. Frishberg, I.V. & Lotsko, D.V. (2009) *Handbook of Non-Ferrous Metal Powders: Technologies and Applications*, 24 February. p. 490. ISBN 9780080559407.
703. Peters, M. & Leyens, C. (2002) Metastabile β-legierungen. In: *Titan und Titanlegierungen*. Wiley-VCH. pp. 23–24. ISBN 978-3-527-30539-1.

704. Lositskii, N.T., Grigor'ev, A.A. & Khitrova, G.V. (1966) Welding of chemical equipment made from two-layer sheet with titanium protective layer (review of foreign literature). *Chemical and Petroleum Engineering*, 2(12), 854–856.

705. Matsui, H., Fukumoto, K., Smith, D.L., Chung, H.M., Witzenburg, W. van & Votinov, S.N. (1996) Status of vanadium alloys for fusion reactors. *Journal of Nuclear Materials*, 233–237(1), 92–99.

706. Hardy, G.F. & Hulm, J.K. (1953) Superconducting silicides and germanides. *Physical Reviews*, 89(4), 884–884.

707. Markiewicz, W., Mains, E., Vankeuren, R., Wilcox, R., Rosner, C., Inoue, H., Hayashi, C. & Tachikawa, K. (1977) A 17.5 Tesla superconducting concentric Nb_3Sn and V^3Ga magnet system. *IEEE Transactions on Magnetics*, 13(1), 35–37.

708. Eriksen, K.M., Karydis, D.A., Boghosian, S. & Fehrmann, R. (1995) Deactivation and compound formation in sulfuric-acid catalysts and model systems. *Journal of Catalysis*, 155(1), 32–42.

709. Abon, M. & Volta, J.-C. (1997) Vanadium phosphorus oxides for n-butane oxidation to maleic anhydride. *Applied Catalysis A: General*, 157(1–2), 173–193.

710. Lide, D.R. (2004) Vanadium. *CRC Handbook of Chemistry and Physics*. CRC Press, Boca Raton. pp. 4–34. ISBN 978-0-8493-0485-9.

711. Manning, T.D., Parkin, I.P., Clark, R.J.H., Sheel, D., Pemble, M.E. & Vernadou, D. (2002) Intelligent window coatings: Atmospheric pressure chemical vapour deposition of vanadium oxides. *Journal of Materials Chemistry*, 12(10), 2936–2939.

712. White, W.B., Roy, R. & McKay, C. (1962) The alexandrite effect: And optical study (PDF). *American Mineralogist*, 52, 867–871.

713. Joerissen, L., Garche, J., Fabjan, C. & Tomazic, G. (2004) Possible use of vanadium redox-flow batteries for energy storage in small grids and stand-alone photovoltaic systems. *Journal of Power Sources* 127(1–2), 98–104.

714. Guan, H. & Buchheit, R.G. (2004) Corrosion protection of aluminium alloy 2024-T3 by vanadate conversion coatings. *Corrosion*, 60(3), 284–296.

715. Sphalerite. Mindat.org.

716. Rosenqvist, T. (1922) *Principles of Extractive Metallurgy*, 2nd edition. Tapir Academic Press. pp. 7, 16, 186. ISBN 82-519-1922-3.

717. Borg, G., Kärner, K., Buxton, M., Armstrong, R. & van der Merwe, S.W. (2003) Geology of the skorpion supergene zinc deposit, Southern Namibia. *Economic Geology*, 98(4), 749. doi:10.2113/98.4.749.

718. Bodsworth, C. (1994) *The Extraction and Refining of Metals*. CRC Press. p. 148. ISBN 0-8493-4433-6.

719. Porter, F.C. (1991) *Zinc Handbook*. CRC Press. ISBN 978-0-8247-8340-2.

720. Gupta, C.K. & Mukherjee, T.K. (1990) *Hydrometallurgy in Extraction Processes*. CRC Press. p. 62. ISBN 0-8493-6804-9.

721. Antrekowitsch, J., Steinlechner, S., Unger, A., Rösler, G., Pichler, C. & Rumpold, R. (2014) 9. zinc and residue recycling. In Worrell, E & Reuter, M. (eds.) *Handbook of Recycling: State-of-the-Art for Practitioners, Analysts, and Scientists*. Elsevier. ISBN: 978-0-12-396459-5.

722. Zinc: World Mine Production (zinc content of concentrate) by Country(PDF). (2009) *Minerals Yearbook: Zinc*. United States Geological Survey, Washington, DC.

723. Besenhard, J.O. (1999) *Handbook of Battery Materials*. Wiley-VCH. ISBN 3-527-29469-4.

724. Wiaux, J.-P. & Waefler, J.-P. (1995) Recycling zinc batteries: An economical challenge in consumer waste management. *Journal of Power Sources*, 57(1–2), 61–65.

725. Culter, T. (1996) A design guide for rechargeable zinc-air battery technology. *Southcon*, 96. Conference Record: 616. ISBN 0-7803-3268-7.

726. Whartman, J. & Brown, I. Zinc air battery-battery hybrid for powering electric scooters and electric buses (PDF). *The 15th International Electric Vehicle Symposium*.

727. Cooper, J.F., Fleming, D., Hargrove, D., Koopman, R. & Peterman, K. A refuelable zinc/air battery for fleet electric vehicle propulsion. *Society of Automotive Engineers Future Transportation Technology Conference and Exposition*.

728. Xie, Z., Liu, Q., Chang, Z. & Zhang, X. (2013) The developments and challenges of cerium half-cell in zinc: Cerium redox flow battery for energy storage. *Electrochimica Acta*, 90, 695–704.

729. Cooper, J.F., Fleming, D., Hargrove, D., Koopman, R. & Peterman, K. A refuelable zinc/air battery for fleet electric vehicle propulsion. *Society of Automotive Engineers Future Transportation Technology Conference and Exposition*.

730. Apelian, D., Paliwal, M. & Herrschaft, D.C. (1981) Casting with zinc alloys. *Journal of Metals*, 33(11), 12–19.

731. Davies, G. (2003) *Materials for Automobile Bodies*. Butterworth-Heinemann. p. 157. ISBN 0-7506-5692-1.

732. Samans, C.H. (1949) *Engineering Metals and Their Alloys*. Macmillan Co. Stuttgart, Germany.

733. Porter, F. (1994) Wrought zinc. In: *Corrosion Resistance of Zinc and Zinc Alloys*. CRC Press. pp. 6–7. ISBN 978-0-8247-9213-8.

734. McClane, A.J. & Gardner, K. (1987) *The Complete Book of Fishing: A Guide to Freshwater, Saltwater & Big-Game Fishing*. Gallery Books. ISBN 978-0-8317-1565-6.

735. Katz, J.I. (2002) *The Biggest Bangs*. Oxford University Press, Oxford. p. 18. ISBN 0-19-514570-4.

736. Zhang, X.G. (1996) *Corrosion and Electrochemistry of Zinc*. Springer. p. 93. ISBN 0-306-45334-7.

737. Weimer, A. (2006) Development of solar-powered thermochemical production of hydrogen from water (PDF). *U.S. Department of Energy*, 17 May.

738. Blew, J.O. (1953) Wood Preservatives (PDF). *Department of Agriculture, Forest Service, Forest Products Laboratory*. hdl:1957/816.

739. Frankland, E. (1849) Notiz über eine neue Reihe organischer Körper, welche Metalle, Phosphor u. s. w. enthalten. *Liebig's Annalen der Chemie und Pharmacie (in German)*, 71(2), 213–216.

740. Paschotta, R. (2008) *Encyclopedia of Laser Physics and Technology*. Wiley-VCH. p. 798. ISBN 3-527-40828-2.

741. Emsley, J. (2001) *Nature's Building Blocks*. Oxford University Press, Oxford. pp. 506–510. ISBN 0-19-850341-5.

742. Lide, D.R. (ed.) (2007–2008) Zirconium. In: *CRC Handbook of Chemistry and Physics*, Volume 4. CRC Press, New York. p. 42. ISBN 978-0-8493-0488-0.

743. Considine, G.D. (ed.) (2005) Zirconium. In: *Van Nostrand's Encyclopedia of Chemistry*. Wylie-Interscience, New York. pp. 1778–1779. ISBN 0-471-61525-0.

744. Peterson, J. & MacDonell, M. (2007) Zirconium. In *Radiological and Chemical Fact Sheets to Support Health Risk Analyses for Contaminated Areas (PDF)*. Argonne National Laboratory. Argonne, Il, USA. pp. 64–65.

745. Nielsen, R. (2005) Zirconium and zirconium compounds. In: *Ullmann's Encyclopedia of Industrial Chemistry*. Wiley-VCH, Weinheim.

746. Stwertka, A. (1996) *A Guide to the Elements*. Oxford University Press, Oxford. pp. 117–119. ISBN 0-19-508083-1.

747. Brady, G.S., Clauser, H.R. & Vaccari, J.A. (2002) *Materials Handbook: An Encyclopedia for Managers, Technical Professionals, Purchasing and Production Managers, Technicians, and Supervisors*. McGraw-Hill Professional, 24 July. p. 1063. ISBN 978-0-07-136076-0.

748. Zardiackas, L.D., Kraay, M.J. & Freese, H.L. (2006) *Titanium, Niobium, Zirconium and Tantalum for Medical and Surgical Applications*. ASTM International, 1 January. p. 21. ISBN 978-0-8031-3497-3.

749. Meier, S.M. & Gupta, D.K. (1994) The evolution of thermal barrier coatings in gas turbine engine applications. *Journal of Engineering for Gas Turbines and Power*, 116, 250.

750. Lee, D.B.N., Roberts, M., Bluchel, C.G. & Odell, R.A. (2010) Zirconium: Biomedical and nephrological applications. *American Society for Artificial Internal Organs Journal*, 56(6), 550–556.

751. Hammond, C.R. (2004) *The Elements, in Handbook of Chemistry and Physics*, 81st edition. CRC Press, Boca Raton, FL, US. pp. 4–1. ISBN 0-8493-0485-7.

752. Papon, P., Leblond, J. & Meijer, P.H.E. (2006) *The Physics of Phase Transitions*. Springer. p. 82. ISBN 978-3-540-33390-6.

753. Tracy, G.R., Tropp, H.E. & Friedl, A.E. (1974) *Modern Physical Science*. p. 268. ISBN 978-0-03-007381-6.

754. Tribe, A. (1868) IX: Freezing of water and bismuth. *Journal of the Chemical Society*, 21, 71.

755. Llewellyn, D.T. & Hudd, R.C. (1998) *Steels: Metallurgy and Applications*. Butterworth-Heinemann. p. 239. ISBN 978-0-7506-3757-2.

756. Davis & Associates, J.R. & Handbook Committee, ASM International (1993) *Aluminium and Aluminium Alloys*. ASM International. p. 41. ISBN 978-0-87170-496-2.

757. Maile, F.J., Pfaff, G. & Reynders, P. (2005) Effect pigments: Past, present and future. *Progress in Organic Coatings*, 54(3), 150.

758. Pfaff, G. (2008) *Special Effect Pigments: Technical Basics and Applications*. Vincentz Network GmbH. p. 36. ISBN 978-3-86630-905-0.

759. Carlin, J.F., Jr. (2010) USGS Minerals Yearbook: Bismuth (PDF). United States Geological Survey.

760. Hammond, C.R. (2000) *The Elements, in Handbook of Chemistry and Physics (PDF)*, 81st edition. CRC Press. ISBN 0-8493-0481-4.

761. Woodhead, J.A. et al. (1991) The metamictization of zircon: Radiation dose-dependent structural characteristics. *American Mineralogist*, 76, 74–82.

762. van Arkel, A.E. & de Boer, J.H. (1925) Darstellung von reinem Titanium-, Zirkonium-, Hafnium- und Thoriummetall. *Zeitschrift für anorganische und allgemeine Chemie (in German)*, 148(1), 345–350.

763. Hammond, C.R. (2004) *The Elements, in Handbook of Chemistry and Physics*, 81st edition. CRC Press. ISBN 0-8493-0485-7.

764. Rancourt, J.D. (1996) *Optical Thin Films: User Handbook*. SPIE Press. p. 196. ISBN 0-8194-2285-1.

765. McKetta, J.J. (1996) *Encyclopedia of Chemical Processing and Design: Thermoplastics to Trays, Separation, Useful Capacity*. CRC Press. p. 81. ISBN 0-8247-2609-X.

766. Uranium. *Encyclopaedia Britannica*.

767. Nilgiriwala, K.S., Alahari, A., Rao, A.S. & Apte, S.K. (2008) Cloning and overexpression of alkaline phosphatase PhoK from sphingomonas sp. strain BSAR-1 for bioprecipitation of uranium from alkaline solutions. *Applied and Environmental Microbiology*, 74(17), 5516–5523.

768. Gupta, C.K. & Mukherjee, T.K. (1990) *Hydrometallurgy in Extraction Processes*, Volume 1. CRC Press. pp. 74–75. ISBN 0-8493-6804-9.

769. (1982) Uranium. *The McGraw-Hill Science and Technology Encyclopedia*, 5th edition. The McGraw-Hill Companies, Inc. ISBN 0-07-142957-3.

770. Uranium. Royal Society of Chemistry.

771. Tungsten. Royal Society of Chemistry.

772. Gold. Royal Society of Chemistry.

773. Zhiling Cao, Henry Gu Cao; "Unified theory and topology of nuclei" *International Journal of Physics*, 2014 2(1) 15–22. Uranium. *Columbia Electronic Encyclopedia*, 6th edition. Columbia University Press.

774. Whitney, D.L. (2002) Coexisting andalusite, kyanite, and sillimanite: Sequential formation of three Al_2SiO_5 polymorphs during progressive metamorphism near the triple point, Sivrihisar, Turkey. *American Mineralogist*, 87(4), 405–416.

775. Kyanite (PDF). Handbook of Mineralogy. 2001.

776. Klein, C. & Hurlbut, C.S., Jr. (1985) *Manual of Mineralogy*, 20th edition. Wiley, p. 380. ISBN 0-471-80580-7.

777. Villalba, G., Ayres, R.U. & Schroder, H. (2008) Accounting for fluorine: Production, use, and loss. *Journal of Industrial Ecology*, 11, 85–101. |access-date= requires |url= (help).

778. Henderson and Marsden (1972) *Lamps and Lighting*. Edward Arnold Ltd. ISBN 0-7131-3267-1.

779. Salvi, S. & Williams-Jones, A. (2004) Alkaline granite-syenite deposits. In Linnen, R.L. & Samson, I.M. (eds.) *Rare Element Geochemistry and Mineral Deposits*. Geological Association of Canada, St. Catharines, ON. pp. 315–341. ISBN 1-897095-08-2.

780. Haxel, G., Hedrick, J. & Orris, J. (2006) *Rare Earth Elements Critical Resources for High Technology*. United States Geological Survey, Reston, VA. USGS Fact Sheet: 087-02.

781. Alleman, J.E. & Mossman, B.T. (1997) Asbestos revisited (PDF). *Scientific American*, 277(July), 54–57.

782. Hearst Magazines (1935) Popular mechanics. *Hearst Magazines*, July. p. 62. ISSN 0032-4558.

783. Jones, D. (2011) *Cambridge English Pronouncing Dictionary*, 18th edition, Roach, P., Setter, J. & Esling, J. (eds.). Cambridge University Press, Cambridge. ISBN 978-052-115255-6.

784. Walker, I.C. (1998) *Marketing Challenges for Canadian Bitumen (PDF)*. International Centre for Heavy Hydrocarbons, Tulsa, OK.

785. Sörensen, A. & Wichert, B. (2009) Asphalt and bitumen. In: *Ullmann's Encyclopedia of Industrial Chemistry*, Wiley-VCH, Weinheim.

786. Boden, T. & Tripp, B. (2012) *Gilsonite Veins of the Uinta Basin, Utah*, Utah Geological Survey, Salt Lake City. UT, Special Study 141.

787. Hayatsu, R., Scott R.G., Studier, M.H., Lewis, R.S. & Auders, E. (1980) Carbynes in meteorites: Detection, low-temperature origin, and implications for intersteller molecules. *Science, 209*(4464):1515–1518.

788. Kim, Kyoung-Sook and Yang, Jong-Mann (1998) "Carbon isotope analysis of individual hydrocarbon molecules in bituminous coal, oil shale, and murchison meteorite". *Journal of Astronomy and Space Sciences*, 15(1), 163–174.

789. Mohd, M.J. & Singh, D.K. (2013) Cashew nutshell liquid resin (PDF). *IJRREST: International Journal of Research Review in Engineering Science and Technology*, 2(1 March), 60–65.

790. Hesp, S.A.M. & Shurvell, H.F. (2010) X-ray fluorescence detection of waste engine oil residue in asphalt and its effect on cracking in service. *International Journal of Pavement Engineering*, 11(6), 541–553.

791. Hurlbut, C.S. & Klein, C. (1985) *Manual of Mineralogy*, 20th edited. Wiley, ISBN 978-0-471-00042-6.

792. Gettens, R.J. & Fitzhugh, E.W. (1993) Azurite and blue verditer. In *Artists' Pigments: A Handbook of Their History and Characteristics*, Volume 2, Roy, A. (ed.). Oxford University Press, Oxford. pp. 23–24.

793. Dana, J.D. & Ford, W.E. (1915) *Dana's Manual of Mineralogy for the Student of Elementary Mineralogy, the Mining Engineer, the Geologist, the Prospector, the Collector, Etc.*, 13th edition. John Wiley & Sons, Inc, Hoboken, NJ. pp. 299–300.

794. Hanor, J. (2000) Barite-celestine geochemistry and environments of formation. In: *Reviews in Mineralogy*, Volume 40. Mineralogical Society of America, Washington, DC. pp. 193–275. ISBN 0-939950-52-9.

795. Rubin, A.E. (1997) Mineralogy of meteorite groups. *Meteoritics & Planetary Science*, 32(2, March), 231–247.

796. Michael, M. & Barite, M. (2009) *Minerals Yearbook*. US Geological Survey, Reston, VA.

797. Kastner, M. (1999) Oceanic minerals: Their origin, nature of their environment, and significance. *Proceedings of the Nationall Academy Sciences USA*, 96(7, 30 March), 3380–3387.

798. Edwards, K.J., Bach, W. & Rogers, D.R. (2003) Geomicrobiology of the ocean crust: A role for chemoautotrophic fe-bacteria. *Biological Bulletin*, 204(April), 180–185.

799. Potassium bentonite. McGraw-Hill Dictionary of Scientific and Technical Terms.

800. Odom, I.E. (1984) Smectite clay minerals: Properties and uses. *Philosophical Transactions of the Royal Society A: Mathematical, Physical and Engineering Sciences*, 311(1517), 391.

801. Hosterman, J.W. & Patterson, S.H. (1992) Bentonite and Fuller's earth resources of the United States. *U.S. Geological Survey Professional Paper 1522*. United States Government Printing Office, Washington, DC, USA.

802. Karnland, O., Olsson, S. & Nilsson, U. (2006) Mineralogy and sealing properties of various bentonites and smectite-rich clay materials. SKB Technical Report TR-06–30. Stockholm, Sweden.

803. Lagaly, G. (1995) Surface and interlayer reactions: Bentonites as adsorbents. In: Churchman, G.J., Fitzpatrick, R.W. & Eggleton, R.A. (eds.) *Clays Controlling the Environment: Proceedings of the 10th International Clay Conference, Adelaide, Australia.* CSIRO Publishing, Melbourne. pp. 137–144. ISBN 0-643-05536-3.

804. Robertson, R.H.S. (1986) *Fuller's Earth: A History of Calcium Montmorillonite.* Volturna Press, Hythe, UK. ISBN 0-85606-070-4.

805. Atkins, P. et al. (2010) *Inorganic Chemistry*, 5th edition. Oxford University Press, Oxford. p. 334. ISBN 9780199236176.

806. Mendham, J., Denney, R.C., Barnes, J.D. & Thomas, M.J.K. (2000) *Vogel's Quantitative Chemical Analysis*, 6th edition. Prentice Hall, New York. p. 316. ISBN 0-582-22628-7.

807. Greenwood, N.N. & Earnshaw, A. (1997) *Chemistry of the Elements*, 2nd edition. Butterworth-Heinemann. p. 205. ISBN 0080379419.

808. Schubert, D. (2011) Boron oxides, boric acid, and borates. In: *Kirk-Othmer Encyclopedia of Chemical Technology.* John Wiley & Sons. pp. 1–68. ISBN 9780471238966.

809. Smith, R.A. (2000) Boric oxide, boric acid, and borates. *Ullmann's Encyclopedia of Industrial Chemistry,* 7th edition. Wiley-VCH. ISBN: 978352732943-4.

810. Hettipathirana, T.D. (2004) Simultaneous determination of parts-per-million level Cr, As, Cd and Pb, and major elements in low level contaminated soils using borate fusion and energy dispersive X-ray fluorescence spectrometry with polarized excitation. *Spectrochimica Acta Part B: Atomic Spectroscopy,* 59(2), 223–229.

811. USGS (2013) 2011 Minerals yearbook, Boron. US Geological Survey, Reston, VA.

812. Kerr, P.F. (1952) Formation and occurrence of clay minerals. *Clays and Clay Minerals,* 1(1), 19–32.

813. Dana, J.D., Klein, C. & Hurlbut, C.S. (1985) *Manual of Mineralogy*, 20th edition. John Wiley and Sons, New York. pp. 395–396. ISBN 0-471-80580-7.

814. Klein, C. (2002) *The Manual of Mineral Science*, 22nd edition. John Wiley & Sons, Inc, Hoboken, NJ. ISBN 0-471-25177-1.

815. Cybulski, A., Moulijn, J.A. (eds.) (2005) *Structured Catalysts and Reactors*, 2nd edition. CRC Press. p. 35. ISBN 978-0-8247-2343-9.

816. Anthony, J.W., Bideaux, R.A., Bladh, K.W. & Nichols, M.C. (ed.) (1997) Corundum. In: *Handbook of Mineralogy* (PDF), Halides, Hydroxides, Oxides, Volume 3. Mineralogical Society of America, Chantilly, VA, US. ISBN 0962209724.

817. Hurlbut, C.S. & Klein, C. (1985) *Manual of Mineralogy*, 20th edition. Wiley. pp. 300–302. ISBN 0-471-80580-7.

818. Root, A.I. & Root, E.R. (2005) *The ABC and XYX of Bee Culture.* Kessinger Publishing, 1 March. p. 387. ISBN 978-1-4326-2685-3.

819. Fields, P., Allen, S., Korunic, Z., McLaughlin, A. & Stathers, T. (2002) Standardized testing for diatomaceous earth (PDF). Proceedings of the *Eighth International Working Conference of Stored-Product Protection,* July. Entomological Society of Manitoba, York, UK.

820. Lartigue, E. del C. & Rossanigo, C.E. (2004) Insecticide and anthelmintic assessment of diatomaceous earth in cattle. *Veterinaria Argentina,* 21(209), 660–674.

821. Fernandez, M.I., Woodward, B.W. & Stromberg, B.E. (1998) Effect of diatomaceous earth as an anthelmintic treatment on internal parasites and feedlot performance of beef steers. *Animal Science,* 66(3), 635–641.

822. Faulde, M.K., Tisch, M. & Scharninghausen, J.J. (2006) Efficacy of modified diatomaceous earth on different cockroach species (Orthoptera, Blattellidae) and silverfish (Thysanura, Lepismatidae). *Journal of Pest Science,* 79(3), August, 155–161

823. Flynn, T.M. (2005) Cryogenic equipment and cryogenic systems analysis. In: *Cryogenic Engineering.* Boca Raton [etc.: CRC, Print].

824. Ferraz et al. (2011) Manufacture of ceramic bricks using recycled brewing spent kieselguhr. *Materials and Manufacturing Processes,* 26(10), 1319–1329.

825. Kay, D.B. & Azam, F. (1999) Accelerated dissolution of diatom silica by marine bacterial assemblages. *Nature*, 397, 508–512.

826. Spencer, L.J. (1911) Epidote. In: Chisholm, H. (ed.) *Encyclopædia Britannica*, Volume 9, 11th edition. Cambridge University Press, Cambridge. p. 689.

827. Blatt, H. & Tracy, R.J. (1996) *Petrology*, 2nd edition. Freeman. pp. 206–210. ISBN 0-7167-2438-3.

828. Apodaca, L.E. Feldspar and nepheline syenite, USGS 2008, Minerals yearbook.

829. Przibram, K. (1935) Fluorescence of luorite and the bivalent Europium ion. *Nature*, 135(3403), 100.

830. Aigueperse, J., Mollard, P., Devilliers, D., Chemla, M., Faron, R., Romano, R. & Cuer, J.P. *Ullmann's Encyclopedia of Industrial Chemistry: Fluorine Compounds, Inorganic*. Wiley-VCH Verlag GmbH & Co. KGaA, 2000.

831. Critical raw materials for the EU, 2010.

832. Miller, M.M., Fluorspar, USGS 2009 Minerals Yearbook, Capper, P. (2005) *Bulk Crystal Growth of Electronic, Optical & Optoelectronic Materials*. John Wiley and Sons, Hoboken, NJ. p. 339. ISBN 0-470-85142-2.

833. Capper, P. (2005) *Bulk Crystal Growth of Electronic, Optical & Optoelectronic Materials*. John Wiley and Sons. p. 339. ISBN 0-470-85142-2.

834. Rost, F.W.D. & Oldfield, R.J. (2000) *Photography with a Microscope*. Cambridge University Press, Cambridge. p. 157. ISBN 0-521-77096-3.

835. Ray, S.F. (1999) *Scientific Photography and Applied Imaging*. Focal Press. pp. 387–388. ISBN 0-240-51323-1.

836. Sutphin, D.M. & Bliss, J.M. (1990) Disseminated flake graphite and amorphous graphite deposit types: An analysis using grade and tonnage models. *CIM Bulletin*, 83(940), August, 85–89.

837. Wainwright, A. (2005) *A Pictorial Guide to the Lakeland Fells, Western Fells*. Frances Lincoln, London. ISBN 0-7112-2460-9.

838. Klein, C. & Hurlbut, C.S., Jr. (1985) *Manual of Mineralogy*, 20th editon. John Wiley & Sons, Hoboken, NJ. pp. 352–353. ISBN 0-471-80580-7.

839. Bock, E. (1961) On the solubility of anhydrous calcium sulphate and of gypsum in concentrated solutions of sodium chloride at 25°C, 30°C, 40°C, and 50°C. *Canadian Journal of Chemistry*, 39(9), 1746–1751.

840. García-Ruiz, J.M., Villasuso, R., Ayora, C., Canals, A. & Otálora, F. (2007) Formation of natural gypsum megacrystals in Naica, Mexico (PDF). *Geology*, 35(4), 327–330.

841. Cockell, C.S. & Raven, J.A. (2007) Ozone and life on the Archaean Earth. *Philosophical Transactions of the Royal Society A*, 365(1856), 1889–1901.

842. Deer, W.A., Howie, R.A. & Zussman, J. (1966) *An Introduction to the Rock Forming Minerals*. Longman, London. p. 469. ISBN 0-582-44210-9.

843. Oster, J.D. & Frenkel, H. (1980) The chemistry of the reclamation of sodic soils with gypsum and lime. *Soil Science Society of America Journal*, 44(1), 41–45.

844. Astilleros, J.M., Godelitsas, A., Rodríguez-Blanco, J.D., Fernández-Díaz, L., Prieto, M., Lagoyannis, A. & Harissopulos, S. (2010) Interaction of gypsum with lead in aqueous solutions. *Applied Geochemistry*, 25(7), 1008.

845. Rodriguez, J.D., Jimenez, A., Prieto, M., Torre, L. & Garcia-Granda, S. (2008) Interaction of gypsum with As(V)-bearing aqueous solutions: Surface precipitation of guerinite, sainfeldite, and $Ca_2NaH(AsO_4)_2 \cdot 6H_2O$, a synthetic arsenate. *American Mineralogist*, 93(5–6), 928.

846. Rodríguez-Blanco, J.D., Jiménez, A. & Prieto, M. (2007) Oriented overgrowth of pharmacolite ($CaHAsO_4 \cdot 2H_2O$) on gypsum ($CaSO_4 \cdot 2H_2O$). *Crystal Growth and Design*, 7(12), 2756–2763.

847. Pohl, W.L. (2011) *Economic Geology: Principles and Practice: Metals, Minerals, Coal and Hydrocarbons: Introduction to Formation and Sustainable Exploitation of Mineral Deposits*. Wiley-Blackwell, Chichester, West Sussex. p. 331. ISBN 978-1-4443-3662-7.

848. Perry, D.L. (2011) *Handbook of Inorganic Compounds*. Taylor & Francis. ISBN 978-1-4398-1461-1.

849. Bellotto, M., Gualtieri, A., Artioli, G. & Clark, S.M. (1995) Kinetic study of the kaolinite-mullite reaction sequence: Part I: kaolinite dehydroxylation. *Physics and Chemistry of. Minerals*, 22(4), 207–214.

850. Meijer, E.L. & Plas, L. van der (1980) Relative stabilities of soil minerals. *Mededelingen Landbouwhogeschool Wageningen*, 80(16), 18p.

851. Millot, G. (1970) *Geology of Clays*. Springer Verlag, New York. 429p.

852. Caillère, S. & Hénin, S. (1947) Formation d'une phyllite du type kaolinique par traitement d'une montmorillonite. *Comptes Rendus des Séances de l'Académie des Sciences* (Paris), 224, 53–55.

853. Norton, F.H. (1939) Hydrothermal formation of clay minerals in the laboratory. *American Mineralogist*, 24, 1–17.

854. Roy, R. & Osborn, E.F. (1954) The system alumina-silica-water. *American Mineralogist*, 39, 853–885.

855. Hawkins, D.B. & Roy, R. (1962) Electrolytic synthesis of kaolinite under hydrothermal conditions. *Journal of the American Ceramic Society*, 45, 507–508.

856. Tomura, S., Shibasaki, Y., Mizuta, H. & Kitamura, M. (1985) Growth conditions and genesis of spherical and platy kaolinite. *Clays and Clay Minerals*, 33, 200–206.

857. Satokawa, S., Osaki, Y., Samejima, S., Miyawaki, R., Tomura, S., Shibasaki, Y. & Sugahara, Y. (1994) Effects of the structure of silica-alumina gel on the hydrothermal synthesis of kaolinite. *Clays and Clay Minerals*, 42, 288–297.

858. Huertas, F.J., Fiore, S., Huertas, F. & Linares, J. (1999) Experimental study of the hydrothermal formation of kaolinite. *Chemical Geology*, 156, 171–190.

859. Brindley, G.W. & DeKimpe, C. (1961) Attempted low-temperature syntheses of kaolin minerals. *Nature*, 190, 254.

860. DeKimpe, C.R. (1969) Crystallization of kaolinite at low temperature from an alumina-silicic gel. *Clays and Clay Minerals*, 17, 37–38.

861. Bogatyrev, B.A., Mateeva, L.A., Zhukov, V.V. & Magazina, L.O. (1997) Low-temperature synthesis of kaolinite and halloysite on the gibbsite: Silicic acid solution system. *Transactions (Doklady) of the Russian Academy of Sciences/Earth Science Sections*, 353A, 403–405.

862. Ciullo, P.A. (1996) *Industrial Minerals and Their Uses: A Handbook and Formulary*. William Andrew. pp. 41–43. ISBN 978-0-8155-1408-4.

863. Diamond, J.M. (1999) Evolutionary biology: Dirty eating for healthy living. *Nature*, 400(6740), 120–121.

864. Leiviskä, T., Gehör, S., Eijärvi, E., Sarpola, A. & Tanskanen, J. (2012) Characteristics and potential applications of coarse clay fractions from Puolanka, Finland. *Central European Journal of Engineering*, 2(2), 10 April, 239–247.

865. Speyer, R. (1993) *Thermal Analysis of Materials*. CRC Press. p. 166. ISBN 0-8247-8963-6.

866. Gaft, M., Nagli, L., Panczer, G., Rossman, G.R. & Reisfeld, R. (2011) Laser-induced time-resolved luminescence of orange kyanite Al_2SiO_5. *Optical Materials* (sciencedirect.com), 33(10), August, 1476–1480.

867. Hurlbut, C.S. & Klein, C. (1985) *Manual of Mineralogy*, 20th edition. Wiley. ISBN 0-471-80580-7.

868. Nechamkin (1968) *The Chemistry of the Elements*. McGraw-Hill, New York.

869. Yoshioka, S. & Kitano, Y. (1985) Transformation of aragonite to calcite through heating. *Geochemical Journal*, 19, 24–249.

870. Thompson, D.W., Devries, M.J., Tiwald, T.E. & Woollam, J.A. (1998) Determination of optical anisotropy in calcite from ultraviolet to mid-infrared by generalized ellipsometry. *Thin Solid Films*, 313–314: 341.

871. Leitmeier, H. (1916) Einige Bemerkungen über die Entstehung von Magnesit und Sideritlagerstätten. *Mitteilungen der Geologischen Gesellschaft in Wien*, 9, 159–166.

872. Brown, T.J., Hobbs, S.F., Mills, A.J., Petavratzi, E., Raycraft, E.R., Shaw, R.A. & Bide, T. European Mineral Statistics 2007–2011, British Geological Survey, 2013. National Environment Research Council (NERC), Keyworth, Nottingham, UK.

873. Hurlbut, C.S., W. Edwin Sharp & Edward Salisbury Dana (1998) *Dana's Minerals and How to Study Them*. John Wiley and Sons, Hoboken, NJ. p. 96. ISBN 0-471-15677-9.

874. Harrison, R.J., Dunin-Borkowski, R.E. & Putnis, A. (2002) Direct imaging of nanoscale magnetic interactions in minerals. *Proceedings of the National Academy of Sciences*, 99(260), 16556–16561 MID 12482930.

875. Oeters, F. et al. (2006) Iron. In: *Ullmann's Encyclopedia of Industrial Chemistry*. Wiley-VCH, Weinheim. doi:10.1002/14356007.a14_461.pub2.

876. Appl, M. (2011) Ammonia, 2: Production processes. In: *Ullmann's Encyclopedia of Industrial Chemistry*. Wiley-VCH. Verlag GmbH & Co. KGaA.

877. Deer, W.A., Howie, R.A. & Zussman, J. (1966) *An Introduction to the Rock Forming Minerals*. Longman. ISBN 0-582-44210-9.

878. Dolley, T.P. (2008) Mica In: *USGS 2008 Minerals Yearbook*. US Geological Survey, Reston, VA.

879. Mica, *USGS mineral commodity summaries* 2011. US Geological Survey, Reston, VA.

880. Wallace, S.W. (1953) The petrology of the Judith Mountains, Fergus County, Montana (Report). U.S. Geological Survey.

881. Sutherland, D.S. (ed.) (1982) *Igneous Rocks of the British Isles,* Vol. 46, Issue 340. John Wiley & Sons, Chichester and New York. p. 211.

882. Eby, G.N. (2012) The Beemerville alkaline complex, northern New Jersey. In: Harper, J.A. (ed.) *Journey Along the Taconic unconformity*. Guidebook, Northeastern Pennsylvania, NJ, and Southeastern New York. 77th Annual Field Conference of Pennsylvania Geologists, Shawnee on Delaware, PA. pp. 85–91.

883. *Shorter Oxford English Dictionary*. Oxford University Press, Oxford, 2002.

884. *The Random House College Dictionary*, revised edition (1980) Any of a class of natural earths, mixtures of hydrated oxides of iron and various earthy materials, ranging in color from pale yellow to orange and red, and used as pigments: A color ranging from pale yellow to reddish-yellow. New York City, USA.

885. Roelofs, I. (2012) *La couleur expliquée aux artistes*. Groupe Eyrolles. ISBN 978-2-212-134865. p. 30.

886. London, D. & Kontak, D.J. (2012) Granitic pegmatites: Scientific wonders and economic bonanzas. *Elements*, 8(4), 3 September, 257–261.

887. Deer, W.A. (2004) *Framework Silicates: Silica Minerals, Feldspathoids and the Zeolites*, 2 edition. Geological Society, London. p. 296. ISBN 1-86239-144-0.

888. Gnandil, K., Tchangbedjil, G., Killil, K., Babal, G. & Abbel, E. (2006) The impact of phosphate mine tailings on the bioaccumulation of heavy metals in marine fish and crustaceans from the coastal zone of Togo. *Mine Water and the Environment*, 25(1), March, 56–62.

889. Joyce A. Ober, Mineral Commodity Summaries; Potash. *USGS 2008 Minerals Yearbook*. US Geological Survey, Reston, VA.

890. Davy, H. (1808) On some new phenomena of chemical changes produced by electricity, in particular the decomposition of the fixed alkalies, and the exhibition of the new substances that constitute their bases, and on the general nature of alkaline bodies. *Philosophical Transactions of the Royal Society of London*, 98, 32.

891. Alikhan, I. (2014) *Management of Agricultural Inputs*. Agrotech Publishing Academy. ISBN 9789383101474.

892. Jasinski, S.M. Potash. (2011) In: *USGS*. UG Geological Survey, Reston, VA.

893. Holleman, A.F., Wiberg, E. & Wiberg, N. (1985) Potassium. *Lehrbuch der Anorganischen Chemie*, in German, 91st–100rd edition. Walter de Gruyter. ISBN 3-11-007511-3.

894. Greenwood, N.N. & Earnshaw, A. (1997) *Chemistry of the Elements*, 2nd edition. Butterworth-Heinemann, Oxford. p. 69. ISBN 0-08-037941-9. Cite uses deprecated parameter |coauthor= (help).

895. Mehta, P.K. (1987) Natural pozzolans: Supplementary cementing materials in concrete. *CANMET Special Publication*, 86, 1–33.

896. Snellings, R., Mertens, G. & Elsen, J. (2012) Supplementary cementitious materials. *Reviews in Mineralogy and Geochemistry*, 74, 211–278.

897. Chappex, T. & Scrivener, K. (2012) Alkali fixation of C-S-H in blended cement pastes and its relation to alkali silica reaction. *Cement and Concrete Research*, 42, 1049–1054.

898. Jackson, J.A., Mehl, J. & Neuendorf, K. (2005) *Glossary of Geology*. American Geological Institute, Alexandria, Virginia. 800pp. ISBN 0-922152-76-4.

899. McPhie, J., Doyle, M. & Allen, R. (1993) *Volcanic Textures A Guide to the Interpretation of Textures in Volcanic Rocks*, Centre for Ore Deposit and Exploration Studies. University of Tasmania, Hobart, Tasmania. 198pp. ISBN 9780859015226.

900. Venezia, A.M., Floriano, M.A., Deganello, G. & Rossi, A. (1992) Structure of pumice: An XPS and 27Al MAS NMR study. *Surface and Interface Analysis*, 18(7), July, 532–538.

901. De Vantier, L.M. (1992) Rafting of tropical marine organisms on buoyant coralla (PDF). *Marine Ecology Progress Series*, 86, 301–302

902. Levy, R. & Przyborski, P. Salt Ponds, South San Francisco Bay. NASA Visible Earth. NASA.

903. Flörke, O.W. et al. (2008) Silica. In: *Ullmann's Encyclopedia of Industrial Chemistry*. Wiley-VCH, Weinheim.

904. Greenwood, N.N. & Earnshaw, A. (1984) *Chemistry of the Elements*. Oxford: Pergamon Press. pp. 393–99. ISBN 0-08-022057-6.

905. Nandiyanto, A.B.D., Kim, S.G., Iskandar, F. & Okuyama, K. (2009) Synthesis of spherical mesoporous silica nanoparticles with nanometer-size controllable pores and outer diameters. *Microporous and Mesoporous Materials*, 120(3), 447.

906. Morgan, D.V. & Board, K. (1991) *An Introduction to Semiconductor Microtechnology*, 2nd edition. John Wiley & Sons, Chichester, West Sussex, England. p. 27. ISBN 0471924784.

907. Ojovan, M.I. (2004) Glass formation in amorphous SiO_2 as a percolation phase transition in a system of network defects. *Journal of Experimental and Theoretical Physics Letters*, 79(12), 632–634.

908. Elliott, S.R. (1991) Medium-range structural order in covalent amorphous solids. *Nature*, 354(6353), 445–452.

909. Shriver and Atkins (2010) *Inorganic Chemistry*, 5th edition. W. H. Freeman and Company, New York. p. 354.

910. Doering, R. & Nishi, Y. (2007) *Handbook of Semiconductor Manufacturing Technology*. CRC Press. ISBN 1-57444-675-4.

911. Lee, S. (2006) *Encyclopedia of Chemical Processing*. CRC Press. ISBN 0-8247-5563-4.

912. Morgan, D.V. & Board, K. (1991) *An Introduction to Semiconductor Microtechnology*, 2nd editon. John Wiley & Sons, Chichester, West Sussex, England. p. 72. ISBN 0471924784.

913. Clow, A. & Clow, N.L. (1952) *Chemical Revolution*. Ayer Co. Pub. pp. 65–90. ISBN 0-8369-1909-2.

914. Deer, H. & Zussman, J. (1963) *Rock Forming Minerals*, Volume 2. Chain Silicates, Wiley. pp. 92–98.

915. Anthony, J.W., Bideaux, R.A., Bladh, K.W. & Nichols, M.C. (1990) *Handbook of Mineralogy*. Mineral Data Publishing, Tucson, Arizona.

916. Deer, W.A., Howie, R.A. & Zussman, J. (1992) *An Introduction to the Rock-Forming Minerals*, 2nd edition. Prentice Hall. ISBN 0-582-30094-0.

917. Webb, T.H. (1824) Letter to the editor: "New localities of tourmalines and talc". *American Journal of Science, and Arts*, 7(55), 55.

918. Wollastonite (2017) *USGS Mineral Commodity Summaries*. USGS, Reston, Virginia, USA.

919. Deer, H. & Zussman, J. (1997) *Rock Forming Minerals: Single Chain Silicates*, Volume 2A, 2nd edition. The Geological Society, London.

920. Andrews, R.W. (1970) *Wollastonite*. Her Majesty's Stationery Office, London.

921. Rollmann, L.D. & Valyocsik, E.W. (1995) *Zeolite Molecular Sieves: Inorganic Syntheses*, Inorganic Syntheses, Volume 30. pp. 227–234.

922. Bhatia, S. (1989) *Zeolite Catalysts: Principles and Applications*. CRC Press, 21 December. ISBN 9780849356285.

923. Virta, R.L. (2010) Zeolites. *USGS 2009 Minerals Yearbook*, October.

924. Ventura, G. & Risegari, L. (2007) *The Art of Cryogenics: Low-Temperature Experimental Techniques*, Surendra Kumar (ed.), 26 November. p. 17. ISBN 978-0-08-044479-6.

925. Wei, L., Kuo, P.K., Thomas, R.L., Anthony, T. & Banholzer, W. (1993) Thermal conductivity of isotopically modified single crystal diamond. *Physical Review Letters*, 70(24), 3764–3767.

926. Wissner-Gross, A.D. & Kaxiras, E. (2007) Diamond stabilization of ice multilayers at human body temperature (PDF). *Physical Review E.*, 76, 020501

927. Fujimoto, A., Yamada, Y., Koinuma, M. & Sato, S. (2016) Origins of sp^{3}C peaks in C$_{1s}$ X-ray photoelectron spectra of carbon materials. *Analytical Chemistry*, 88, 6110.

928. Walker, J. (1979) Optical absorption and luminescence in diamond (PDF). *Reports on Progress in Physics*, 42(10), 1605–1659.

929. Shirey, S.B. & Shigley, J.E. (2013) Recent advances in understanding the geology of diamonds. *Gems & Gemology*, 49(4), 1 December.

930. Shigley, J.E., Abbaschian, R. & Shigley, J.E. (2002) Gemesis laboratory created diamonds. *Gems & Gemology*, 38(4), 301–309.

931. Shigley, J.E., Shen, A.H.-T., Breeding, C.M., McClure, S.F. & Shigley, J.E. (2004) Lab grown colored diamonds from chatham created gems. *Gems & Gemology*, 40(2), 128–145.

932. Yarnell, A. (2004) The many facets of man-made diamonds. *Chemical and Engineering News*, 82(5), 26–31.

933. Werner, M. & Locher, R. (1998) Growth and application of undoped and doped diamond films. *Reports on Progress in Physics*, 61(12), 1665.

934. O'Donoghue, M. & Joyner, L. (2003) *Identification of Gemstones*. Butterworth-Heinemann, Great Britain. pp. 12–19. ISBN 0-7506-5512-7.

935. Erlich, E.I. & Hausel, W.D. (2002) *Diamond Deposits: Origin, Exploration, and History of Discovery*. Society for Mining, Metallurgy, and Exploration, Littleton, CO. ISBN 0-87335-213-0.

936. Barnard, A.S. (2000) *The Diamond Formula*. Butterworth-Heinemann. p. 115. ISBN 0-7506-4244-0.

937. Zhang, W., Ristein, J. & Ley, L. (2008) Hydrogen-terminated diamond electrodes, II: Redox activity. *Physical Review E*, 78(4), 041603.

938. Pierson, H.O. (1993) *Handbook of Carbon, Graphite, Diamond, and Fullerenes: Properties, Processing, and Applications*. William Andrew. p. 280. ISBN 0-8155-1339-9.

939. James, D.S. (1998) *Antique Jewellery: Its Manufacture, Materials and Design*. Osprey Publishing. pp. 82–102. ISBN 0-7478-0385-4.

940. Prelas, M.A., Popovici, G. & Bigelow, L.K. (1998) *Handbook of Industrial Diamonds and Diamond Films*. CRC Press. pp. 984–992. ISBN 0-8247-9994-1.

941. Tozlukov, V. Gem cutting: Popular mechanics. *Hearst Magazines*, 74(5), 760–764, 1940. ISSN 0032-4558.

942. Hurlbut, C.S., Jr. & Kammerling, R.C. (1991) *Gemology*. John Wiley & Sons, New York. p. 203. ISBN 0-471-52667-3.

943. Wise, R.W. (2001) *Secrets of the Gem Trade: The Connoisseur's Guide to Precious Gemstones*. Brunswick House Press. p. 108. ISBN 0-9728223-8-0.

944. Bauer, M. Precious Stones, vol. 1 (Dover Jewelry and Metalwork) Dover Publications (1968). ISBN 9780486219103

945. Milisenda, C.C. (2005) Rubine mit bleihaltigen Glasern gefullt. *Zeitschrift der Deutschen Gemmologischen Gesellschaft* (in German) (Deutschen Gemmologischen Gesellschaft), 54(1), 35–41.

946. McCure, S.F., Smith, C.P., Wang, W. & Hall, M. (2006) Identification and durability of lead glass-filled rubies. *GIA Global Dispatch, Gems & Gemology*, 42(1), 22–34. Gemology Institute of America.

947. Maiman, T.H. (1960) Stimulated optical radiation in ruby. *Nature*, 187(4736), 493–494.

948. (2009) Ruby and sapphire products and distribution: A quarter century of change. *Gems & Gemology,* 45(4), 236–259. Gemology Institute of America.

949. Heaton, N. (1912) *The Production and Identification of Artificial Precious Stones.* Annual Report of the Board of Regents of the Smithsonian Institution, 1911. Government Printing Office, USA. p. 217.

950. Scheel, H.J. & Fukuda, T. (2003) *Crystal Growth Technology* (PDF). J. Wiley, Chichester, West Sussex. ISBN 0-471-49059-8.

951. Dobrovinskaya, E.R., Lytvynov, L.R. & Pishchik, V. (2009) *Sapphire: Materials, Manufacturing, Applications.* Springer. p. 3. ISBN 0-387-85694-3.

952. Starr, C. (2005) *Biology: Concepts and Applications.* Thomson Brooks/Cole. p. 94. ISBN 0-534-46226-X.

953. Bianchi, M.P. (2002) Gallium nitride collector grid solar cell. U.S. Patent 6, 447, 938. Northrop Grumman Systems Corp.

954. Hyndman, D.W. & Alt, D.D. (2002) *Roadside Geology of Oregon,* 18th edition. Mountain Press Publishing Company, Missoula, Montana. p. 286. ISBN 0-87842-063-0.

955. Rudler, F.W. (1911) Amethyst. In: Chisholm, H. (eds.) *Encyclopædia Britannica,* Volume 1, 11th edition. Cambridge University Press, Cambridge. p. 852.

956. Greenwood, N.N. & Earnshaw, A. (1997) *Chemistry of the Elements,* 2nd edition. Butterworth-Heinemann. ISBN 0080379419.

957. Lameiras, F.S., Nunes, E.H.M. & Vasconcelos, W.L. (2009) Infrared and chemical characterization of natural amethysts and prasiolites colored by irradiation. *Materials Research,* 12(3), 315–320.

958. O'Donoghue, M. (2006) *Gems,* 6th edition. Butterworth-Heinemann. ISBN 978-0-7506-5856-0.

959. Wise, R.W. (2005) *Secrets of the Gem Trade: The Connoisseur's Guide to Precious Gemstones,* Brunswick House Press, Lenox, MA. ISBN 0-9728223-8-0.

960. Rossman, G.R. (1994) Ch.13: Colored varieties of the silica minerals. In: Heaney, P.J., Prewitt, C.T. & Gibbs, G.V. (eds.) *Silica: Physical Behavior, Geochemistry, and Materials Applications.* Reviews in Mineralogy, Volume 29. Mineralogical Society of America. pp. 433–468. ISBN 0-939950-35-9.

961. O'Donoghue, M. (1997) *Synthetic, Imitation, and Treated Gemstones.* Taylor & Francis. pp. 124–125. ISBN 978-0-7506-3173-0.

962. George, O. (1957) The amateur lapidary. *Rocks & Minerals,* 32(9–10), 471.

963. CIT (2009) Color in the beryl group. *Mineral Spectroscopy Server.* California Institute of Technology. Minerals.caltech.edu

964. Ibragimova, E.M., Mukhamedshina, N.M. & Islamov, A.Kh. (2009) Correlations between admixtures and color centers created upon irradiation of natural beryl crystals. *Inorganic Materials,* 45(2), 162.

965. Viana, R.R., Da Costa, G.M., De Grave, E., Stern, W.B. & Jordt-Evangelista, H. (2002) Characterization of beryl (aquamarine variety) by Mössbauer spectroscopy. *Physics and Chemistry of Minerals,* 29(1), 78.

966. Blak, A.R., Isotani, S., Watanabe, S. (1983) Optical absorption and electron spin resonance in blue and green natural beryl: A reply. *Physics and Chemistry of Minerals.* 9(6), 279.

967. Nassau, K. (1976) The deep blue Maxixe-type color center in beryl (PDF). *American Mineralogist,* 61, 100.

968. Klein, C., Dutrow, B., Dana, J.D. (2007) *The Manual of Mineral Science.* (after James D. Dana), 23rd edition. J. Wiley, Hoboken, NJ. ISBN 0-471-72157-3.

969. CIT (2009) Color in the beryl group. *Mineral Spectroscopy Server.* California Institute of Technology. Minerals.caltech.edu

970. Ibragimova, E.M., Mukhamedshina, N.M., & Islamov, A.Kh. (2009) Correlations between admixtures and color centers created upon irradiation of natural beryl crystals. *Inorganic Materials,* 45(2), 162.

971. Ege, C. (2002) What gemstone is found in Utah that is rarer than diamond and more valuable than gold?. *Survey Notes: Utah Geological Survey*, 34(3), September.

972. Klein, C., Hurlbut, C.S., Jr. (1985) *Manual of Mineralogy*, 20th edition. Wiley, New York. ISBN 0-471-80580-7.

973. Clark, D. Is synthetic Alexandrite Real Alexandrite? *International Gem Society*.

974. Weinberg, D. & Farnell, K. Alexandrite synthetics and imitations. in Alexandrite Tsarstone Collectors Guide. (2006).

975. Mitchell, T.E. & Marder, J.M. (1982) Precipitation in cat's-eye chrysoberyl. *Electron Microscopy Soc.* Proceedings.

976. Hoover, D.B., Williams, B., Williams, C. & Mitchell, C. (2008) Magnetic susceptibility, a better approach to defining garnets. *The Journal of Gemmology*, 31(3/4), 91–103.

977. Joyce, E. (1987) [1970] *The Technique of Furniture Making*, 4th edition, Peters, A. (ed.). Batsford, London. ISBN 071344407X.

978. Murdoch, J. (1957) Crystallography and X-ray measurement of howlite from California. *American Mineralogist*, 42, 521–524.

979. Tay Thye Sun. (2018) The changing face of Jade (PDF). *Alumni Newsletter No. 3.* pp. 5–6.

980. Liu, L. (2003) The products of minds as well as hands: Production of prestige goods in neolithic and early state periods of China. *Asian Perspectives*, 42(1), 1–40. p. 2.

981. Duda, R., Rejl, L. (1990) *Minerals of the World*. Arch Cape Press. ISBN 0-517-68030-0.

982. Willing, M.J. & Stocklmayer, S.M. (2003) A new chrome chalcedony occurrence from Western Australia. *Gems & Gemology*, 23, 265–279.

983. Cairncross, B. (2005) *Field Guide to Rocks & Minerals of Southern Africa*. Struik. ISBN 1-86872-985-0.

984. Hey, M.H. & Embrey, P.G. (1974) Twenty-eighth list of new mineral names (PDF). *Mineralogical Magazine*, 39(308), 903–932.

985. Hey, M.H. (1970) Twenty-sixth list of new mineral names (PDF). *Mineralogical Magazine*, 37(292), 954–967.

986. O'Donoghue, M. (2006) *Gems*. Butterworth-Heinemann. ISBN 0-7506-5856-8.

987. Insider gemologist: What are the identifying characteristics of the different varieties of green chalcedony? *GIA Insider (Gemological Institute of America)*, 7(13), 2005.

988. Heaney, P.J. (1994) Structure and chemistry of the low-pressure silica polymorphs. *Reviews in Mineralogy and Geochemistry*, 29, 1–40.

989. Lule-Whipp, C. (2006) Chromium chalcedony from Turkey and its possible archeological connections (PDF). *Gems & Gemology: Proceedings of the 4th International Gemological Symposium & GIA Gemological Research Conference, San Diego, California 42.* p. 115.

990. Thoresen, L. Ancient Glyptic art: Gem engraving and gem carving.

991. Sanders, J.V. (1964) Colour of precious opal. *Nature*, 204(496), 1151–1153.

992. Sanders, J.V. (1968) Diffraction of light by opals. *Acta Crystallographica*, A24, 427–434.

993. Klein, C. & Hurlbut, C.S. (1998) *Manual of Mineralogy*, 21st edition. Wiley. ISBN 0-471-80580-7.

994. Astratov, V.N., Bogomolov, V.N., Kaplyanskii, A. A., Prokofiev, A.V., Samoilovich, L.A., Samoilovich, S.M. & Vlasov, Y.A. (1995) Optical spectroscopy of opal matrices with CdS embedded in its pores: Quantum confinement and photonic band gap effects. *Il Nuovo Cimento D*, 17(11–12), 1349–1354.

995. Sima, P. (1990) *Evolution of Immune Reactions*. Taylor & Francis. ISBN 0849365937.

996. Neil, H.L. et al. (2001) *Pearls: A Natural History*. Harry Abrams, Inc. ISBN 0-8109-4495-2.

997. Scarratt, K. (1999) *The Pearl and the Dragon*, 1st edition. Houlton. ISBN 0-935681-07-8.

998. Deer, W.A., Howie, R.A. & Zussman, J. (1966) *An Introduction to the Rock Forming Minerals*. Longman. pp. 340–355. ISBN 0-582-44210-9.

999. Anthony, J.W., Bideaux, R.A., Bladh, K.W. & Nichols, M.C. (eds.) (2004) Quartz. *Handbook of Mineralogy* (PDF). III (Halides, Hydroxides, Oxides). Mineralogical Society of America, Chantilly, VA, US. ISBN 0962209724.

1000. Briggs, P. & McIntyre, C. (2013) *Tanzania Safari Guide: With Kilimanjaro, Zanzibar and the Coast.* Bradt Travel Guides. p. 104. ISBN 978-1-84162-462-4.

1001. James, Y. Study of Heat Treatment. *Yourgemologist.com.* International School of Gemology.

1002. Wise, R.W. (2005) *Secrets of the Gem Trade: The Connoisseur's Guide to Precious Gemstones.* Brunswick House Press, 31 December. pp. 35, 220. ISBN 978-0-9728223-8-1.

1003. Weldon, R. An introduction to gem treatments. gia.edu. Gemological Institute of America, Inc.

1004. Roskin, G. (2005) Natural-color tanzanite. *JCK Magazine,* February.

1005. King, R.M. (2014) Featured gemstone: Tanzanite. *GIA Library.* Gemological Institute of America, Inc. Carlsbad, California, USA.

1006. Hawthorne, F.C. & Henry, D.J. (1999) Classification of the minerals of the tourmaline group. *European Journal of Mineralogy,* 11, 201–215.

1007. Nassau, K. (1984) *Gemstone Enhancement: Heat, Irradiation, Impregnation, Dyeing, and Other Treatments.* Butterworth Publishers. Oxford, UK.

1008. Hubert, J.F. (1962) A zircon-tourmaline-rutile maturity index and the interdependence of the composition of heavy mineral assemblages with the gross composition and texture of sandstones. *Journal of Sedimentary Research,* 32(3), September.

1009. ASTM, C18. C119-08 Standard terminology relating to dimension stone. *ASTM,* 2008. p. 8. ISBN 0-8031-4118-1.

1010. ASTM, C18. C119-06 Standard terminology relating to dimension stone. *ASTM,* 2007. pp. 11–13. ISBN 0-8031-4104-1| Types of Stones by Group|.

1011. Mead, L. & Austin, G.S. (2005) Dimension stone. In: *Industrial Minerals and Rocks,* 7th edition. AIME-Society of Mining Engineers, Littleton CO. pp. 907–923. ISBN 0-87335-233-5.

1012. Blatt, H. & Tracy, R.J. (1997) *Petrology,* 2nd edition. Freeman, New York. p. 66. ISBN 0-7167-2438-3.

1013. Kearey, P. (2001) *Dictionary of Geology.* Penguin Group, London and New York. p. 163. ISBN 978-0-14-051494-0.

1014. Doronina, N.V., TsD, L., Ivanova, E.G. & Trotsenko, IuA. (2005) Methylophaga murata sp. nov.: A haloalkaliphilic aerobic methylotroph from deteriorating marble. *Mikrobiologiia,* 74(4), 511–519.

1015. Cappitelli, F., Principi, P., Pedrazzani, R., Toniolo, L. & Sorlini, C. (2007) Bacterial and fungal deterioration of the Milan Cathedral marble treated with protective synthetic resins. *The Science of the Total Environment,* 385(1–3), 172–181.

1016. Cappitelli, F., Nosanchuk, J.D., Casadevall, A., Toniolo, L., Brusetti, L., Florio, S., Principi, P. Borin, S. & Sorlini, C. (2007) Synthetic consolidants attacked by melanin-producing fungi: Case study of the biodeterioration of Milan (Italy) cathedral marble treated with acrylics. *Applied Environmental Microbiology,* 73(1), 271–277.

1017. Serpentine, Encyclopedia Britannica, 1911, s:1911 Encyclopædia Britannica/Serpentine (mineral).

1018. Hunter, Sir W.W. & Burn, Sir R. (1907) *The Imperial Gazeteer of India,* Volume 3. Clarendon Press, Henry Frowde Publishers, Oxford, England. p. 242.

1019. Marshak, S. *Essentials of Geology,* 5th edition. W.W. Norton & Co. ISBN 97803936011.

1020. Bhan, K.K., Vidale, M. & Kenoyer, J.M. (2002) Some important aspects of the Harappan technological tradition. In: Settar, S. & Korisettar, R. (eds.) *Indian Archaeology in Retrospect.* Manohar Press, New Delhi.

1021. Sassaman, K.E. (1993) *Early Pottery in the Southeast: Tradition and Innovation in Cooking Technology.* University Alabama Press, 30 March. ISBN 978-0-8173-0670-0.

1022. Rosendahl, E. (1987) *The Vikings.* The Penguin Press. London, UK. p. 105.

1023. Blander, M. (1989) *Calculations of the Influence of Additives on Coal Combustion Deposits* (PDF). Argonne National Laboratory. Argonne, Il, USA. p. 315.

1024. Taylor, T.N., Taylor, E.L. & Krings, M. (2009) *Paleobotany: The Biology and Evolution of Fossil Plants.* Academic Press. ISBN 978-0-12-373972-8.

1025. Nunn, J., Cottrell, A., Urfer, A., Wibberley, L. & Scaife, P. (2003) *A Lifecycle Assessment of the Victorian Energy Grid*. Cooperative Research Centre for Coal in Sustainable Development, February. (CCSD) Pullenvale, Quensland, Australia. p. 7.

1026. Rosenkranz, J. & Wichtmann, A. (2005) *Balancing Economics and Environmental Friendliness: The Challenge for Supercritical Coal-Fired Power Plants with Highest Steam Parameters in the Future* (PDF). Siemens AG. Munich, Germany.

1027. Cziesla, F., Bewerunge, J. & Senzel, A. (2009) Siemens, A.G. Lünen: State-of-the art ultra supercritical steam power plant under construction (PDF).

1028. Chisholm, H. (ed.) (1911) Coke. In: *Encyclopædia Britannica*, Volume 6, 11th edition. Cambridge University Press, Cambridge. p. 657.

1029. Speight, J.G. (2008) *Synthetic Fuels Handbook: Properties, Process, and Performance*. McGraw-Hill Professional. pp. 9–10. ISBN 978-0-07-149023-8.

1030. Lee, S. (1996) *Alternative Fuels*. CRC Press. pp. 166–198. ISBN 978-1-56032-361-7.

1031. Cleaner Coal Technology Programme (1999) Technology status report 010: coal liquefaction (PDF). Department of Trade and Industry, October.

1032. Boyajian, G. Electricity from natural gas. Introduction to Primus' STG+ Technology Primus Green Energy.

1033. Campbell, L., Phillips, F., Lague, J. & Broekhuijsen, J. Composition of natural gas. World Bank, GGFR partners unlock value of wasted gas. *World Bank*, Washington, DC, 14 December 2009.

1034. US Geological Survey, Organic origins of petroleum; Roberts, Ken. Modular design of smaller-scale GTL plants (PDF). *Petroleum Technology Quarterly*.

1035. Tobin, J., Shambaugh, P. & Mastrangelo, E. (2006) Natural gas processing: The crucial link between natural gas production and its transportation to market (PDF).

1036. Zimmerman, B.E. & Zimmerman, D.J. (1995) *Nature's Curiosity Shop*. Contemporary Books, Lincolnwood, Chicago, IL. p. 28. ISBN 978-0-8092-3656-5.

1037. The World Factbook. LaMonica, M. (2012) Natural gas tapped as bridge to biofuels. *MIT Technology Review*, 27 June.

1038. Hugron, S., Bussières, J. & Rochefort, L. (2013) *Tree Plantations within the Context of Ecological Restoration of Peatlands: Practical Guide* (PDF) (Report). Peatland Ecology Research Group (PERG), Laval, Québec, Canada.

1039. Keddy, P.A. (2010) *Wetland Ecology: Principles and Conservation*, 2nd edition. Cambridge University Press, Cambridge, UK. 497 p. Chapter 1.

1040. Keddy, P.A. (2010) *Wetland Ecology: Principles and Conservation*, 2nd edition. Cambridge University Press, Cambridge, UK. pp. 323–325, 497.

1041. World Energy Council (2007) Survey of energy resources 2007 (PDF).

1042. Keddy, P.A. 2010. *Wetland Ecology: Principles and Conservation*, 2nd edition. Cambridge University Press, Cambridge, UK. 497p. Chapter 7.

1043. Gorham, E. (1957) The development of peatlands. *Quarterly Review of Biology*, 32, 145–166.

1044. Kopp, O.C. Lignite. In: *Encyclopædia Britannica*. Encyclopaedia Britanica, Inc. USA.

1045. Ghassemi, A. (2001) *Handbook of Pollution Control and Waste Minimization*. CRC Press. p. 434. ISBN 0-8247-0581-5.

1046. George, A.M. (1975) State electricity Victoria. Petrographic report No 17; Perry, G.J. & Allardice, D.J. *Coal Resources Conference*, NZ, 1987, Proc.1, Sec. 4. Paper R4.1.

1047. Blatt, H., Middleton, G. & Murray, R. (1972) *Origin of Sedimentary Rocks*. Prentice-Hall Inc., NJ. ISBN 0-13-642702-2.

1048. Norman, J.H. (2001) *Nontechnical Guide to Petroleum Geology, Exploration, Drilling, and Production*, 2nd edition. Penn Well Corp., Tulsa, OK. pp. 1–4. ISBN 087814823X.

1049. Ollivier, B. & Magot, M. (2005) *Petroleum Microbiology*. American Society of Microbiology, Washington, DC, 1 January. ISBN 9781555817589.

1050. Speight, J.G. (1999) *The Chemistry and Technology of Petroleum*, (3rd edition, rev. and expanded ed. Marcel Dekker, New York. pp. 215–216, 543. ISBN 0824702174.

1051. Guerriero, V. et al. (2012) A permeability model for naturally fractured carbonate reservoirs. *Marine and Petroleum Geology*. Elsevier, 40, 115–134.

1052. Guerriero, V. et al. (2011) Improved statistical multi-scale analysis of fractures in carbonate reservoir analogues. *Tectonophysics*. Elsevier, 504, 14–24.

1053. EC (2008) The raw materials initiative: Meeting our critical needs for growth and jobs in Europe. COM 699.

1054. OECD (2009) A sustainable materials management case study: Critical metals and mobile devices. OECD, Paris, France.

1055. EC (2010) Critical Raw Materials at EU Level European Commission, 2011. *Critical Raw Materials for the EU*. 85p.

1056. EC (2014) Report on critical materials for the EU. Report of the Ad hoc Working Group on defining critical raw materials. European Commission, Brussels, Belgium.

1057. European Commission (2014a) *Report on Critical Raw Materials for the EU*. 41p.

1058. European Commission (2014b) *Annexes to the Report on Critical Raw Materials for the EU*. 38p.

1059. COM2017. List of critical raw materials for the EU, 2017. European Commission, Brussels, Belgium.

1060. COM2017. Explanatory Memorandum on the 2017 list of Critical Raw Materials for the EU, 2017. European Commission, Brussels, Belgium.

1061. US Department of Energy (2011) *Critical Materials Strategy*. Office of Policy and International Affairs, 190p. Webster's New Collegiate Dictionary, 1975. Webster's New Collegiate Dictionary. Thomas Allen & Son Limited, Toronto, 1536p.

1062. Ayres, R. & Peiró, L. (2013) Material efficiency: Rare and critical metals. Phil Trans R Soc A 371: 20110563. The Royal Society, London.

1063. Simandl, G.J., Akam, C. & Paradis, S. (2015) Which materials are "critical" and "strategic". In: Simandl, G.J. & Neetz, M. (eds.) *Symposium on Strategic and Critical Materials Proceedings, November 13–14, 2015, Victoria, British Columbia*. British Columbia Ministry of Energy and Mines, British Columbia Geological Survey Paper 2015–3. pp. 1–4.

1064. Dutch economy, Preliminary results, Centre for Policy Related Statistics, Critical materials in the Publisher Statistics Netherlands, Henri Faasdreef 312 2492 JP The Hague Prepress Statistics Netherlands: Grafimedia ISSN: 1877–3036. © Statistics Netherlands, The Hague/Heerlen, 2010.

1065. Economic Stockpiling in Foreign Countries: Appendix C discusses economic stockpiling in the nine-nation European Economic Community (EEC) in general, and then specifically examines the stockpiling policies of three foreign countries, Japan, France, and Sweden.

1067. US Department of Defense (2013) Strategic and critical materials 2013 report on stockpile requirements. *Office of the Under Secretary of Defense for Acquisition, Technology and Logistics*. 189p.

1068. Strategic and Critical Materials. Operations report to Congress, Operations under the Strategic and Critical, Materials Stock Piling Act during Fiscal Year 2016. *Office of the Under Secretary of Defense for Acquisition, Technology, and Logistics*, January, 2017.

1069. Priorities for critical materials for a circular economy, European Academies Science Advisory Council (EASAC) ISBN 978-3-8047-3679-5. © German National Academy of Sciences Leopoldina, 2016.

1070. EASAC (2015) *Circular Economy: Commentary from the Perspectives of Natural and Social Sciences*. European Academies' Science Advisory Council. Halle (Saale), Germany.

1071. *The Role of Mining in National Economies*, 2nd edition. Mining's Contribution to Sustainable Development. International Council on Mining and Metals (ICMM), London, UK, October, 2014. ISBN: 978-1-909434-11-0.

1072. Krautkraemer, J. (2005) The economics of natural resource scarcity, the state of the debate. Chapter 3 in scarcity and growth revisited. Resources for the Future. doi:10.22004/ag.econ.10562; Handle: RePEc:rffdps:10562.

1073. EC COM (2011) Tackling the challenges in commodity markets and on raw materials. COM 25.

1074. COM (2012) 82 final. Making raw materials available for Europe's future wellbeing proposal for a European innovation partnership on raw materials.

1075. Deloitte (2015) *Study on Data for a Raw Material System Analysis: Roadmap and Test of the Fully Operational MSA for Raw Materials*. Prepared for the European Commission, DG GROW.

1076. Levinson, A.A., Gurney, J.J. & Kirkley, M.B. (1992) Diamond sources and production: Past, present, and future. *Gems and Gemology*, 28(4), 234–254.

1077. Teal, L. & Jackson, M. (1997) Geologic overview of the Carlin trend gold deposits and descriptions of recent deep discoveries. In: Vikre, P., Thompson, T.B., Bettles, K., Christensen, O. & Parratt, R. (eds.) *Carlin-Type Gold Deposits Field Conference: Society of Economic Geologists Guidebook Series*, Littleton, Colorado, USA. Volume, 28. pp. 3–38.

1078. U.S. Bureau of Mines (1953–93) Minerals yearbook for 1950, 1955, 1960, 1965, 1970, 1975, 1980, 1985, and 1990. In: *Metals and Minerals*, Volume 1. Bureau of Mines, Washington, DC, US, variously paged.

1079. U.S. Geological Survey (1997 and 2002) Minerals tearbook for 1995 and 2000. *Metals and Minerals*, Volume I. Reston, VA, U.S. Geological Survey, variously paged.

1080. British Geological Survey (2012) *Risk List 2012: An Update to Supply Risk Index for Elements or Element Groups That Are of Economic Value.*

1081. Hagelüken, C. (2012) Recycling and substitution: Opportunities and limits to reduce the net use of critical materials in hi-tech applications. In: *Proceedings of the Critical Materials Conference, Brussels, Belgium, March.*

1082. Hagelüken, C. (2014) Recycling of critical metals. In: *Critical Metals Handbook*. John Wiley & Sons, Hoboken, NJ.

1083. Ndlovu, S., Simate, G.S. & Matinde, E. *Waste Production and Utilization in the Metal Extraction Industry*. CRC Press, Boca Raton. (ISBN 9781498767293).

1084. New technologies and strategies for fuel cells and Hydrogen Technologies in the phase of recycling and dismantling, Grant No. 700190: HyTechCycling, WP2 Regulatory analysis, critical materials and components identification and mapping of recycling technologies, D2.1 Assessment of critical materials and components in FCH technologies. Brussels, Belgium.

1085. Hagelüken, C. & Meskers, C. (2010) Complex life cycles of precious and special metals. In: Graedel, T.E. & van der Voet, E. (eds.) *Linkages of Sustainability*. Strüngmann Forum Report, The MIT Press. Cambridge, MA, USA.

1086. Wellmer, F. & Hagelüken, C. (2015) The feedback control cycle of mineral supply, increase of raw material efficiency, and sustainable development. *Minerals*, 5, 815–836.

1087. UNEP (2011) Assessing mineral resources in society: Metal stocks and recycling rates. Summary booklet based on the two reports of the Global Metal Flows Group "Metal Stocks in Society: Scientific Synthesis" and "Recycling Rates of Metals: A Status Report". UNEP, Geneva, Switzerland.

1088. UNEP (2013) *Metal Recycling: Opportunities, Limits, Infrastructure*. United Nations Environmental Programme. International Resource Panel, M.A. Hudson et al. UNEP, Geneva, Switzerland.

1089. Schuler, D. et al. (2011) *Study on Rare Earths and Their Recycling.*

1090. Drexler, K.E. (1986) *Engines of Creation: The Coming Era of Nanotechnology*. Doubleday. ISBN 0-385-19973-2.

1091. Drexler, K.E. (1992) *Nanosystems: Molecular Machinery, Manufacturing, and Computation*. John Wiley & Sons, New York. ISBN 0-471-57547-X.

1092. Hochella Jr., M.F., Lower, S.K., Maurice, P.A., Penn, R.L., Sahai, N. Sparks, D.L. & Twining, B.S. (2008) Nanominerals, mineral nanoparticles, and earth systems. *Science*, 319(5870), 21 March, 1631–1635.

1093. Hennebert, P., Anderson, A. & Merdy, P. (2017) Mineral nanoparticles in waste: Potential sources, occurrence in some engineered nanomaterials leachates, municipal sewage sludges and municipal landfill sludges. *J Biotechnol Biomater*, 7, 261.

1094. Froggett, S.J., Clancy, S.F. & Boverhof, D.R. & Canady, R.A. (2014) A review and perspective of existing research on the release of nanomaterials from solid nanocomposites. *Part Fibre Toxicol*, 11, 17.

1095. OECD (Organisation for Economic Co-operation and Development) (2015b) Working party on resource productivity and waste-recycling of waste containing nanomaterials, Paris, France. p. 17.

1096. OECD (Organisation for Economic Co-operation and Development) (2015d) Working party on resource productivity and waste: The fate of engineered nanomaterials in sewage treatment plants and agricultural applications, Paris, France. p. 17.

1097. Kaegi, R., Ulrich, A., Sinnet, B., Vonbank, R. & Wichser, A. et al. (2008) Synthetic TiO2 nanoparticle emission from exterior facades into the aquatic environment. *Environmental Pollution*, 156, 233–239.

1098. Mann, S. (2006) Nanotechnology and construction (PDF). *Nanoforum.org*. European Nanotechnology Gateway, 31 October.

1099. Yang, S., Pelton, R., Raegen, A., Montgomery, M. & Dalnoki-Veress, K. Nanoparticle flotation collectors: Mechanisms behind a new technology. Langmuir, 27(17), 10438–10446.; Copyright © 2011 American Chemical Society.

1100. Yang, S., Pelton, R., Montgomery, M. & Cui, Y. (2012) Nanoparticle flotation collectors III: The role of nanoparticle diameter. *ACS Applied Materials Interfaces*, 4(9), 4882–4890. Copyright © 2012 American Chemical Society.

1101. Yang, S., Marzieh Razavizadeh, B.B., Pelton, R. & Bruin, G. (2013) Nanoparticle flotation collectors: The influence of particle softness. *ACS Appl. Mater. Interfaces*, 5(11), 4836–4842. doi:10.1021/am4008825. Publication Date (Web), 21 May 2013. Copyright © 2013 American Chemical Society.

1102. RS & RAE, Nanoscience and nanotechnologies: Opportunities and uncertainties. Royal Society and Royal Academy of Engineering. July 2004. ISBN 0854036040

1103. ACS, (2003) Nanotechnology: Drexler and Smalley make the case for and against "molecular assemblers". *Chemical & Engineering News* (American Chemical Society), 81(48), 37–42.

1104. Allhoff, F., Lin, P., Moore, D. (2010) *What Is Nanotechnology and Why Does It Matter?: From Science to Ethics*. John Wiley and Sons, Hoboken, NJ. pp. 3–5. ISBN 1-4051-7545-1.

1105. Prasad, S.K. (2008) *Modern Concepts in Nanotechnology*. Discovery Publishing House. pp. 31–32. ISBN 81-8356-296-5.

1106. Eric, D.K. (1992) *Nanosystems, Molecular Machinery, Manufacturing, and Computation*. John Wiley & Sons, New York. ISBN 9780471575474.

1107. SCENIHR (2009) Scientific committee on emerging and newly identified health risks, risk assessment of products of nanotechnologies. *European Commission, DG Health & Consumers*, Brussels.

1108. Bolea, E., Laborda, F., Castillo, J.R. (2010) Metal associations to microparticles, nanocolloids and macromolecules in compost leachates: Size characterisation by assymetrical flow field-flow fractionation coupled to ICP-MS. Anal Chim Acta 661. pp. 206–214.

1109. Narayan, R.J., Kumta, P.N., Sfeir, C.H., Lee, D.-H., Choi, D. & Olton, D. (2004) Nanostructured ceramics in medical devices: Applications and prospects. *JOM*, 56(10), 38–43.

1110. Schwab, K. The fourth industrial revolution. What It Means and How to Respond. SNAPSHOT December 12, 2015; wef.ch/4IRbook.

1111. Moavenzadeh, J. The 4th industrial revolution: Reshaping the future of production. *World Economic Forum*. DHL Global Engineering & Manufacturing Summit. 7 October 2015. Amsterdam.

1112. MyWave: GmcBride, London. The Fourth Industrial Revolution, April 2016.

1113. Reckhenrich-Studie, World Economic Forum: Skill: 4th Industrial Revolution, The future of jobs.

1114. WEF_AM16 Report, Davos-Klosters, Mastering the Fourth Industrial Revolution, 2016.

1115. Port, J., Murphy, D.J. (2017) Mesothelioma: Identical routes to malignancy from asbestos and carbon nanotubes (PDF). *Current Biology*, 27(21), R1156–R1176.

1116. Chernova, T., Murphy, F.A., Galavotti, S., Sun, X.-M., Powley, I.R., Grosso, S., Schinwald, A., Zacarias-Cabeza, J., Dudek, K.M. & Dinsdale, D. et al. (2017) Long-fiber carbon nanotubes replicate asbestos-induced mesothelioma with disruption of the tumor suppressor gene Cdkn2a (Ink4a/Arf) (PDF). *Current Biology*, 27(21), 3302–3314.

1117. Maret, W., Moulis, J.-M. (2013) Chapter 1: The bioinorganic chemistry of cadmium in the context of its toxicity. In: Sigel, A., Sigel, H., Sigel, R.K.O (eds.) *Cadmium: From Toxicology to Essentiality*. Metal Ions in Life Sciences, Volume 11. Springer, London. pp. 1–30.

1118. Hayes, A.W. (2007) *Principles and Methods of Toxicology*. CRC Press, Philadelphia. pp. 858–861. ISBN 978-0-8493-3778-9.

1119. Navas-Acien, A. (2007) Lead exposure and cardiovascular disease: A systematic review. *Environmental Health Perspectives*, 115(3), 472–482.

1120. McFarland, R.B. & Reigel, H. (1978) Chronic mercury poisoning from a single brief exposure. *Journal of Occupational Medicine and Toxicology*. 20(8), 532.

1121. UNEP & WHO. *Mercury, Environmental Health Criteria Monograph No. 001*. World Health Organization, Geneva, 1976. ISBN 92-4-154061-3. (UNEP: United Nations Environmental Programme).

1122. UNEP & WHO, Inorganic mercury *Environmental Health Criteria Monograph No. 118*. World Health Organization, Geneva, 1991. ISBN 92-4-157118-7. (UNEP: United Nations Environmental Programme).

1123. Sandenbergh, R., Fauconnier, C. & Baxter, R. (2009) Mining as a catalyst for development in underdeveloped regions: An African perspective. Addis Ababa, Ethiopia.

1124. Tumafor, L. (2009) African mining vision and the continent's development. *3rd World Network Africa* Available from: www.twnafrica.org.

1125. Walker, M. & Jourdan, P. (2003) Resource-based sustainable development: An alternative approach. Raw Materials Report. https://doi.ort 10.1080/140410403010019435.

1126. UNECA. (2011) Minerals and Africa's development: International study group report on Africa's mineral regimes to industrialisation in South Africa, minerals and energy. United Nations Economic Commission for Africa (UNECA) Addis Ababa, Ethiopia.

1127. Wright, G. & Czelusta, J. (2003) Mineral resource and economic development. Paper prepared for the *Conference on Sector Reform in Latin America*, Standford Centre for International Development.

1128. Felix, W., Gibbs, J., Schute, C., Joost, W. & Ollila, S. Materials technology: overview. June 2016.

1129. Eddy, M.D. (2008) *The Language of Mineralogy: John Walker, Chemistry and the Edinburgh Medical School 1750–1800*. CFRS-ICSU. Scientific relations between academia and industry: Building on a new era of interactions for the benefit of society. Freedom and Responsibility in the conduct of science, Sigtuna, Sweden, 2011.

1130. African Union Commission (AUC). (2008) Plan of action for African acceleration of Industrialization-Promoting Resource-Based Industrialization: A Way Forward.

1131. Lahti, A. (2007) Resource-based exporting and industrialisation: How to learn from the Nordic countries?, Addis Ababa.

1132. Lydall, M. (2009) Backward linkage development in the South African PGM industry: A case study. *Resources Policy*, 34(3).

1133. Lydall, M. (2010) Getting the basics right: Towards optimizing mineral-based linkages in Africa, contribution to the ISG's review of African Mining Regimes.

1134. San Cristobal, J.R. & Biezma, M.V. (2006) Mining industry in the European Union: Analysis of inter-industry linkages using input-output analysis, Resources Policy.

1135. Katumba, G. et al. (2008) Optical, thermal and structural characteristics of carbon nanoparticles embedded in ZnO and NiO as selective solar absorbers. *Solar Energy Materials & Solar Cells*, 92, 12850–1292.

1136. Johnson, C.S., Li, N., Lefief, C. & Thackeray, M.M. (2007) Anomalous capacity and cycling stability of $xLi_2MnO_3.(1-x)LiMO_2$ electrodes (M = Mn, Ni, Co) in lithium batteries at 50°C. *Electrochemistry Communications*. 9, 787–795.

1137. Mudder, T.I. & Botz, M.M. (2004) Cyanide and society: A critical review. *European Journal of Mineral Processing and Environmental Protection*. 4, 62–74.

1138. Bambagioni, V., Biachini, C., Marchionni, A., Filippi, J., Vizza, F., Teddy, J., Serp, P. & Zhiani, M. (2009) Pd and Pt-Ru anode electrocatalyst supported on multi-walled carbon nanotubes and their use in passive and active direct alcohol fuel cells with an anion-exchange membrane. (alcohol = methanol, ethanol, glycerol). *Journal of Power Sources*, 190, 241–251.

1139. Fleming, C.A., Mezei, A., Bourricaudy, E., Canizeres, M. & Ashbury, M. (2011) Factors influencing the rate of gold cyanide leaching and adsorption on activated carbon, and their impact on the design of CIL and CIP circuits. *Minerals Engineering*, 24, 484–494.

1140. The African Diaspora Programme (2007) Mobilizing the African Diaspora for development. Geneva, Switzerland.

1141. World Bank (2002) *Constructing Knowledge Societies: New Challenges for Tertiary Education*, World Bank, Washington, DC.

1142. Inter Academy Council (2004) *Inventing a Better Future: A Strategy for Building Worldwide Capacities in Science and Technology*. Amsterdam. ICTP, Trieste, Italy. p. 1.

1143. The Sustainable Development Goals (SDGs) (or Global Goals for Sustainable Development) are a collection of 17 global goals set by the United Nations Development Programme.

1144. Commission for Africa (2005) Our common interest: Report of the commission for Africa, March. UNECA, Addis Ababa, Ethiopia. p. 139.

1145. Tracy, B., Cloete, N. & Pillay, P.(2012)*Universities and Economic Development: Case Study*. Uganda and Makerere University, CHE, Kampala, Uganda.

1146. African Union (2005) 4th ordinary session of the African Union Assembly, Abuja, Nigeria.

1147. Mugabe, R.G. Speech on the occasion of his inauguration ceremony as President of the Republic of Zimbabwe. [accessed 22 August 2013].

1148. Parliamanent of Zimbabwe, Cap. 25:33 Pan-African Minerals University of Science and Technology Act, No. 9/2016.

1149. Hodge, G.A. & Greve, C. (2007) Public: Private partnerships: An international performance review. *Public Administration Review*,. 67(3), 545–558.

1150. Caldwell, N., Roehrich, J.K. & George, G. (2017) Social value creation and relational coordination in public-private collaborations. *Journal of Management Studies*, 54(6), 906–928.[1]

1151. Hart, O. (2003) Incomplete contracts and public ownership: Remarks, and an application to public-private partnerships. *Economic Journal*, 113, C69–C76.

1152. Iossa, E. & Martimort, D. (2012) Risk allocation and the costs and benefits of public-private partnerships. *RAND Journal of Economics*, 43, 442–474.

1153. NOIE: Productivity and Organisational Transformation: Optimising investment in ICT, (Ovum) Canberra, 2003.

1154. Asabere, N.Y. & Kusi-Sarpong, S. (2013) The relationship and role of information & communication technology (ICT) in the mining industry: An analysis of Supply Chain Management (SCM): A case study. *International Journal of Engineering Research and Application (IJERA) ISS*, 3, 2248–9622.

1155. Fasanghari, M., Roudsari, F.H. & Chaharsooghi, S.K. (2008) Assessing the impact of information technology on supply chain management. *World Applied Sciences Journal*, 4(1), 87–93.

1156. Asabere, N.Y., Oppong, D. & Kusi-Sarpong, S. (2012) A review of the roles and importance of information and communication technologies (ICTs) in supply chain management of organisations and companies. *International Journal of Computer Science and Network*, 1(4), 70–80.

1157. Shavazi, R.A., Abzari, M. & Mohammadzadeh, A. (2009) A research in relationship between ICT and SCM. *World Academy of Science, Engineering and Technology*, 50, 92–101.

1158. Lancioni, R., Schau, H.J. & Smith, M.F. (2003) Internet impacts on supply chain management. *Industrial Marketing Management*, 32, 173–175.

1159. Gek, W.T., Shawand, M.J. & Fulkerson, B. (2000) Web-based supply chain management. *Information Systems Frontiers*, 2(1), 41–55.

1160. Campbell, A.J. & Wilson, D. (1996) Managed networks: Creating strategic advantage. In: Iacobucci, ýD (ed.) *Networks Marketing*. Sage, London, England.

1161. Evangelista, P. (2002) The role of ICT in the logistics integration process of shipping lines. *Pomorski Zbornik*(1), 61–78.

1162. Sweeney, E. (2007) *Re-Engineering the Supply Chain: Making SCM Work for You in Perspectives on Supply Chain management and Logistics: Creating Competitive Organisations in the 21st Century*, Sweeney, E. (ed.). Blackhall Publishers, Dublin, Chapter 16. pp. 295–306.

1163. The InterAcademy Partnership (IAP) is a global network linking academies of sciences, medicine and engineering with more that 130 national and regional member academies working together to support the vital role of science and its effort to seek solutions that address the world's most challenging problems.

1164. African Academy of Sciences. AAS history: Laying the foundation. *AAS*.

1165. African Academy of Sciences. AAS history: Inauguration and establishment: Phase 1 (1985–1988). *AAS*.

1166. NASAC, Network of African Science Academies (NASAC) established in 2001 with Headquarters in Nairobi, Kenya.

1167. AMASA, Annual Meetings of African Science Academies.

1168. Akyeampong, E. (2000) Africans in the Diaspora: The Diaspora and Africans. *African Affairs*, 99(395), 183–215.

1169. CIDO Directorate. The Diaspora Division. Statement: The Citizens and Diaspora Organizations Directorate (CIDO).

1170. Nkiru, I. and Nzegwu, O. (2009) *The New African Diaspora*. Indiana University Press, Blooming, IN, USA.

1171. UNCTAD. UN Conference on Trade and Development: Trade Misinvoicing in Primary Commodities in Developing Countries (2016) (PDF) Geneva, Switzerland.

1172. Portugal-Perez, A. & Wilson, J.S. (2008) Trade costs in Africa: Barriers and opportunities for reform, policy research working paper, no. 4619, World Bank, Development Research Group. Washington DC, USA.

1173. Valiante, D. & Egenhofer, C. (2013) *Price Formation in Commodity Markets*. European Capital Markets Institute/Centre for European Studies, Brussels. p. 129.

1174. Hoekman, B. & Martin, W. (2012) Reducing distortions in international commodity markets an agenda for multilateral cooperation In: Evenett, S.J. & Jenny, F. (eds.) *Trade, Competition, and the Pricing of Commodities*. Centre for Economic Policy Research (CEPR) and Consumer Unity & Trust Society (CUTS International), London.

1175. Humphreys, D. (2011) Mineral pricing regimes and the distribution of rents in the value chain. POLINARES Internal Working Document.

1176. Fliess, B. & Mård, T. (2012) Taking stock of measures restricting the export of raw materials: Analysis of OECD inventory data. *OECD Trade Policy Papers*, No. 140, OECD Publishing Paris, France.

1177. OECD (2014) *OECD conference on trade in raw materials*. Paris, 3 November.

1178. London Metal Exchange/Europe Economics (2013) Executive summary. *Warehouse Study*.

1179. Cavalcanti, T., Mohaddes, K. & Raissi, M. (2011) Commodity price volatility and the sources of growth (PDF). Cambridge Working Papers in Economics. *Journal of Applied Econometrics*, 2015, vol. 30(6) pp. 852–873.

1180. Connor, J. (2012) Price effects of international cartels in markets for primary products In: Evenett, S.J. & Jenny, F. (eds.) *Trade, Competition, and the Pricing of Commodities*. Centre for Economic Policy Research (CEPR) and Consumer Unity & Trust Society (CUTS International), London.

1181. African Development Bank Group (2009a) Africa and the global economic crisis: Strategies for preserving the foundations of long-term growth. Working Paper # 98. African Development Bank, Abidjan.

1182. African Development Bank Group (2009b) Impact of the crisis on African economies: Sustaining growth and poverty reduction: African perspectives and recommendations to the G20. African Development Bank, Abidjan.

1183. Menzie, W., Soto-Viruet, Y., Bermúdez-Lugo, O., Mobbs, P., Perez, A., Taib, M. & Wacaster, S. (2013) Review of selected global mineral industries in 2011 and an outlook to 2017. U.S. Geological Survey Open-File Report 2013–1091. USGS, Reston, Virginia, USA.

1184. Arkhipov, I. & Wild, F. (2013) Russia, South Africa seek to create OPEC-style platinum bloc. *Bloomberg*, 27 March.

1185. Suslow, V. (2005) The changing international status of export cartel exemptions. *American University International Law Review*, 20(4), 785–828.

1186. Martyniszyn, M. (2012) Export cartels: Is it legal to target your neighbour? Analysis in Light of Recent Case Law. *Journal of International Economic Law*, 15(1), 181–222.

1187. Larsson, V. & Ericsson, M. (2014) E&MJ's Annual Survey of Global Metal-mining Investment. *E&MJ Engineering and Mining Journal*, January, 26–31.

1188. Hathaway, N. (1997) Privatisation and the cost of capital. *Agenda*, 4(1), 1–10.

1189. Borcherding, T., Pommerehne, W.W. & Schneider, F. (1982) Comparing the efficiency of private and public production: The evidence from five countries. *Zeitschrift fur Nationalokonomie Journal of Economics*(2), 127–156.

1190. Otto, J., Andrews, C., Cawood, F., Doggett, M., Guj, P., Stermole, F. & Stermole, J. (2006) *Mining Royalties: A Global Study on Their Impact on Investors, Government, and Civil Society*, Directions in Development, Energy and Mining, World Bank, Washington, DC.

1191. ShriSahoo, L.K. Energy efficiencyEnergy efficiency in Mining Sector, Central Institute of Mining & Fuel Research (CIMFR), CSIR.

1192. Hedden, S., Moyer, J.D. & Rettig, J. (2013) *Fracking for Shale Gas in South Africa: Blessing or Curse?* Institute for Security Studies and Pardee Centre for International Futures. Pretoria, South Africa and Denver, Colorado, USA.

1193. Maest et al. (2006) Predicted versus actual water quality at hardrock mine sites: Effect of inherent geochemical and hydrologic characteristics. Poster presented at the 7th International Conference on Acid Mine Drainage (ICARD), St. Louis MO, USA.

1194. Jung, M.C. & Thornton, I. (1996) Heavy metals contamination of soils and plants in the vicinity of a lead-zinc mine, Korea. *Applied Geochemistry*, 11, 53–59.

1195. Hoostal, M.J., Bidart-Bouzat, M.G. & Bouzat, J.L. (2008) Local adaptation of microbial communities to heavy metal stress in polluted sediments of Lake Erie. *FEMS Microbiology Ecology*, 65, 156–168.

1196. Kaushik, A. & Kaushik, C.P. (2010) *Basics of Environment and Ecology*. New Age International (p) Limited, Publishers, New Delhi.

1197. Kimura, S., Bryan, C.G., Hallberg, K.B. & Johnson, D.B. (2011) Biodiversity and geochemistry of an extremely acidic, low-temperature subterranean environment sustained by chemolithotrophy. *Environmental Microbiology*, 13(8), 2092–2104.

1198. Mummey, D.L., Stahl, P.D. & Buyer, J.S. (2002) Soil microbiological properties 20 years after surface mine reclamation: Spatial analysis of reclaimed and undisturbed sites. *Soil Biology and Chemistry*, 34, 1717–1725.

1199. Niyogi, D.K., William, M.L., Jr., McKnight, D.M. (2002) Effects of stress from mine drainage on diversity, biomass, and function of primary producers in mountain streams. *Ecosystems*(5), 554–567.

1200. Nuss, P. & Eckelman, M.J. (2014) Life cycle assessment of metals: A scientific synthesis. *PLoS One*, 9(7), e101298.

1201. Fang-Mei Tai, Shu-Hao Chuang. Published online September 2014 in Scientific Research Publishing Inc. (SciRes).

1202. Chen, C.H. (2011) The major components of corporate social responsibility. *Journal of Global Responsibility*, 2, 85–99.

1203. Wells Jr., L.T. (1977) Negotiating with third world governments. *Harvard Business Review*, 55(1), January–February.

1204. Bird, F. & Smucker, J. (2007) The social responsibilities of international business firms in developing areas. *Journal of Business Ethics*, 73, 1–9.

1205. UNIDO (2007) UNIDO initiative aims to develop CSR awareness among European SMEs. *Ethics World.*

1206. OECD (2001) Corporate responsibility: Private initiatives and public goals.

1207. Lockwood, K. (2016) Analysis of investment in mining TIPS annual forum, June.

1208. Dunning, J.H. (1997) Re-evaluating the benefits of foreign direct investment. In: Dunning, J.H. *Alliance Capitalism and Global Business*. Routledge, London.

1209. OECD Global Forum ON International Investment Conference on Foreign Direct Investment and the Environment Lessons to be Learned from the Mining Secto 7–8 February 2002.

1210. Jaspers, F. & Mehta, I. (2007) Developing SMEs through business linkages: MozLink experience. A manual for companies, NGOs and government entities version 1.0. International Finance Corporation (IFC), World Bank Group. Washington, D.C., USA.

1211. Humphreys, M., Sachs, J. & Stiglitz, J. (2007) *Escaping the Resource Curse*. Columbia University Press, New York.

1212. ICMM, World bank & UNCTAD (2008) *Resource Endowment Toolkit: The Challenge of Mineral Wealth*, London: ICMM.

1213. Van Wijnbergen, S. (1984) The "Dutch disease": A disease after all?. *The Economic Journal*, 94(373)m 41.

1214. Sachs, J. & Warner, A. (1999) The big rush, natural resource booms and growth. *Journal of Development Economics*, 59, 43–76.

1215. Ben Hammouda, H., Karingi, S., Njuguna, A. & Sadni-Jallab, M. (2006) Diversification: Towards a new paradigm for Africa's development. *ATPC Work in Progress No. 35*, African Trade Policy Centre. United Nations Economic Commission for Africa, UNECA, Addis Ababa, Ethiopia.

1216. Collier, P. & Venables, A.J. (2007) Rethinking trade preferences: How Africa can diversify its exports. Center for Economic and Policy Research Discussion Papers 6262, Center for Economic and Policy Research Discussion Papers, Washington, DC.

1217. Dennis, A. & Shepherd, B. (2007) Trade costs, barriers to entry, and export diversification in developing countries. Policy Research Working Paper Series 4368, World Bank, Washington, DC.

1218. Ang, A., Gorovyy, S. & van Inwegen, G.B. (2011) Hedge fund leverage. *Journal of Financial Economics*, 102(1), 102–126.

1219. Stulz, R. (2007) Hedge funds past, present, and future. *Journal of Economic Perspectives*, 21(2), 175–194. CiteSeerX 10.1.1.475.3895.

1220. Stowell, D. (2010) *An Introduction to Investment Banks, Hedge Funds, and Private Equity*, The New Paradigm. Academic Press. Cambridge, Massachusetts, USA.

1221. Boyson, N.M., Stahel, C.W., Stulz, R.M. (2010) Hedge fund contagion and liquidity shocks. *The Journal of Finance*, 65, 1789–1816.

1222. Christopherson, R. & Gregoriou, G.N. (2004) *Commodity Trading Advisors: Risk, Performance Analysis, and Selection*. John Wiley & Sons, Hoboken, NJ. pp. 377–384. ISBN 0-471-68194-6.

1223. Duc, F. & Schorderet, Y. (2008) *Market Risk Management for Hedge Funds: Foundations of the Style and Implicit Value-at-Risk*. Wiley. pp. 15–17. ISBN 0470722991.

1224. PrivatisatioN and Nationalisation Jean-Pierre Dupuis. National accounts and financial statistics statistics directorate Organisation for Economic Co-operation and Development (OECD). Paper presented at the *Fourth Meeting of the Task Force on Harmonization of Public Sector Accounting (TFHPSA) Hosted by the International Monetary Fund Washington, DC, 3–6 October 2005*.

1225. Aluga, M. & Mwaanga, P. Impact of privatization of mining industry in developing nations on African engineering education: A case study of Zambia. The 6th African Engineering Education Conference, CUTS, at Bloemfentein, FS, South Africa, 2016.

1226. Forrest, S.A., Rowley, H., Trickey, S. & Underwood, J. (2010) Strategic implications for the platinum group metals: Privatization vs. nationalization in South Africa. The 4th International

Platinum Conference, Platinum in transition 'Boom or Bust', The Southern African Institute of Mining and Metallurgy. Keynote address: Strategic implications for the platinum group metals: Privatization vs. nationalization in South Africa. Forrest, S.A., Rowley, H., Trickey, S. & Underwood, J., SFA (Oxford) Ltd, UK.

1227. Iheduru, O. (2004) Black economic power and nation-building in post-apartheid South Africa. *Journal of Modern African Studies*, 42(1), 1–30.

1228. Chiume, L. & Kingston, M.L. (2006) The role of an investment bank in broad-based BEE transactions: A banker's perspective. *New Agenda: South African Journal of Social and Economic Policy*, 22, 25–29.

1229. Cargill, J. (2005) Black corporate ownership: Complex codes can impede change. In: *Conflict and Governance: Economic Transformation: Audit 2005*. Institute for Justice and Reconciliation, Cape Town.

1230. Fauconnier, A. & Mathur-Helm, B. (2008) Black economic empowerment in the South African mining industry: A case study of Exxaro Limited. *S. Afr. J. Bus. Manage.*, 39(4).

1231. Byrnes, P., Grosskopf, S. & Hayes, K. (1986) Efficiency and ownership: Further evidence. *Review of Economics and Statistics*, 68, May, 337–341.

1232. Mitchell, P. (2009) Taxation and investment issues in mining: The role of tax in investment decisions. In *Advancing the Extractive Industries Transparency Initiative (EITI) in the Mining Sector*. pp. 27–31. Transparency Initiative

1233. Henderson Global Investors (2005) *Responsible Tax*. Henderson Global Investors ltd., London, UK.

1234. Otto, J., Andrews, C., Cawood, F., Doggett, M., Guj, P., Stermole, J. & Tilton, J. (2006) *Mining Royalties: A Study of Their Impact on Investors, Government, and Civil Society*. World bank, Washington, DC.

1235. Wittendorff, J. (2010) *Transfer Pricing and the Arm's Length Principle in International Tax Law*. Kluwer Law International. ISBN 90-411-3270-8.

1236. OECD: Transfer pricing guidelines for multinational enterprises and tax administrations. OECD Publishing, Paris. Organization for Economic Cooperation & Development, July 2010.

1237. Cooper, J., Fox, R., Loeprick, J., Mohindra, K. (2016) *Transfer Pricing and Developing Economies: A Handbook for Policy Makers and Practitioners*. World Bank, Washington, DC. pp. 18–21. ISBN 978-1-4648-0970-5.

1238. Humphreys, D. (2013) New mercantilism: A perspective on how politics is shaping world metal supply. *Resources Policy*, 38(3), 341–349.

1239. Evenett, S.J. & Jenny, F. (eds.) (2012) *Trade, Competition, and the Pricing of Commodities*. Centre for Economic Policy Research (CEPR) and Consumer Unity & Trust Society (CUTS International), London.

1240. Global Corruption Report (2001) Transparency international. The Global Anti-Corruption Coalition: Transparency.org.

1241. Sanderson, H. (2014) Lawsuit alleges manipulation of precious metals benchmark. *Financial Times*, 26 November.

1242. Otto, J. & Cordes, J. (2002) *Regulation of Mineral Enterprise: A Global Perspective on Economics, Law and Policy*. Rocky Mountain Mineral Law Foundation. Westminster, Colorado, USA.

1243. Kelsall, G.H. & Wood, D.J.D. (1982) Cassiterite electrochemistry: Hydrometallurgical routes to Tin. In: Osseo-Assare, K. & Miller, J.D. (eds.) *Hydrometallury: Research, Development and Plant Practice*. A.I.M.E., New York.

1244. Gudyanga, F.P. (1988) *Electrohydrometallurgical Reduction of Cassiterite (SnO₂) Associated with Sulphide Minerals*. PhD Thesis, Imperial College, University of London.

1245. Wills, B.A. (1988) *Mineral Processing Technology: An Introduction to the Practical Aspects of Ore Treatment and Mineral Recovery*, 4th edition, International Series on Materials and Technology, Volume 41, Pergamon Press. ISBN 0-08-034936-6.

1246. Mosher, J.B. (2016) Communition circuits for gold ore processing. In: Adams, M.D. (ed.) *Gold Ore Processing: Project Development and Operations*, 2nd edition. Elsevier. ISBN 978-0-444-63658-4.

1247. Swaziland National Trust Commission and Swaziland National Museum: Ref. 5421: Ngwenya Mine. Wikipedia.

1248. Zimba, J., Henwood, D., Simbi, D.J. & Navara, E. (1999) A three dimensional diffusion model for the austenitisation of ferric spheroidal graphite irons. *Journal of Material Science and Technology*, 15, 1024–1030, September.

1249. Deschenes, G., Pratt, A., Riveros, P., Fulton, M. (2002) Reactions of gold and sulfide minerals in cyanide media. *Mineral and Metallurgical Processing Journal*, 19(4), 169–177.

1250. Clep, O., Alp, L., Deveci, H. (2011) Improved gold and silver extraction from refractory antimony ores by pretreatment with alkaline sulphide leach. *Hydrometallurgy*, 105, 234–239.

1251. Rare Earth Elements and Selected Examples of Usages. Available from: https://en.wikipedia.org/wiki/Rare-earth element

1252. van der Berg, N.G., Malherbe, J.B., Botha, A.J., Friedland, E. & Jesser, W.A. (2009) *Surface Interface Anal*, 42, 1156.

1253. Prasad, P.N. (2004) *Nanophotonics*, John Wiley & Sons, Hoboken, NJ.

1254. Tuovinen, O.H. & Kelly, D.P. (1974) Use of micro-organisms for the recovery of metals. *International Metals Review*, 19(No. 179), 21.

1255. Dai, X. & Breuer, P.L. (2013) Leaching and electrochemistry of gold, silver, and gold-silver alloys in cyanide solutions: Effect of oxidant and lead (II) ions. *Hydrometallurgy*, 133, 139–148.

1256. Habashi, F. (1971) Pressure hydrometallurgy: Key to better and non-polluting processes. *Engineering and Mining Journal*, 172, 96–100.

1257. Gudyanga, F.P., Mahlangu, T., Roman, R.J., Mungoshi, J. & Mbeve, K. (1999) An acidic pressure oxidation pretreatment of refractory gold concentrates from the Kwekwe roasting plant, Zimbabwe. *Minerals Engineering*, 12(8).

1258. Liddell, K.S., Adams, M.D. (2012) Kell hydrometallurgical process for extraction of platinum group metals and base metals from flotation concentrates. *Journal of South African Institute of Mining and Metallurgical Transaction*, 112, 31–36.

1259. El-Sayed, M.A. (2004) Small is different: Shape-, and composition-dependent properties of some colloidal semiconductor nanocrystals. *Accounts of Chemical Research*, 37, 326.

1260. Welham, N.J., Kelsall, G.H. & Diaz, M.A. (1993) Thermodynamics of Ag-Cl-H_2O and Ag-I-H_2O systems at 298 K. *Journal of Electroanalytical Chemistry*, 361(1–2), 39–47.

Index

Note: Numbers in bold indicate tables on the corresponding page.

Printed in the United States
by Baker & Taylor Publisher Services